STRESS: A Reference Manual

A Problem-Oriented Computer Language for Structural Engineering

STRESS: A Reference Manual

A Problem-Oriented Computer Language for Structural Engineering

The Department of Civil Engineering
Massachusetts Institute of Technology

Steven J. Fenves
Robert D. Logcher
Samuel P. Mauch

The M.I.T. Press
Massachusetts Institute of Technology
Cambridge, Massachusetts, and London, England

PREFACE

Work on STRESS began in the Fall of 1962 under the direction of Professor S. J. Fenves of the University of Illinois, who was a visiting member of the M.I.T. faculty during the year 1962-1963. The project staff included Professor R. D. Logcher, Professor S. P. Mauch, Mr. K. F. Reinschmidt, and Mr. R. L. Wang. In the Fall of 1963 the project was placed under the general supervision of Professor J. M. Biggs with Professors Logcher and Mauch directly in charge of the programming effort. During the ensuing period important contributions were made to the debugging of the system by Professor Z. M. Elias, Mr. R. V. Goodman, Miss S. C. Finkelstein, Mr. S. G. Mazzotta, Mr. J. R. Roy, and Mr. A. C. Singhal.

The STRESS project has been partially supported from a major grant for the improvement of engineering education made to M.I.T. by the Ford Foundation. Additional support was provided by Project MAC, an M.I.T. research program, sponsored by the Advanced Research Project Agency, Department of Defense, under Office of Naval Research, Contract No. Nonr-4102(01). Reproduction in whole or in part is permitted for any purpose of the U.S. Government. The work was done in part at the M.I.T. Computation Center, and the aid and support of the Center and its personnel are gratefully acknowledged.

Although the STRESS processor has been extensively tested by M.I.T. personnel, no warranty is made regarding the accuracy and reliability of the program and no responsibility is assumed by M.I.T. in this connection.

S. J. Fenves
R. D. Logcher
S. P. Mauch

January 1965

CONTENTS

STRESS: A Reference Manual

INTRODUCTION

STRESS (STRuctural Engineering System Solver) is a programming system for the solution of structural engineering problems on digital computers. This manual contains the description of the STRESS processor system as presently (April 1964) implemented on the IBM 7094 computer at the M. I. T. Computation Center. The purpose of the manual is to provide the information necessary for the implementation of the STRESS processor on other computer systems, and for the inclusion of additional capabilities in processors already implemented.

This manual serves as a complement to STRESS: A User's Manual.[1] A thorough familiarity with the User's Manual is recommended before attempting to implement or expand the system.

It is assumed that the person planning to implement or change the system is an experienced FORTRAN programmer. Knowledge of assembly language programming and monitor operation will be helpful in implementing STRESS for special monitors and other machines.

One of the primary objectives in the development of the STRESS system has been the ease of expansion and modification of the system. Expansions can be accomplished by the incorporation of additional input statements and corresponding subroutines, so that the system can be dynamically maintained with components added, modified, or replaced to reflect the individual requirements of a particular organization. It has been intended that such changes need not require the services of experienced systems programmers but can be readily accomplished by programmers now engaged in the development and maintenance of conventional special-purpose programs.

This manual consists of seven chapters. Chapter 1 contains a general description of the system in terms of the organization of the processor and usage of memory. Chapter 2 describes in detail the dynamic memory allocation algorithm used by the system. The method of input translation is discussed in Chapter 3.

---

[1]S. J. Fenves, R. D. Logcher, S. P. Mauch, and K. F. Reinschmidt, STRESS: A User's Manual, The M. I. T. Press, Cambridge, Mass., 1964.

The general method of indeterminate analysis is described in Chapter 4. Chapter 5 gives the function of all parameters and arrays in the system. Chapter 6 describes the complete system as currently implemented. Finally, in Chapter 7 guidelines and examples are presented for the incorporation of possible additional components of the system.

Chapter 1

GENERAL DESCRIPTION OF SYSTEM

This chapter gives an introductory description of the STRESS processor system necessary for the understanding of the details described in the succeeding chapters. Specific aspects covered are the organization of the system, the internal functioning of the processor, the use of memory by the system, and the method of storing data.

1.1 Organization of the System

The organization of the STRESS Language, described in the User's Manual, reflects one of the initial specifications set by the authors, namely, that the language used should closely resemble the accepted terminology of structural engineering. The organization of the system, that is, the processor which acts on the input statements to produce the specified results, was dictated by two additional specifications: first, that the system be highly flexible in order to handle the widest possible range of structural problems; and second, that the system be easily understood, so that modifications on the system could be easily performed by persons who are not primarily systems programmers. Only experience in actual use will determine how successfully these specifications have been met.

With the last of these specifications in mind, the most widely available programming language, namely, FORTRAN II, has been used almost exclusively in the development of the system, even where considerably greater sophistication could have been achieved through symbolic programming. Symbolic coding has been restricted primarily to a function subprogram used for input decoding, and to the subroutines dealing with dynamic memory allocation. It is believed that these subroutines need not be changed for any IBM 709-7090-7094 system. However, they must be reprogrammed for any computer that differs from the 7094 in word format or length, internal representation of alphanumeric characters, or access to secondary storage (tapes or disks), and for

3

use with any other compiler.

In planning the STRESS system, the authors have intended to restrict themselves to the development portion alone, and to leave the implementation of the system for particular machine and monitor requirements to interested outside parties. This decision is reflected in the organization of the system and the contents of this manual in two ways.

First, the system as described in this manual was not intended to be ready to run on any arbitrary large-scale digital computer. The system was written for the particular equipment configuration directly available to the authors, namely, the IBM 7094 computer at the M.I.T. Computation Center operating under the MIT-FMS monitor. The system does not possess the general flexibility in terms of interaction with a monitor, tape units assignments, etc., that would be necessary for a fully implemented system. Provisions have been made for certain flexibility, but such provisions were not considered a primary objective in the development.

Second, the system described in this manual is the end product of the development phase, and no attempt has been made to "clean up" the programs. As a consequence, the program listings contain certain inconsistencies in nomenclature, as well as a few logical branches, calling parameters, etc., which have become obsolete, but which have not been removed. Finally, there are many provisions for potential extensions of the system that are not currently implemented.

This manual should therefore be considered essentially a progress report on the current status of the project. The authors intend to continue developing the system, and hope that interested organizations will do the same.

1.2 Processor Organization

As presently implemented, STRESS contains the core of a comprehensive system for the solution of structural engineering problems, and there is ample room for the development of components which precede, parallel, or follow those presently in the system. The capabilities and limitations of the present version of STRESS are described in Chapter 1 of the User's Manual.

The general flow of the process will be described in detail in Chapter 6. Essentially, the processor consists of three phases:

1. Phase Ia performs the input functions, that is, the decoding of statements and the storage of parameters, data, and logical flags. This phase handles the input both for an initial problem description and for modifications on subsequent reruns.

2. Phase Ib performs the editing, diagnostic, and compiling
    functions.
3. Phase II performs the actual execution of the operations
    specified or implied by the input statements.

Phase Ia is terminated after a SOLVE or SOLVE THIS PART
statement is decoded. This statement also initiates Phase Ib.
If the diagnostic of Phase Ib is successful, Phase II is automatically initiated. Upon completion of Phase II, control again trans-
fers to Phase Ia. At this point, several possibilities exist, depending on the order of the statements following the SOLVE THIS
PART statement:

1. a FINISH statement terminates processing of a job. The
    system is ready to accept the next job in sequence.
2. a SELECTIVE OUTPUT and one or more PRINT statements
    cause the output of the data specified by the statements in
    an interpretive fashion.
3. a MODIFICATION statement initiates action for the accept-
    ance of modifications to the problem previously analyzed.

The normal process, then, consists of cyclical alternations of
Phases Ia, Ib, and II until all the modifications have been processed. After the FINISH statement, the system again returns to
Phase Ia. Thus, an arbitrary number of separate problems can
be handled in sequence.
    The internal organization of the processor is built upon the
CHAIN subroutine of FORTRAN II. Phases Ia and Ib, as well as
the procedures for handling all normal and error returns from
Phase II comprise the first link. The various subroutines of the
execution phase are grouped into the subsequent links according
to their logical order. Wherever a bifurcation of paths exists,
the link currently in high-speed storage determines which chain
link is to be called in next. The detailed description of the system components making up the various links is given in Chapter 6.

1.3 Memory Usage

    The usage of memory adopted for the STRESS processor was
dictated by three considerations: (a) that there should be no arbitrary size limitations on the problems to be solved, (b) that core
memory be used as efficiently as possible, and (c) that large
problems be conveniently handled without sacrificing the efficiency
of solving small problems.
    To illustrate these points, consider the storage of the stiffness
matrices of the members in the structure. A FORTRAN statement DIMENSION ST(6, 6, 100) would reserve 3600 locations to

store the stiffness matrices of 100 space frame members. However, if the flexibility method were specified, this storage would have to be EQUIVALENCEd to another array, or would be unused. Furthermore, if a particular problem were a plane truss, only 4 of the 36 elements of a matrix would be involved, yet there would be no way of using the remaining 32 elements per matrix, or increasing the number of members that could be handled. Finally, if the system were designed primarily for large problems — with, say, only one stiffness matrix in high-speed storage at a time — smaller problems, which could fit all at once in core, would be heavily penalized.

The dynamic memory allocation algorithm was incorporated in the system to overcome these objections. In this scheme, all data pertaining to the problem being run are organized into arrays. The size of the arrays is completely determined from the input data for the problem. In the example just cited, the array of stiffness matrices is of size NB*JF*JF, where NB, the number of members, is specified by the NUMBER OF MEMBERS statement, and JF, the number of degrees of freedom, is determined from the TYPE statement.

The entire high-speed memory, with the exception of a small area used for parameters and the area occupied by the current program link, is considered as a pool available for the storage of arrays. Furthermore, the secondary storage (scratch tapes) is considered as a logical extension of the high-speed storage. Thus, the programmer writing a subroutine only has to <u>define</u> the size of the arrays he will generate, <u>allocate</u> the arrays he will need at any particular stage, and, <u>prior</u> to returning control to the calling program, <u>release</u> the arrays no longer needed. The allocation system will insure that all allocated arrays are in core, and supply a reference for addressing the elements of an array.

The dynamic memory allocation system is described in detail in Chapter 2. Admittedly, the use of this system removes a major programming facility of FORTRAN, that is, direct and multiple subscripting of arrays. However, the versatility and efficiency gained more than compensate for the small amount of additional programming.

The organization of memory is illustrated schematically in Figure 1.1. Two of the areas, "Current Program Link" and "Pool," have been described in Sections 1.2 and 1.3. The modified CHAIN program used in the system automatically sets the address of the bottom of the pool to the location immediately above the highest location used by the program link. The four areas at the top of memory are

1. <u>Working area.</u> These 108 locations can be used as working storage by the subroutines. In general, no information is transmitted through these locations from one subroutine to the next.

2. <u>System parameters.</u> This area of memory holds all parameters that indicate the status of the system, and that are essentially independent of the size and type of problem being handled.

3. <u>Problem parameters.</u> These parameters completely describe the problem being handled, and are used by essentially all subroutines.

4. <u>Codewords.</u> These parameters are associated with the data arrays described previously. Each array has an associated codeword, which completely describes the status of the array.

Working area 108_{10}
System parameters 50_{10}
Problem parameters 42_{10}
Array codewords 100_{10}
Pool
Current program link

Figure 1.1. Schematic layout of memory.

The function of all parameters and the makeup of each array are described in detail in Chapter 5.

Chapter 2

DYNAMIC MEMORY ALLOCATOR

This chapter describes the memory allocation procedure used in the STRESS system. The procedure is designed both to provide the flexibility needed in operating on a multitude of data arrays of varying sizes and to enable the programmer to gain access to the arrays with relative ease.

2.1 General Features of the Memory Allocator

The principal features of the memory control scheme are as follows:

1. Only the amount of storage needed for an array is used, as specified by the data.
2. All use of tape is automatic and need not concern the programmer.
3. Tape is used only when memory requirements exceed the available core space.
4. Space used by arrays not needed at any stage during processing is automatically available for other use.

In order to have this flexibility, the programmer must accept the following disadvantages:

1. Numerous calls to the memory control routines are needed to provide and control arrays.
2. Use of reference words is needed to gain access to the arrays.
3. Some sacrifice in multiple and automatic subscripting is required.

Memory control consists of automatic arrangement of programs and data in primary and secondary storage. The arrangement of programs in STRESS is made at development or compilation time by lumping series of subroutines into blocks. These blocks are called into use under program control. The IBM-FMS CHAIN subroutine is used to load program links in core memory. The

8

special data-control routine that has been written for STRESS
uses in a flexible manner the rest of core for data arrays.

2.2 Program Allocation

All data which are used by more than one subprogram, or are a
function of problem size, are placed in COMMON storage, that is,
outside the part of core that contains programs. The core storage available for data consists of all the space that is not used by
program. Figure 2.1 shows the arrangement of core in detail.

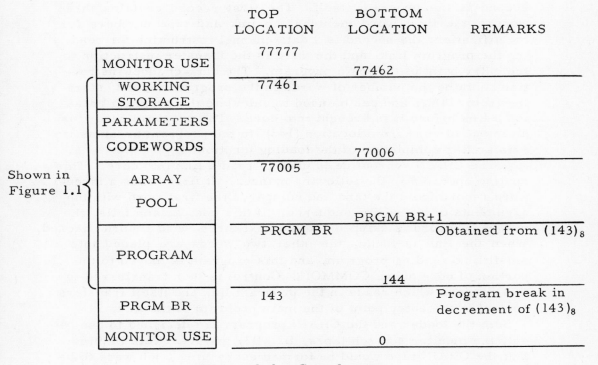

Figure 2.1. Core layout.

Locations 0 to $(143)_8$ are used by the FORTRAN monitor, as well
as $(77462)_8$ to $(77777)_8$. Programs are loaded upward from $(144)_8$,
and COMMON storage is allocated downward from $(77461)_8$. It
should be recalled that FORTRAN II stores arrays downward,
needing the highest location word address as a reference (base
address).

The highest location used by programs, called the program
break (PRGM BR), is a variable function of the programs in core
at any particular time. At running time the program blocks are
brought into core from tape with the CHAIN subroutine. Each

chain link is an executable set of programs, with a main program, and library subroutines; the chain link is logically terminated either with a call to EXIT or to CHAIN for loading another link. Parts of core not containing program are not destroyed as a result of reading a new program link into core.

At setup time the program blocks are placed on tape by the IBM BSS loader. CHAIN control cards, giving the link number and tape drive, delineate the links. All the programs after a CHAIN card are loaded into core together with their required library routines, thereby satisfying all transfer vector requirements. The loader then writes two records per link on the tape. The first record is for identifying and reading the link, and the second is the program itself. The first record contains three words: the first word includes the link and tape numbers for identification; the second is a data channel instruction for reading the program link; and the third is the transfer instruction to the entry point of the main program. The data-channel instruction contains the number of words in the program record, starting from $(143)_8$, and can be used to calculate the program break for a link before it is brought into core. The program record is an image of core from location $(143)_8$ to the program break as it exists after completion of the loading function.

The CHAIN subroutine is used to read a link into core. This routine operates in the following manner. It first reads a three-word record from the tape and compares the first word with the arguments of its calling sequence. If this comparison fails, the program record is skipped and the next three-word record checked. When the link is found, the other two words are placed into a small link-reading program, and this program is moved to the portion of core above COMMON. Control is then transferred to this reader, which reads in the program link and itself transfers control to the entry point of the main program of the link.

Both the loader and the CHAIN program are designed to use only the monitor scratch tapes, B2, B3, and A4. This implies that the CHAIN tape would be formed every time a job were to be run. This would, however, waste more than 3 minutes of 7094 time for each run. Forming the tape infrequently and reusing the CHAIN tape would therefore become a necessity. Reusing this tape on a monitor tape drive might inconvenience or restrict usage, so that modifications to the CHAIN subroutine would have to be made to enable a tape to be read from another tape drive. It was considered feasible to alter the CHAIN subroutine, but not the loader. Therefore, the CHAIN tape is formed on one drive (A4) using the loader, and it is executed on another drive (A5) employing a special version of CHAIN which disregards the tape number in the calling sequence when selecting the drive, using only the drive designated for execution (A5). A very short starter program, calling CHAIN for the first link, and a copy of the special version of CHAIN are sufficient to start execution on drive A5. No chain control cards are used with this starter deck.

Noting the contents of the first word of the identifying record on the CHAIN tape, the user must be cautioned in writing his calls to the CHAIN program, both within the links and in the starter program. The identifying word is formed by the loader and therefore contains the drive number on which the link was formed (A4). The calls to CHAIN for link N must be CALL CHAIN(N,A4) even though another tape drive will be used.

The dynamic memory allocation routine makes it possible to use all available core for data arrays. A problem exists in assuring that data arrays are not overwritten by program when bringing in a link longer than the previous link. There are two system parameters set and used by the allocation routines that are related to this condition. These are NT, the bottom of the pool for the link in core, and NL, the next location down from the top of the pool to be assigned for an array. (See Section 5.2.2.) A new value of NT must be found for the new block of programs, and this value checked against NL. If the new NT is greater than NL, the next link will overlap data. A memory reorganization is needed to eliminate the overlap by compacting memory towards the top of core. (See Section 2.5.)

Since a check for overlap of program and data must be done before a program link is read into core, and the size of the link is not known until its identification record is read into core, this check must be performed in the CHAIN subroutine. This modification has been added to the subroutine and is needed in all links. It is not needed, however, in the starter version since memory has not yet been used.

During debugging reuse of the tape on another drive may not be important. By making only the data-overlap modification to CHAIN, STRESS will operate using drive A4. This tape may also be restarted with a main program of CALL CHAIN(1,A4), taking the subroutine from the library. The subroutine FILES (see Section 2.6) normally uses tapes B1, B2, B3, and A4 for scratch tapes. For this use FILES must then be changed for execution with the chain tape on A4, say by changing the scratch tape from A4 to A5.

2.3 Memory Use for Data

If a program is to use core memory in a flexible manner so as not to restrict the program to a fixed arrangement of memory, the programming system must resort to indirect addressing and control functions. In the operating programs indirect addressing and control must be done with FORTRAN statements. It is this specification which determines the manner of memory use.

The control function in FORTRAN can be specified by CALL statements to the memory supervisor. In order to use an array the programmer must

1. Define necessary parameters about an array.
2. Specify a need for an array.
3. Signify completion of the need for an array.

These operations can be put in any order, so long as there is a define operation first. The FORTRAN calls are described in Section 2.4.

The only means of indirect addressing available with FORTRAN is in the use of subscripts. An array position can be specified by a subscript referenced to a fixed location (base address). The subscript is understood to refer always to the fixed location, so that its use is independent of what preceded or follows its reference. This subscript is a variable that is controlled by the STRESS allocator. Its value may change as a result of the memory-control process, but its treatment as a variable reference does not alter the array referencing. This subscript, however, requires direct referencing, so that it must also be assigned to a particular (fixed) location.

In order that individual subprograms be independent of changes in memory usage, their references (base addresses) must not change location because of additional references in other subprograms. For this reason the subscript reference is given a fixed location independent of other referencing, namely, the top of COMMON. A single-subscripted array is given this reference (U equivalenced with IU) and any array is referenced by its subscript as belonging to U (or IU for integer quantities). The array references are assigned a particular fixed location for each array in an area near the top of COMMON which will not be used for the arrays themselves. For a newly added array, a reference word is added to the COMMON statements of subsequent subprograms and to the list of COMMON usage (see Chapter 5). The subscripts of U do not start with 1, but with some variable, set by an initiation CALL START in order to locate all arrays below this fixed-assignment location area of COMMON.

The fixed location area at the top of COMMON is referred to in almost all subroutines in the processor. Thus use of this area must be consistent. For convenience, the area has been separated into blocks of words with each block containing words used for different purposes. It must be noted, however, that these are not restrictions on groups of words, since any use can be made of a location as long as it is consistent throughout all the programs. Figure 2.1 shows the division of storage.

The first 108 words are set aside as a working storage for

temporary use by small components of the processor. This area
can be used, for example, for small multiple-subscripted arrays.
The COMMON statements correspond to the usage for a particular
subroutine and are different in other subroutines. This area does
not have the same computational meaning in different parts of the
processor and therefore its value cannot be considered available
between the parts.

The remaining fixed locations in COMMON have the same meaning between the parts, as enumerated in Chapter 5. Although
separated into three parts — system parameters, problem parameters, and codewords — intermixing is permissible. The only
functional difference is that restoration of an initial problem
specification (reading a MODIFICATION OF FIRST PART statement) returns the information in the initial problem specification
to the locations from TOP $(77272)_8$ to the value of NL at a stage
near the end of Phase Ib. A few system parameters then are not
restored to the first-part status.

2.4 Codewords and Array Words

Array usage and control involves more than merely a subscript.
The word layout of FORTRAN II, however, allows the use of the
reference word for more than just storing subscripts. Therefore,
the reference word, designated a <u>codeword,</u> is used to store almost all the information about an array. All FORTRAN subscripting is performed using only the decrement of the word. All
computation of subscripts for an array uses only one codeword
and positive integer constants and variables. The only requirement on the arrangement of the codeword outside the decrement
would then be that the sign be plus when the codeword is used for
subscript computations.

s	1	2	3 17	18	19	20	21 35
p	q		F	r	u	v	N

Figure 2.2. Codeword schematic.

Each codeword consists of one memory word utilizing the full
36 bits of the word, and each contains almost all the necessary
information about the array that it references. Figure 2.2
shows the location of the information packed into the codeword.
The symbols indicate variables set by one or more of the memory organization processes. The letter F is used for subscripting
and may be altered as a result of the memory-control process.
The value of F is set by the allocator so that if the codeword is

s	1	2	3 17	18-20	21 35
s	u	v	N if M = 0		M

Figure 2.3. Array word schematic.

added to the subscript of an element in the array the result will be the subscript of that element in U or IU. The U(F) is the word before the start of the array.

Table 2.1. Meaning of Memory-Control Variables

Symbol	Word	Bit Position	Meaning
N	Codeword	21-35	Array size, number of storage locations
p	Codeword	s	Position: + when in core − when not in core (either on tape or not yet used)
q	Codeword	1	Redefinition indicator: 1 if array was redefined while on tape 0 if not
r	Codeword	18	Array content: 0 if data 1 if codewords
s	Array word	s	Release status: + needed − released
u	Codeword Array word	19 1	Core retention priority: 0 low 1 high
v	Codeword Array word	20 2	Erase status: 0 can be erased 1 must be saved
F	Codeword	3-17 2-17	Reference: Location-core (p = +) array subscript Location-tape (p = −) array file number
M	Array word	21-35	Backreference: Address of the codeword of associated array
J			Subarray number: J = 0 corresponds to codeword array

For the memory-reorganization process it is necessary to supply a referencing from the array to the codeword as well. This is termed the backreference and is placed in a word, called the array word, at U(F). It exists only for an array in core or, if an array has been moved, for a hole in memory. Figure 2.3 shows the form of the array word. If M = 0, no backreference exists. Then the decrement contains the size of the array that was in the hole. Thus, for an array of size N, N + 1 words are used in core with the array word occupying the highest location, U(F), and the Ith word, U(F + I). Table 2.1 defines all symbols used in the codewords and array words.

For many types of data, the number of arrays needed is a function of the problem size. These arrays are then really subarrays of a single array. All the subarrays together could be considered as a single array, but this implies that all subarrays would be in core when one of them is needed. This would overly restrict capacity. Alternatively, each subarray could be assigned a codeword and referenced individually. Since the fixed-location area of COMMON is not appropriate for this, the codewords for the subarrays must themselves constitute an array. This corresponds to a second level of indirect addressing and is the limit provided for with the present allocator.

2.5 Reorganization Process

As space for arrays are requested by programs using calls to the control routines, arrays are placed in sequence down from the top of the array data pool (N1) with NL, the next available space, updated each time. The symbol NT is the bottom of the pool. At the time an array of size N is needed, if NT + N > NL, then that array will not fit into core with its present arrangement. A memory reorganization must take place in order to fit the array into core.

The form of reorganization is the simplest consistent with the form of data, that is, a compaction of all needed (not released) arrays toward the top of the pool. All low-priority arrays not needed are put on tape or erased. High-priority arrays will be put on tape only if their space is needed to fit the array desired. Since the compaction process overwrites the top of the pool, it is ordered to proceed from the top of the pool downward, dealing with the arrays in the order they exist in core. This requires that the reorganization process deal with the arrays first and reference the codewords since the codewords are altered by the process. The array word provides the reference to the codeword and contains information used only for an array in core, namely, the release status s.

2.6 Programming with the Allocator

The FORTRAN CALL statements that are used in the programs
to control memory can now be described. For examples we will
consider two codewords, NAME and MEMB, where NAME refers
to a single array and MEMB to a group of N arrays, that is, second-level arrays. Because of the integer computations performed,
all codewords must have FORTRAN integer names. In usage, the
codeword and the array become synonymous.

Initialization of the allocator is accomplished with the entry

CALL START(TN, NCW, NMAX)

(TN and NMAX are no longer meaningful), where NCW is the number of words contained in the fixed-location area, that is, above
the array data pool. The word START is presently called whenever
the STRUCTURE statement is translated and NCW is set at 300.
These 300 words still contain about 60 unused words. Initialization consists of zeroing the NCW words, which process effec-
tively zeroes the data pool by destroying references, and setting
the top (N1), bottom (NT), and next available location (NL = N1)
of the pool. The word NT is set from PRGM BR in location $(143)_8$.

Definition of the array is accomplished with the order,

CALL DEFINE(NAME, N, r, u, v)

where NAME is the array name appearing in COMMON
N is the array size
r, u, v are either 0 or 1 as defined in Table 2.1.

If the codeword is zero, this array has not been used before in
the problem. For this case, p is set to minus and N, r, u, and v
are put into the codeword. If the codeword is not zero, one of
three processes is carried out: If p is minus, q is set to 1 and
N, r, u, and v are put in the codeword. If p is plus and N is unchanged, r, u, v are put into the codewords. If p is plus and N is
changed, the array is moved to the current bottom of the pool
(NL); appropriate references are changed; N, r, u, and v are put
into the codeword; and the array is either truncated or extended,
with the extension zeroed. A single exception is that if the array
is the bottom array in the pool, it need not be moved. It is important to note that redefinition may change F and, since moving
is a function of NT and NL, may even cause a memory reorganization, changing F in many codewords.

Array need is specified by the call

CALL ALOCAT (NAME)

If p is minus and F in the codeword NAME is zero, the array has not been used. In this case, space is allocated, the array word formed with a plus sign, F placed in the codeword, and the array zeroed. If p is plus, the array is in core, so only s in the sign of the array word is set to plus. If p is minus and F not zero, the array is on tape and is automatically brought from tape into core and the references established. If q was not zero, the array is read from tape differently and either truncated or the extension zeroed. If there is insufficient space in core for the array, a reorganization takes place before the array is allocated.

The array is now ready for use in core. To set the Ith element of the array equal to A or IA, the following statements are required:

$$IJ = NAME + I$$
$$U(IJ) = A \text{ or } IU(IJ) = IA \qquad (2.1)$$

When an array is temporarily or permanently not needed, this information can be specified with the call

CALL RELEAS(NAME)
                or
CALL RELEAS(NAME, u, v)

With this call the sign of the array word is set to minus and u and v placed in the array word. If u and v are given in the calling sequence, the codeword is also updated.

For second-level arrays, referencing must still be done through fixed-location codewords at the first level. Control operations must be performed on the first-level or codeword array as well as the data subarrays. The variable J indicates the subarray number, referenced by the Jth codeword in the codeword array. Operations on the codeword array are indicated by J = 0. The corresponding calls are

CALL DEFINE(MEMB, J, N, r, u, v)
CALL ALOCAT(MEMB, J)
CALL RELEAS(MEMB, J)
                or
CALL RELEAS(MEMB, J, u, v)

In order to define a subarray MEMB, J, the codeword array MEMB, 0 must have been previously defined with $N \geq J$ and allocated since the subarray definition operates on the words of the codeword array. Note that r = 1 can only be used for J = 0.

When the codeword array and a desired subarray are in core, (allocated), subscript references for the Ith element in the Jth subarray are computed as follows:

$$IJ = MEMB + J$$
$$IK = IU(IJ) + I \qquad (2.2)$$
$$A = U(IK) \text{ or } IA = IU(IK)$$

As shown in Statements 2.1, IJ is the subscript of the word in the first-level array, or, in this case, the subscript of the code-word for the data array. The codeword for the Jth data array is then IU(IJ), so that the second and third of Statements 2.2 become equivalent to 2.1.

It must be noted that any array allocation or redefinition may require a memory reorganization at the time the allocator is called. Since F's for all arrays in core change during a reorganization, all needed subscripts must be recomputed after a CALL ALOCAT or a redefining CALL DEFINE. In other words, subscripts cannot be carried beyond these call statements. Efficient organization can compensate for this restriction.

Additional restrictions implicit in the memory process are

1. The entire operation of altering codewords must be done only by calls to the allocator routines. This restriction arises from the interrelation of codewords and arrays. A codeword, for example, may not be set equal to zero to erase an array in core since the codeword is referenced during a memory reorganization from the array.

2. All calls to the allocator routines must reference the code-word directly, not a dummy temporary variable. This is necessary because the allocator routines alter the code-words and need the codeword address from the calling sequence for backreferencing.

Although a data array might be zeroed by setting each of its elements equal to zero, it might also be zeroed as follows:

CALL DEFINE(NAME, 0, 0, 0, 0),

and if previously released, then,

CALL ALOCAT(NAME)
CALL DEFINE(NAME, N, 0, 0, 1)

Redefinition for an array on tape involves changing the codeword and placing the flag q used for reading the array into core. If the array is on tape, the second of two successive redefinitions overwrites the first, which therefore has no explicit effect on the array elements. If the array is in core, both redefinitions provide for explicit actions, the first truncating the array, the second extending it and zeroing the extension. The allocation is to assure that the array is in core, so that the two redefinitions have explicit effects. Second-level data arrays can also be zeroed by either re-definition or by setting their elements to zero, but since a code-word implies referencing outside itself, a codeword array can only be zeroed by redefinition. The redefinition process will eliminate the references. It should be noted that such a zeroing of a code-word not only zeroes the second-level arrays but requires their redefinition.

An additional routine helpful for debugging will give an image of an array in all internal forms: octal, floating point, FORTRAN integer, and BCD. It is entered with

CALL DUMP(NAME)

or

CALL DUMP(MEMB, J)

Subprograms must, however, be recompiled after debugging to eliminate the intermediate dumps.

All tape handling is done in a single small FORTRAN subroutine called FILES. FILES contains logical tape numbers used for scratch tapes and computes the tape to be used from the file number. Some attempt is made to keep from having very short records on the tapes: through a form of buffering in the subroutine, files less than 254 words are combined into files of approximately this length.

In order to compute tape numbers and record numbers on the tape from the file number, a maximum number of files per record had to be set (at present it is 20). With four scratch tapes, the formulas for computing an internal tape number and record number for file L are

$$\left. \begin{aligned} NIT &= \frac{L-1}{20} + 1 - 4\left(\frac{L-1}{80}\right) \\ NIR &= \frac{L-1}{80} + 1 \end{aligned} \right\} \qquad (2.3)$$

where NIT is an internal tape number and NIR a FORTRAN record number on the tape. If the number of scratch tapes are to be changed, the only changes in FILES would involve the two arithmetic statement functions (Equations 2.3). For N scratch tapes, Equations 2.3 become

$$\left. \begin{aligned} NIT &= \frac{L-1}{20} + 1 - N\left(\frac{L-1}{N \times 20}\right) \\ NIR &= \frac{L-1}{N \times 20} + 1 \end{aligned} \right\} \qquad (2.4)$$

using integer arithmetic. The internal-external tape correspondence (NTAPE) would also have to be altered, and the limit on the tape initiation loop changed.

2.7 M. I. T. FMS Conventions

Logical-physical correspondence of tapes differs widely with installations. Logical tape numbers are used in only two subprograms in STRESS: FILES and DUMIO. The tape correspond-

ence in the present system is shown in Table 2.2. Scratch tapes
are B1, B2, B3, and A4. Input is A2 and output, A3. These can
be easily changed to suit any monitor system.

Table 2.2. M. I. T. Tape Correspondence

Logical	2	4	5	6	7	8	9
Physical	A3	A2	B1	B2	B3	A4	A5

   All STRESS BCD output is programmed with FORTRAN PRINT
statements, and input with the equivalent of the READ statement.
Output occurs from numerous programs, but input occurs only
from subroutine MATCH. (See Chapter 3.) The M. I. T. library
routines interpret the PRINT and READ statements as WRITE
OUTPUT TAPE 2 and READ INPUT TAPE 4. To avoid extensive
changes of source programs and recompilation for use with the
IBM library routines, a dummy routine DUMIO is added which
contains the entries to the PRINT and READ subroutines. The
action for each entry is to place the appropriate logical unit number in the accumulator and transfer to the corresponding tape
routine. This method also has the advantage of localizing tape
references.

Chapter 3

LANGUAGE TRANSLATION

The form of data input to the STRESS processor differs greatly
from most other computer programs. Format and ordering restrictions have been almost eliminated but are replaced by the
need to program explicitly the language translation. This chapter
describes the method of data input and translation used in STRESS
and gives some examples of how input flexibility can be programmed.

3.1 Nature of Input Data

The input technique used in the STRESS processor treats input
in a logical rather than physical form yet allows input in its general sense to be programmed in FORTRAN. For this purpose a
single FAP subroutine has been written to process input fields
logically and provide information about the input that can be
used by the FORTRAN program to control further processing and
data storage.

The nature of input does not require the rigidity of the FORMAT
statements needed with FORTRAN. It is more logical to have the
input form follow the same rules that govern natural language.
The logical rather than the physical composition of the data can
be interpreted during input as an alternative to FORMAT statements. The fact that a group of characters starts with a letter
is sufficient to recognize a word. Similarly, a number indicates
numerical data; a decimal point distinguishes a floating-point
number from an integer; and a blank or a comma after a group
of characters indicates the end of the group.

Input to STRESS deals with three types of information: BCD
words, floating-point numbers, and integers. Each of these
requires a separate form of conversion from card form to internal representation. Each implies different types of actions by
its presence. A number is a data item to be stored, and an integer normally signifies where the number is to be stored. A word

can be data to be stored, information about the type of data, or
more generally information about the input process. In most
cases, a word in BCD form has no meaning to an algebraic processor, and its meaning must be determined from some dictionary.
The simplest form of dictionary, used in STRESS, consists of a
list of acceptable words. The position of a word in a dictionary
characterizes its identity and is a logical quantity which can be
used for further computations.

With the input process controlled by decoding input fields,
statements may be processed in an arbitrary order. The amount
of data used for control is generally small. The input process
has, in its simplest form, a tree structure, with decoding of
statements proceeding up the tree. In this simple form each
branch is programmed separately to perform its operations,
process the input, and continue up the tree.

Character manipulation required by the free-field format can
be done only in a symbolic language. A single subroutine, written
in FAP, performs all input functions to the processor. The program reads a full card (72 columns) into a buffer in BCD form.
This buffer can then be searched for logical fields, that is, groups
of characters separated by a blank, comma, or the start or end
of the card. For numbers, the appropriate conversion from BCD
to binary can then be performed. The BCD word itself can be returned or it can be compared to a dictionary.

3.2 Specifications for Subprogram MATCH

The function-type subprogram MATCH is designed to be easily
usable with FORTRAN for operations on logical input fields. The
translation of a field identifies its form and meaning. Branching
on translated words can then be accomplished in FORTRAN with
the "computed GO TO" statement, with the control variable determined from the position of the translated word in the dictionary
list.

The subprogram MATCH is used by a FORTRAN translation program with the following form and argument definitions:

I=MATCH(J, K, M)

where:

M is the operation code.
I is the result code.

For

M = 0, read next logical field.
M = 1, start a new card, read first logical field.
M = 2 or 3, read the next number, skipping letters.
M = 4, skip one logical field (word).

M = 5, determine buffer reference address.

For M = 0, 1, or 4, J is a dictionary list name. The next logical field is isolated and if the field is a word, it is compared with the words in list J for correspondence. If the field is a number, it is converted. For M = 5, J is the subscript of U locating the buffer. The Ith word in the 12-word buffer can then be referenced as U(J+I) for printing or saving.

The symbol K represents the location where the results are stored. The form of the result depends on the value of I, that is, the form of the logical field, as follows:

I	Form of Field	Form of Results in K
1	6 consecutive blanks	unchanged
2	integer number	FORTRAN integer form
3	floating number	floating-point form
4	word, not found in list J	BCD form of first 6 characters of word
5	word, found in list J	FORTRAN integer for word location in list J
6	the word DO or DITTO	unchanged

For I = 3 and 4, floating-point FORTRAN names are needed for operations on the results, so that K must be equivalenced with a floating-point variable, say BK. If a word is found, only the first six characters are used.

Two fixed-location symbols are provided to designate two different card types. An "*" in card column 1 designates a comment card, which is echo printed, but otherwise ignored. A "$" in card column 1 designates the card as a continuation of the previous card. The usual test for the end of a statement is a check for I = 1. If the logical fields on a card carry beyond column 66, I = 1 is determined by finding no $ in column 1 of the next card. On the other hand, if M = 1 (read a new statement) is specified and a $ is found in column 1, this signifies a continuation error. To mark this error, I is set to 4 and K contains in BCD the word "$ERROR."

The form of the lists can be specified most efficiently in FAP. Their form, however, is sufficiently easy to understand, so that a programmer should have no difficulty altering them. The form of a typical list in FAP is

```
          ENTRY      LISTN
LISTN     DEC        10              (number of words in list)
          BCI        1, 0WORD1
          BCI        1, 0WORD2
           .          .
           .          .
           .          .
          BCI        1, WORD10
          END
```

For words of length six letters or more in the list, the word in
the list is started following the comma. For words of less than
six characters, leading zeroes are used so that the sum of zeroes
and characters equals six. Since the list name is a subroutine
name, and it is used as an argument in the calling sequence to
MATCH, the list name must appear on a FORTRAN F card in the
subroutine using the list. The F card performs the function of
placing the named variable location in the transfer vector list
rather than in variable storage.

   MATCH gives an echo print of each card as it is read. Error
messages associated with the translation of a statement can be
printed when detected and will therefore appear directly below
the statement in error for easy association.

3.3 Translation with MATCH

   Although MATCH provides a general input capability, versatile
translation requires extensive logical programming. Virtually
any type of input can be performed. Input to the STRESS processor can be separated into three types: words to direct the
translation process, words to identify the data, and finally the
data. The last two types might also perform the function of the
first. The data might be a code number constructed as a function
of the input words, or integer decimal numbers.

   The translation process consists of identifying a logical field,
operating on it according to its type and value, and branching
to some appropriate point to continue. For example, a series of
about 15 statements at the start of subroutine PHAS1A read the
first field in a new statement. The input specifications allow a
series of decisions depending on the outcome. If six consecutive
blanks are detected, the next logical field in the statement is
examined as if it were the first. This follows from the specification that a statement can be started anywhere on a card. In order
that blank cards be allowed, it is necessary to count the number
of words of blanks. Integer numbers as a first word are acceptable only if the statement follows a tabular header or another
table entry. A table indicator (ITABLE) can then be used for
branching. Starting a statement with a decimal number is an
error, as is a word not contained in a list LIST1 of acceptable
first words. A word found in the list is used for branching. The
tabular mode is terminated as soon as a word is found starting
a statement. The word position is saved for possible later reference, as detection of DO or DITTO causes branching on the
last branch saved. In the following program illustrating this input specification for starting statements, statement numbers in
the 90's are used for printing error messages.

```
C        START NEW CARD
   10 J=0
         IX=0
         M=1
C        READ FIRST FIELD
   11 I1=MATCH(LIST1,K,M)
         GO TO (12,14,97,94,16,17), I1
C        BLANK FIELD, TRY AGAIN
   12 IX=IX+1
         M=0
         IF(IX-12) 11,10,10
C        INTEGER–SAVE AS JOINT OR MEMBER NUMBER
   14 J=K
C        IF IN TABULAR MODE, GO TO JOINT OR
C        MEMBER INPUT
         IF(ITABLE-1) 98,100,200
C        STOP TABULAR MODE
   16 ITABLE=0
C        SAVE BRANCH NUMBER
         K1=K
C        BRANCH TO CONTINUE PROCESSING
   17 GO TO (100,200,...), K1
```

The first word in a statement defines at least partially the statement type. The branch on K1 goes to program parts that decode the appropriate types of statement. In order to keep the "computed GO TO" statements from getting too long, the dictionary is divided into several lists according to the association of possible words in a translation process. All permissible first words appear in LIST1. There is another list for possible third words in the NUMBER statement. The second word in the NUMBER statement is skipped using M=4 in the call to MATCH.

The subroutine READ is used to read much of the numeric and alphabetic member and joint data. Alphabetic information is formed into a code computed from the positions of the words in the list used. Whenever a set of words representing a code is translated, the code is stored in an array LABL and a storage counter tabulating the number of codes incremented. When a decimal number is read, it is temporily stored in an array BETA and another storage counter incremented. The entry in the array LABL identifies the numeric data for the corresponding elements in the array BETA. The elements in LABL can be used for storing the corresponding elements of BETA in a data array. The label numbers in LABL(K) are computed to give the position in the data array for storage of the Kth element of BETA, which is stored in an array, for example, NDUM, by

J=NDUM+LABL(K)
U(J)=BETA(K)

If labels are not given, the array LABL is filled with a sequential set of numbers, thus giving the alternative of fixed-order or identified data (see Section 2.7 in STRESS: A User's Manual). For alphabetic data, the codes in the array LABL are the input results, which can then be further processed.

As an example, consider the translation of the data or labels FORCE X, FORCE Y, FORCE Z, MOMENT X, MOMENT Y, and MOMENT Z. Assume LIST6 is made up as follows:

```
        ENTRY  LIST6
LIST6   DEC    5
        BCI    1, 00000X
        BCI    1, 00000Y
        BCI    1, 00000Z
        BCI    1, 0FORCE
        BCI    1, MOMENT
        END
```

The following set of statements will place code numbers 1 through 6 for the labels in the order just given above:

```
10   IJ=0
     L=0
12   I1=MATCH(LIST6, K, 0)
     GO TO (30, 100, 200, 300, 14, 95), I1
14   GO TO (20, 20, 20, 16, 18), K
16   IJ=0
     GO TO 12
18   IJ=3
     GO TO 12
20   L=L+1
     LABL(L)=IJ+K
     GO TO 12
30   Subsequent processing
```

Note that a series of 6 blanks is used to determine completion. Then L shows the number of labels formed. In addition, flexibility in input form is produced: the words FORCE and MOMENT need not be repeated to form labels, so the user could write FORCE X Y Z. Obviously, the preceding program segment places in sequence the code numbers for any arbitrary sequence of labels.

3.4 User Modification of Language

Any modification of the STRESS processor by users will most likely involve modifications of the input phase. A series of possible types of modifications is briefly discussed. Sample extensions of the processor are given in Chapter 7.

The easiest type of modification consists of changes in the vocabulary. If a firm is used to particular terms and wishes to retain them, only the appropriate entries in the dictionary lists need be changed. For convenience all the lists are included in a single subroutine with multiple entries. By threading through the input programs in the statement-translation phase, the place can be found where the word to be changed is read. This place will disclose which list is used; the word to be changed can be located in this list. Only the subprogram of the list need be recompiled (reassembled).

To add a new statement type, the first word in the new statement must be added to LIST1 and a branch to its decoding added to the branch on first words. If the decoding will use parts of other decoding routines, appropriate logical constants may have to be introduced for branching. The new decoding portion may consist of a series of calls to MATCH, looking and testing on certain forms of logical fields and storing data for subsequent processing. Existing translation programs can serve as a guide to writing new parts.

If new arrays are needed for data storage or subsequent processing, their names (codewords) must be included in the COMMON statements in the subroutines referring to the names. Locations in COMMON already used are enumerated in Chapter 5. Appropriate array definition and allocation are required before storage of data. The definition function may require an additional statement of the size-descriptor type. The array should be released while not needed.

To add additional branches on existing statement types, the procedure is similar to new statement types, but branching is deferred to later word translation. For additional size descriptors, size-descriptor names have to be added to the list. If the statement form is not different from the other size descriptors, then the translation process is unchanged. At present all branching on processes necessitated by translation of a NUMBER statement is done in the subroutine SIZED. A new branch must then be added in the subroutine SIZED to perform the processes required by the translation.

Chapter 4

METHOD OF SOLUTION

The present STRESS program performs the static, linear analysis of elastic lumped-parameter structures. This structural analysis problem involves the setting up and the solution of a set of linear algebraic equations. The present chapter discusses the mathematical formulation of the structural analysis problem as it is implemented in the STRESS processor. The formulation is presented in matrix notation: only rectangular Cartesian space coordinates are used. A prereading of "A Network Topological Formulation of Structural Analysis" by S. J. Fenves and F. H. Branin, Jr.,[1] is essential to understanding this chapter.

4.1 The Network Concept

At present STRESS performs structural analysis solely by the stiffness method, treating the displacements as unknowns. This is no restriction on the class of problems than can be handled. Since certain structural forms could be analyzed more efficiently by the flexibility method, it is hoped that this alternative method will be implemented in the system as well. Many of the solution steps could be used as presently programmed.

The computational procedure of the structural analysis in STRESS is based on a network interpretation of the governing equations, the principal feature of which is the clear segmentation in processing of the geometrical, mechanical, and the topological relationships and properties of the structure. This allows a concise and systematic computational algorithm that is equally well applicable for different structural types such as plane or space structure, pinned or rigid member connections.

In "A Network Topological Formulation of Structural Analysis,"

---

[1] S. J. Fenves and F. H. Branin, Jr., "A Network Topological Formulation of Structural Analysis," J. Struct. Div., ASCE, ST4, 483 (August 1963).

a detailed description of this formulation is given for the homo-
genous structural type (that is, either all members rigidly connected or all members pinned). Only a summary of important
relationships is therefore reproduced here.

A somewhat more detailed description of the procedure used
to include local force releases (for example, hinges, rollers)
for members and support joints is given in Section 4.2.

One essential difference of procedure used in the STRESS processor from that in Fenves and Branin's paper should be pointed
out: for reasons of numerical accuracy the single global-coordinate system has been replaced by a series of joint-coordinate
systems, all oriented in the same direction but each having its
own origin at a joint. The only effect of this on the general formulation is that the transformation T of stiffnesses from local
to global (joint) coordinates involves only a rotation and that the
entries of the incidence matrix A become $[-\tau], [I], [0]$ rather than
$[-I], [I], [0]$. The submatrix $[\tau]$ is the displacement translation matrix in joint coordinates from the start to the end of the member.The
origin of each member-coordinate system is located at the end
of the member. All quantities referring to member coordinates
are indicated with a superscript asterisk.

We now consider a linear elastic structure consisting of slender
members connecting the joints of the structure.

The statement of the problem is given as follows:

Given: 1. A structure that determines the topological
                        matrix $\overline{A}$ and the geometric transformations
                        T and T^{-1} between member and joint coordi-
                        nates. The branch-node incidence matrix $\overline{A}$
                        is defined as follows: the element (submatrix)
                        $\bar{a}_{ij}$ is $[-\tau]$, $[I]$, or $[0]$, depending on whether
                        the ith branch (member) is positively, nega-
                        tively, or not connected to joint j.
                    2. The primitive-flexibility or stiffness matrices
                        F^* or K^* of all the members in member coordi-
                        nates, or the necessary mechanical and geo-
                        metrical properties of the members, so that
                        F^* and/or K^* can be computed.
                    3. The applied member distortions U^* in mem-
                        ber coordinates and the applied joint loads P'
                        in joint coordinates. STRESS accepts a larger
                        class of input loads, all of which are processed
                        into equivalent contributions to U^* and P'.

Find: The induced member forces R^* and distortions V^*
                    such that in joint coordinates
                    1. $KV = R$ (4.1)
                    2. $A^t R = P'$ (4.2)

where the following relationships hold:

$$u = Au' \qquad (4.3)$$
$$V = U + u \qquad (4.4)$$
$$R = P + p \qquad (4.5)$$

where   U = applied member distortions
               V = induced member distortion
               R = induced member forces
               u' = joint displacements
               u = total member distortion
               P = applied member forces
               p = total member forces

We now introduce the joint displacements u' as variables

$$V = U + Au' \qquad (4.6)$$
$$R = KV = KU + KAu' \qquad (4.7)$$
$$P' = A^t R = A^t KU + A^t KAu' \qquad (4.8)$$

This is the governing set of linear algebraic equations, which is solved for u'

$$u' = (A^t KA)^{-1}(P' - A^t KU) \qquad (4.9)$$

After we have found u', backsubstitution gives the induced member distortions and forces as

$$V = U + Au' = A(A^t KA)^{-1} P' + (I - A(A^t KA)^{-1} A^t K)U$$
$$\qquad (4.10)$$
$$R = KV \qquad (4.11)$$

4.2 Local Releases

The formulation given by Equations 4.1 through 4.11 assumes that the structure is homogenously connected in the sense that all members have rigid connections to the joints in the directions compatible with the structural type. (For a frame this means rigid connections in all directions, and for a truss a rigid axial connection.)

If a structure deviates locally from this standard condition it must be so stated by specifying releases. A release is associated with a particular force component and implies that this component is independent of the deformations of the structure (that is, it is prescribed). A joint release relates to a reaction component at a support joint, and a member release relates to a member-force component at a given point along the axis of the member.

4.2.1 Member releases. The load-deflection relationship for the unreleased member in local-member coordinates is

$$R^* = K^* V^*$$

(4.12)

For the released member we assume that a given set of components of the member-force vector R_r at R is prescribed to be zero (any fixed value could be used, but generally would not be meaningful). We consider here only the cases where the origin of the release system is located either at the start A or at the end B of the member. The procedure described could be extended to include releases within a member. Such cases can, however, always be reduced into the situation where R is either at A or B by introducing a fictitious joint at R.

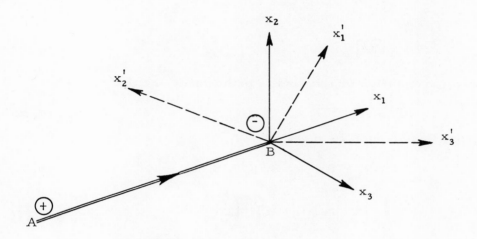

Figure 4.1. Member and member-release coordinates.

The release conditions are assumed to be given in the member-release coordinate system x_1', x_2', x_3' shown in Figure 4.1, which is related to the member coordinate system at B by

$$V^* = T_r V_r$$
$$R^* = (T_r^{-1})^t R_r$$

(4.13)

where T_r is the transformation matrix between the member- and the release-coordinate system. The subscripts r refer to the release system. We derive an equation of the type of Equation 4.12 that is valid for the released member. Equations 4.13 are written in the release systems as

$$R_r = K_r V_r$$

(4.14)

where

$$K_r = T_r{}^t K^* T_r \tag{4.15}$$

we define

$$\tilde{V} = \Lambda V_r = (\Lambda^t)^{-1} V_r \tag{4.16}$$

$$\tilde{R} = \Lambda R_r \tag{4.17}$$

where

$$\tilde{R} = \left[\dfrac{\tilde{R_1}}{0}\right] \tag{4.18}$$

and

$$\Lambda \equiv \left[\dfrac{\Lambda_1}{\Lambda_2}\right] \tag{4.19}$$

is a permutation matrix and is orthogonal; that is,

$$\Lambda^{-1} = \Lambda^t \tag{4.20}$$

from Equations 4.16 and 4.14

$$\tilde{R} = \Lambda R_r = \Lambda K_r V_r = \Lambda K_r \Lambda^t \tilde{V} \tag{4.21}$$

$$\left(\dfrac{\tilde{R_1}}{0}\right) = \left[\dfrac{\Lambda_1}{\Lambda_2}\right] K_r (\Lambda_1{}^t \,|\, \Lambda_2{}^t) \cdot \left(\dfrac{\tilde{V_1}}{\tilde{V_2}}\right) \tag{4.22}$$

$$\left(\dfrac{\tilde{R_1}}{0}\right) = \begin{bmatrix} \tilde{K}_{11} & \tilde{K}_{12} \\ \tilde{K}_{21} & \tilde{K}_{22} \end{bmatrix} \cdot \left(\dfrac{\tilde{V_1}}{\tilde{V_2}}\right) \tag{4.23}$$

The definitions of $K_{11}\ldots K_{22}$ are implied by Equations 4.22 and 4.23. Solving Equation 4.23 for $\tilde{R}_1$ while eliminating $\tilde{V}_2$ gives

$$\tilde{R}_1 = (\tilde{K}_{11} - \tilde{K}_{12}\tilde{K}_{22}{}^{-1}\tilde{K}_{21})\tilde{V}_1 \tag{4.24}$$

If back permutation is effected, we get for R_r

$$R_r = \Lambda_1{}^t (\tilde{K}_{11} - \tilde{K}_{12}\tilde{K}_{22}{}^{-1}\tilde{K}_{21})\Lambda_1 V_r \tag{4.25}$$

Equation 4.25 is the modified load-deflection relationship for the released member in released coordinates. The inverse transformations of Equation 4.13 brings us back to member coordinates.

$$R^* = (T_r{}^{-1})^t \Lambda_1{}^t (\tilde{K}_{11} - \tilde{K}_{12}\tilde{K}_{22}{}^{-1}\tilde{K}_{21})\Lambda_1 T_r{}^{-1} \cdot V^* \tag{4.26}$$

or

$$R^* = \overline{K}^* v^*$$

where

$$\overline{K}^* = (\Lambda_1 T_r^{-1})^t (\widetilde{K}_{11} - \widetilde{K}_{12} \widetilde{K}_{22}^{-1} \widetilde{K}_{21}) \Lambda_1 T_r^{-1} \qquad (4.27)$$

The effective stiffness for the released member is $\overline{K}^*$.

If releases are prescribed at both ends of the member, the procedure just described can be applied serially to the two release groups.

If a number of such release groups is given such that an unstable member is implied, this fact will become automatically apparent, in that the matrix $\widetilde{K}_{22}$ will be singular if the releases that are about to be incorporated would render the member unstable.

4.2.2 Fixed-end forces. The prescribed releases not only affect the member stiffness but also the fixed-end force vectors resulting from applied member loads.

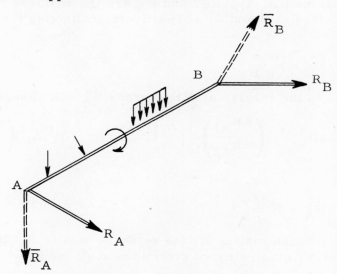

Figure 4.2. Definition of fixed-end force vectors.

We next describe a procedure to compute the modified fixed-end forces. The stiffness of the member without releases is K^* and the corresponding fixed-end force vectors are R_A^* and v_B^*. (Figure 4.2). We assume that R_B^* has been computed from

$$R_B^* = K^* v_B^* \qquad (4.28)$$

where v_B^* is the cantilever deflection vector at end B resulting from the applied loads. From statics

$$R_A^* = F_{CA} - T^{-1}R_B^* \tag{4.29}$$

where F_{CA} is the cantilever force vector at A resulting from the given loads.

It is immediately clear from the definition of K^* and R^* that in the case where there is a release at B only we obtain

$$\bar{R}_B^* = \bar{K}_B^* v_B^* \tag{4.30}$$

$$\bar{R}_A^* = F_{CA} - T^{-1}\bar{R}_B^* \tag{4.31}$$

The case where the member start A is released only can be handled by analogy if $\bar{K}_A^*$ and v_A^* are available.

If both ends of the member have released force components, a special procedure is necessary because then the deflection vector v_B^* does not correspond to the actual support conditions of the member at A.

We assume that $\hat{K}_B^*$, $\hat{R}_B^*$, and $\hat{R}_A^*$ are the stiffness matrix and force vectors at B and A, respectively, reflecting the releases at B but not those at A.

$$\hat{R}_B^* = \hat{K}_B^* v_B^* \tag{4.32}$$

Then due to the release at A the forces $\hat{R}_A^*$ will change by ΔR_A^*:

$$\Delta R_A = \Lambda \Delta R_A^* \Lambda^T = \begin{pmatrix} \Delta R_{A1} \\ \hline \Delta R_{A2} \end{pmatrix} = \begin{bmatrix} \Lambda_1 \\ \hline \Lambda_2 \end{bmatrix} T^{-1}\hat{K}_B^* T^{-1t}(\Lambda_1^t | \Lambda_2^t)\begin{pmatrix} v_{A1} \\ \hline v_{A2} \end{pmatrix} \tag{4.33}$$

Let

$$\hat{K}_A^* = T^{-1}\hat{K}_B^* T^{-1t} \tag{4.34}$$

The permutation matrix for the releases at A is Λ. If v_{A2} and ΔR_{A2} refer to the released directions at A, we have

$$v_{A1} = 0$$

$$\Delta R_{A2} = -\Lambda_2 R_A^* \tag{4.35}$$

The final fixed-end forces become

$$\bar{R}_A^* = \hat{R}_A^* + \Delta R_A^*$$

$$\bar{R}_B^* = F_{CB} - T\bar{R}_A^* \tag{4.36}$$

Rewriting Equation 4.33 in partitioned form and solving it for ΔR_{A1} by elimination of v_{A2}, we find

$$\Delta R_{A1} = -\Lambda_1 \hat{K}_A^* \Lambda_2^{\ t} \left(\Lambda_2 \hat{K}_A^* \Lambda_2^{\ t} \right)^{-1} \Lambda_2 \hat{R}_A^* \tag{4.37}$$

or

$$\Delta R_A^* = \Lambda^t \begin{pmatrix} \Delta R_{A1} \\ \Delta R_{A2} \end{pmatrix} = (\Lambda_1^{\ t} | \Lambda_2^{\ t}) \left[\frac{\Lambda_1 \hat{K}_A^* \Lambda_2^{\ t} \left(\Lambda_2 \hat{K}_A^* \Lambda_2^{\ t} \right)^{-1} \Lambda_2}{-\Lambda_2} \right] \hat{R}_A^* \tag{4.38}$$

If we define

$$I' = \Lambda_1^{\ t} \Lambda_1$$

$$I'' = \Lambda_2^{\ t} \Lambda_2 = I - I' \tag{4.39}$$

It follows that

$$\bar{R}_A^* = I' \left[I - \hat{K}_A^* \Lambda_2^{\ t} \left(\Lambda_2 \hat{K}_A^* \Lambda_2^{\ t} \right)^{-1} \Lambda_2 \right] \hat{R}_A^* \tag{4.40}$$

$\bar{R}_B^*$ can then be found from statics. It follows that

$$\bar{R}_B^* = \bar{K}^* v_B^* + T\Omega F_{CA} \tag{4.41}$$

where

$$\bar{K}^* = T[I' - I' \hat{K}_A^* \Lambda_2^{\ t} \tilde{K}_{22}^{\ -1} \Lambda_2] \hat{K}_A^* T^t \tag{4.42}$$

By using the definition of I' one can prove that Expression 4.42 is equal to Expression 4.42a, which shows that the first term of Equation 4.41 is the same as expression 4.26.

$$\bar{K}^* = T[I' - I' \hat{K}_A^* \Lambda_2^{\ t} \tilde{K}_{22}^{\ -1} \Lambda_2] \hat{K}_A^* I' T^t \tag{4.42a}$$

Also

$$\Omega = I'' + \Lambda_1^{\ t} \tilde{K}_{12} \tilde{K}_{22}^{\ -1} \Lambda_2 \tag{4.42b}$$

where $\tilde{K}_{12}$ and $\tilde{K}_{22}$ are given by

$$\tilde{K}_{12} = \Lambda_1 \hat{K}_A^* \Lambda_2^{\ t} \tag{4.43}$$

$$\tilde{K}_{22} = \Lambda_2 \hat{K}_A^* \Lambda_2^{\ t} \tag{4.44}$$

It is noted that all matrix operations involving multiplication with parts of a permutation matrix can be carried out on the computer by inspection rather than by actual multiplication.

4.2.3 Joint releases. The term joint release as defined here applies to support joints only. It differs from a member release at the same point (if more than one member is incident to this joint) in that the joint release requires only that the sum of all member forces from all the members and external applied loads in the released direction be zero. Each individual member may still carry forces in this direction (Figure 4.3).

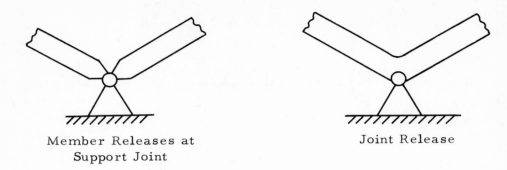

Member Releases at   Joint Release
Support Joint

Figure 4.3.   Distinction between member and joint releases.

The stiffness formulation for the complete structure, including support joints is

$$\begin{bmatrix} K_{11}' & K_{12}' \\ K_{21}' & K_{22}' \end{bmatrix} \begin{pmatrix} u' \\ \bar{\bar{u}}' \end{pmatrix} = \begin{pmatrix} P' \\ \bar{\bar{P}}' \end{pmatrix} \tag{4.45}$$

where $\bar{\bar{u}}'$ and $\bar{\bar{P}}'$ are the displacements and joint forces at the support joints.   The elements of the stiffness matrix are defined as

$$K_{11}' = A^t K A$$
$$K_{21}' = \bar{A}^t K A$$
$$K_{12}' = A^t K \bar{\bar{A}} \tag{4.46}$$
$$K_{22}' = \bar{\bar{A}}^t K \bar{\bar{A}}$$

In Equation 4.46, A refers to the portion of the augmented incidence matrix involving free joints,[2] while $\bar{A}$ refers to the portion

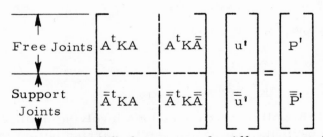

Figure 4.4.   Modified structural stiffness matrix.

involving support joints.   This partitioning of the structural matrix can be represented as is shown in Figure 4.4.

We apply a permutation to Equation 4.45 such that the equations associated with the released directions are moved to the bottom

---

[2] See Fenves and Branin, op. cit.

and also rotate the equations in the coordinate system of the joint releases by

$$\bar{\bar{u}}'_p = (\Lambda R)\bar{\bar{u}}' \tag{4.47}$$

where Λ is the (orthogonal) permutation matrix and R is the rotation matrix R_{CO}, from the release system to the joint-coordinate system. We define

$$Q = (\Lambda R) = \left[\frac{Q_1}{Q_2}\right] \tag{4.48}$$

$$\bar{\bar{u}}'_p = \begin{pmatrix} \bar{\bar{u}}'_{P1} \\ \hline \bar{\bar{u}}'_{P2} \end{pmatrix} \tag{4.49}$$

where u'_{P1} are the prescribed joint displacements at support joints in nonreleased directions.

$$\bar{\bar{P}}' = \begin{pmatrix} \bar{\bar{P}}'_{P1} \\ \hline \bar{\bar{P}}'_{P2} \end{pmatrix} = Q\bar{\bar{P}}' \tag{4.50}$$

Here P'_{P2} is prescribed and represents the applied joint forces at released supports in released directions. We can now further partition Equation 4.45 by introducing Expressions 4.49 and 4.50:

$$\begin{bmatrix} K_{11} & K_{12} & K_{13} \\ K_{21} & K_{22} & K_{23} \\ K_{31} & K_{32} & K_{33} \end{bmatrix} \begin{pmatrix} u' \\ \bar{\bar{u}}'_{P1} \\ \bar{\bar{u}}'_{P2} \end{pmatrix} = \begin{pmatrix} P' \\ \bar{\bar{P}}'_{P1} \\ \bar{\bar{P}}'_{P2} \end{pmatrix} \tag{4.51}$$

where

$$K_{11} = A^t K A$$
$$K_{12} = K_{21}{}^t = A^t K \bar{\bar{A}}(\Lambda_1 R)^t$$
$$K_{13} = K_{31}{}^t = A^t K \bar{\bar{A}}(\Lambda_2 R)^t$$
$$K_{23} = K_{32}{}^t = \Lambda_1 R \bar{\bar{A}}^t K \bar{\bar{A}}(\Lambda_2 R)^t$$
$$K_{33} = \Lambda_2 R \bar{\bar{A}}^t K \bar{\bar{A}}(\Lambda_2 R)^t \tag{4.52}$$

Solving (4.51) for u' by elimination of $\bar{\bar{u}}'_{P1}$, we obtain

$$u' = \mathcal{K}^{-1}\mathcal{P}' \tag{4.53}$$

where

$$\mathcal{K} = K_{11} - K_{13}K_{33}{}^{-1}K_{31} \tag{4.54}$$

$$\mathscr{P}' = P' - K_{13}K_{33}^{-1}P_{P2} - (K_{12} - K_{23}K_{33}^{-1}K_{32})\,\bar{\bar{u}}'_{P1} \qquad (4.45)$$

Replacing $\bar{\bar{P}}'_{P2}$ and $\bar{\bar{u}}'_{P1}$ by the corresponding quantities in joint coordinates, we arrive at

$$\mathscr{P}' = P' - K_{13}K_{33}^{-1}\Lambda_{2}R\bar{\bar{P}}' - (K_{12} - K_{23}K_{33}^{-1}K_{32})\Lambda_{2}R\bar{\bar{u}}' \qquad (4.56)$$

Equation 4.56 describes the effective joint loads $\mathscr{P}'$ on the free joints, including the forces applied in the direction of released support components and the displacements in the direction of unreleased support components.

4.3 Implementation in Processor

The methods outlined in Sections 4.1 and 4.2 are well suited for application to a digital computer program. They allow a general program to be written in concise form and the data to be stored efficiently. The same program can handle different types of structures, and only the minimum of needed information is stored. For example, only the nonzero elements of the incidence matrix $\overline{A}$ or the stiffness matrix $\overline{A}^t K \overline{A}$ are stored, and, where possible, logical submatrices of up to 36 elements are stored only symbolically as a single binary bit. (See also Section 6.5.7.)

The whole solution process thus becomes literally a sequence of file-processing operations; that is, a given array of numbers is operated on to generate a new array, then further operations are done on this new array, etc. The solution process (execution phase) can be summarized in these steps:

1. Generate a table of member stiffness matrices for the nonreleased structure. This is done by computing first the member flexibility F^* (F^* = cantilever deflection due to a unit-force vector at the end of the cantilever) and then obtain the stiffness as

    $$K^* = F^{*-1}$$

    where K^* is stored in KMKST.
2. If there are any member releases, the K^* table is modified to $\overline{K}^*$ according to the procedure given in Section 4.2 for member releases.
3. From the raw data for the loads, for each loading condition the total equivalent force vectors P' and $\bar{P}'$ are generated, comprising the load vectors for the governing joint-equilibrium equations. STRESS accepts five different types of input loads:

a. Joint loads
b. Member-end loads
c. Member loads within the member
d. Member distortions
e. Support-joint displacements

In the kinematically determinate structure (which is considered during step 3) the different load types give contributions to various data arrays kept internally. These arrays are the following: the member-end loads at the start and the end of the members, P^+ and P^- (arrays KPPLS and KPMNS); the member distortions U (array KUV); and the effective joint loads P' and $\bar{\bar{P}}'$ at the free joints and at the support joints, respectively (arrays KPPRI and KPDBP).

The mathematical representation of the processing steps required for the different load types are shown as follows:

a. Joint loads
Store directly in P' or $\bar{\bar{P}}'$.

b. Member-end loads
Convert the given member-load components from member coordinates to joint coordinates.
Add contribution to P^+ and/or P^-
Compute contribution to P' as

$$\bar{P}' = \bar{A}^t P^{\pm}$$

$$\bar{P}' = \left(\frac{P'}{\bar{\bar{P}}'}\right), \quad \bar{A}^t = \left[\frac{A^t}{\bar{A}^t}\right]$$

c. Member loads
Compute equivalent member-end forces and proceed as in b.

d. Member distortions U^*
Rotate to joint coordinates at end joint.
Add contribution to U.
Compute contribution to $P^{\pm}$:

$$P^{\pm} = KU$$

Compute contribution to $\bar{P}'$:

$$\bar{P}' = \bar{A}^t P^{\pm} = \bar{A}^t KU$$

e. Support-joint displacement u'
Compute resulting member distortions

$$U = \bar{A}\left(\frac{0}{\bar{\bar{u}}'}\right) \quad \text{and add these to U.}$$

Contributions to $P^{\pm}$:

$$P^{\pm} = KU = K\overline{A}\left(\frac{0}{\overline{\overline{u}}'}\right)$$

Contributions to $\overline{P}'$:

$$\overline{P}' = \overline{A}^t P^{\pm} = \overline{A}^t KU = \overline{A}^t K\overline{A}\left(\frac{0}{\overline{u}'}\right)$$

4. Rotate all member stiffnesses to joint coordinates of the member-end joints.

    $$K = R\overline{K}^* R^t$$

    where K is stored in table form in the second-level array KSTDB.

5. Generate the structural stiffness $\overline{A}^t K\overline{A}$ from the incidence table $\overline{A}$ and the stiffness table K. $\overline{A}^t K\overline{A}$ is stored in the arrays KDIAG and KOFDG, using the arrays IFDT and IOFDG as bookkeeping arrays. Only nonzero elements of the lower left half of matrix $\overline{A}^t K\overline{A}$ are stored in KOFDG.

6. If there are joint releases, the stiffness tables KDIAG, KOFDG, IOFDG, IFDT ($\overline{A}^t K\overline{A}$), and the load vectors KPPRI (P') are modified according to the procedure of Section 4.2 for joint releases. If there are no free joints, this operation yields all displacement components at the released supports.

7. If there are any free joints, solve the final governing joint-equilibrium equations for the free-joint displacements u', reflecting member as well as joint releases. The resulting u' are stored in KPPRI.

8. Backsubstitution. From the free- and the support-joint displacements compute the induced member distortions and member-end forces according to Equations 4.10 and 4.11 and add them to the contributions in the kinematically determinate structure.

The detailed execution of the individual steps is described in Section 5 of Chapter 6, as well as in the flow charts of the corresponding programs.

All data arrays, such as the displacement vector u' or the member stiffnesses K^* are arranged in storage (either in core or in secondary storage) in table form, either as first-level or second-level data arrays, using the procedures described in Chapter 2. The matrix operations as they appear in the formulation of the problem can thus be done largely by bookkeeping and inspection, rather than formally. For example, the matrix $\overline{A}$ is stored only as two incidence tables giving the start and end joint of each member (JPLS, JMIN). Similarly, the "matrix product" $\overline{A}^t K\overline{A}$ is not formally computed but is formed by bookkeeping.

First, only the nonzero submatrices on the diagonal of the complete matrix K are stored in a second-level data array KSTDB.
The lower left half of the matrix $\overline{A}^t K \overline{A}$ is then formed by means
of one loop on the members, adding the proper member stiffness
once to the diagonal stiffness submatrices and once to the off-
diagonal stiffness submatrix corresponding to the joints which
the member connects. Thus even for structures for which the
storage for the complete matrix $\overline{A}^t K \overline{A}$ would require many times
the capacity of the IBM 7094 core, the actual storage requirement
during the formation of $\overline{A}^t k \overline{A}$, by bookkeeping, never exceeds a few
hundred machine words, and the computer time used for the com-
putation is a fraction of that required by a formal multiplication.
Similar methods are applied in all phases of the total process,
for example, for the operations required due to local-member
or joint releases as well as during the solution of the stiffness
equations and backsubstitution.

Since all operations are performed a member or a joint at a
time, all procedures are applicable to all structural types. Only
the three subroutines dealing with the computation of member
flexibilities, member fixed-end forces, and output actually dis-
tinguish between the structural types. In all other subroutines,
the operations are completely generalized using variable length
loops of the form DO N I=1, JF, where JF is the number of degrees of freedom for the structural type considered.

Chapter 5

DESCRIPTION OF PARAMETERS, CODEWORDS, AND ARRAYS

In this chapter, a detailed description is given of the system parameters, problem parameters, and codewords comprising the fixed storage of the system. As described in Chapter 3, the codewords controlling the data arrays are synonymous with the arrays themselves. Therefore, the makeup of the data arrays is presented in terms of the codewords.

5.1 Organization of Parameters and Codewords

The $(300)_{10}$ locations from TOP+1 through TOP+300 (see Section 5.2) are assigned fixed functions in the STRESS system, as follows:

Number of Words	Absolute Addresses (Octal)[1]	Function
108	77461 − 77306	Working storage
50	77305 − 77224	System parameters
42	77223 − 77152	Problem parameters
100	77151 − 77006	Codewords

The working area is available for direct subscripted operations, and normally does not contain information carried from one subroutine to the next. This portion will not be discussed further.

The remaining 192 locations are reserved for parameters and codewords. Not all of this space is used at the present time, nor are all existing parameters and codewords needed by every subroutine. In order to maintain the proper COMMON allocations, dummy filler arrays SYSFIL, PROFIL, and CODFIL of appropriate dimensions are used in the subroutines.

The actual assignment of parameters and codewords followed the chronological development of the system, and is therefore

---

[1]We assume TOP = $(77462)_8$ to conform with the FMS Monitor on the 7094.

not in a very logical order. In the following detailed description,
an attempt was made to group parameters in a logical order. For
quick reference, the parameters and codewords are tabulated by
location in Appendix C.

5.2 System Parameters

The following parameters describe or control the status of the
system and are essentially independent of problem size.

5.2.1 Parameters pertaining to program status

Name	Address	Description
CHECK	77305	Solution consistency parameter, used in Phase 1. The 36 bits of CHECK are used as consistency flags. Flags not present (bit = 0) at end of Phase IB indicate an input error and cause a diagnostic to be printed. If the flag is present (bit = 1), the consistency is satisfied. The significance of the bits in terms of the error message printed is given in the following list.

Bit position	Code number of error	Message
35	15	NUMBER OF JOINTS NOT SPECIFIED
34	16	NUMBER OF SUPPORTS NOT SPECIFIED
33	17	NUMBER OF MEMBERS NOT SPECIFIED
32	13	NUMBER OF LOADINGS NOT SPECIFIED
31	19	TYPE OF STRUCTURE NOT SPECIFIED
30	20	METHOD OF SOLUTION NOT SPECIFIED
29	21	PRINTING OF RESULTS NOT YET POSSIBLE
28	35	NO LOADS SPECIFIED
27	28	NUMBER OF JOINTS GIVEN NOT EQUAL TO THE NUMBER SPECIFIED
26	30	NUMBER OF MEMBERS NOT EQUAL TO THE NUMBER SPECIFIED

Name (continued)	Address (continued)			Description (continued)
5.2.1 (continued)				
CHECK (continued)				
		Bit position (continued)	Code number of error (continued)	Message (continued)
		25	31	NUMBER OF MEMBER PROPERTIES GIVEN NOT EQUAL TO THE NUMBER SPECIFIED
		24	32	UNACCEPTABLE STATEMENTS PRESENT
		23	33	STRUCTURAL DATA INCORRECT
		22	34	LOADING DATA INCORRECT
		21	8	MODIFICATION NOT ACCEPTABLE
		20	10	JOINT NUMBER GREATER THAN THE NUMBER SPECIFIED
		19	11	MEMBER NUMBER GREATER THAN THE NUMBER SPECIFIED
		18	14	LOADING NUMBER GREATER THAN THE NUMBER SPECIFIED
		17	29	NUMBER OF SUPPORTS GIVEN NOT EQUAL TO THE NUMBER SPECIFIED
		16	18	NUMBER OF LOADINGS GIVEN NOT EQUAL TO THE NUMBER SPECIFIED
		15	26	METHOD SPECIFIED NOT AVAILABLE

Name (continued)	Address (continued)	Description (continued)

5.2.1 (continued)

INORM	77303	Overflow index = 0 normal operating mode = 1 if memory overflow occurred
ISCAN	77301	Mode index = 1 if normal solution mode = 2 if scanning mode
ISUCC	77274	Error indicator = 1 if solution is still successful = 2 if solution has failed
ISOLV	77302	Indicator for the stage of solution success- fully completed = 1 System is entered = 2 PHAS1B = 3 MEMBER = 4 MRELES = 5 LOADPC = 6 TRANS = 7 ATKA = 8 JRELES = 9 FOMOD = 10 SOLVER
IMOD	77277	Mode index for modifications = 1 no modifications = 2 changes = 3 additions = 4 deletions
IRST	77261	Modification entry, structural data (set by Phase Ia, but not used)
IRLD	77260	Modification entry, load data
IRPR	77257	Modification entry, printing
ICONT	77275	Index for stage of modifications = 0 if single solution = N if N^{th} solution (N-1st modification)
IPRG	77262	Print request parameter = 1 if print requested (selective output) = 0 otherwise

5.2.2 Parameters controlling core storage

NMAX	77304	Pool size available in core
TOP	77272	Address contains address of top of COMMON+1, that is, $(77462)_8$
N1	77271	Top of pool = TOP - 301_{10} - In Address
NL	77270	Next available storage location to be assigned

Name (continued)	Address (continued)	Description (continued)

5.2.2 (continued)

NT	77267	Bottom of pool
NREQ	77266	Additional memory space required for last memory request to ALOCAT

5.2.3 Parameters controlling tape storage

NXFIL	77276	Initial data file number
LFILE	77264	Next file number to be used
TN	77265	Scratch-tape logical number last used
NTAPE	77256 to 77252	Five scratch-tape logical numbers
NXFILE	77251 to 77245	Five scratch-tape position numbers, indicating number of files on each scratch tape NTAPE

5.2.4 Miscellaneous

TOLER	77263	Tolerance limit. Relative discrepancy allowed between the member length internally computed from joint coordinates and the sum of lengths of subsegments in a variable member, as specified under member properties.

5.3 Problem Parameters

The parameters in this group describe the problem currently being processed by the system.

5.3.1 Type descriptors

IMETH	77214	Method of solution code = 1 for stiffness method = 2 for flexibility method
ID	77220	Identification for structural type = 1 for plane truss = 2 for plane frame = 3 for plane grid = 4 for space truss = 5 for space frame

5.3.2 Basic size descriptors

NB	77222	Number of bars (members) in the structure
NJ	77223	Number of joints in the structure

Name (continued)	Address (continued)	Description (continued)

5.3.2 (continued)

NDAT	77221	Number of support joints
NFJS	77212	Number of free joints (= NJ-NDAT)
JF	77217	Number of degrees of kinematic freedom at a joint = 2 for plane truss = 3 for plane frame, plane grid, or space truss = 6 for space frame
NCORD	77215	Number of coordinate directions = 2 for planar structures = 3 for spatial structures
NLDS	77213	Number of loading conditions
NLDSI	77164	Number of independent loading conditions
NMR	77206	Total number of member-release components
NJR	77205	Total number of joint-release components

5.3.3 External size descriptors

MEXTN	77167	Highest external member number: MEXTN, JEXTN, and LEXTN control the sizes of all input-oriented arrays associated with members, joints, and loadings, respectively. Initially (during the original problem, ICONT = 0) they are set equal to NB, NJ, and NLDS. On succeeding modifications, they are reset to the highest external number encountered during reading. Every loop operating on input-oriented data must have these parameters as limits and must include a test for non-existing members, joints, and loadings, respectively.
JEXTN	77170	Highest external joint number (see MEXTN)
LEXTN	77166	Highest external loading number (see MEXTN)

5.3.4 Derived size descriptors

NSQ	77216	= JF*JF: size of submatrices
NSTV	77211	= JF*NFJS: length of a vector having JF components at each free joint
NMEMV	77210	= JF*MEXTN: length of vector having JF components per member

Name (continued)	Address (continued)	Description (continued)

5.3.4 (continued)

NDSQ	77203	= NJR*NJR: size of matrix K_{33} in Equation 4.51
NDJ	77202	= JF*NDAT: length of a vector having JF components at each support
NFJS1	77176	= NFJS+1: internal number of first support joint

5.3.5 Counters

JJC	77175	Number of free joints given
JDC	77174	Number of support joints given
JMIC	77173	Number of member-incidence statements given
JMPC	77172	Number of member-property statements given
JLD	77171	Highest loading number
JLC	77165	Current loading condition number (for modifications)
NLDG	77156	Number of loading conditions given

5.3.6 Optional array indices

IPSI	77207	Code for table of Beta angles = 1 if any member has a Beta angle given = 0 if no Beta angles are given
IYOUNG	77163	Young's modulus indicator = 0 if no E values given = 1 if E given for each member $\neq 0 \neq 1$, E value to be used for all members
ISHEAR	77162	Shear modulus indicator, similar to IYOUNG
IEXPAN	77161	Coefficient of thermal expansion indicator, similar to IYOUNG
IDENS	77160	Member density indicator, similar to IYOUNG

5.3.7 Miscellaneous

ISODG	77204	Number of member-force components set in Phase Ia.
IXX	77201	Position of subarray KMEGA for current member with release at start of member
NPR	77200	Current number of subarrays KMEGA set by MRELES, used by FIXM

Name (continued)	Address (continued)	Description (continued)

| NBB | 77177 | Current number of submatrices KOFDG (= NB up to subroutine JRELES, may be larger than NB after JRELES and SOLVER) |
| NBNEW | 77157 | Current defined length of codeword array of KOBDG (see also NBB) |

5.4 Codewords

The codewords in this group control the data arrays used by the system. The makeup and function of the codewords are discussed in Chapter 3. The makeup of each data array is described in this section in connection with the codeword controlling the array. An asterisk after the location indicates a second-level array.

5.4.1 Titles

| NAME | 77151 | 12-word array containing the structure identification name in BCD |
| MODN | 77140 | 12-word array containing the modification name in BCD |

5.4.2 Input joint data

JTYP	77101	Joint type array, JEXTN long. Joint types are as follows: = 0 joint does not exist = 1 free joint = 2 support = 3 deleted (changed to zero in PHAS1B)
JEXT	77106	External joint table NJ long. JEXT(J) contains the external joint number corresponding to the internal joint number J.
JINT	77105	Internal joint table, JEXTN long. JINT(J) contains the internal joint number corresponding to the external joint J.
KXYZ	77150	Joint-coordinate table, 3*JEXTN long. Coordinates XYZ of each joint stored in order by external joint number.
KJREL	77147	Joint-release codes. First word: number of joints released, NR. Succeeding groups of 5 words per released support joint: external joint number, release

Name (continued)	Address (continued)	Description (continued)

5.4.2 (continued)

KJREL (continued)   code, 3 angles in decimal degrees.

1	NR	number of released joints
2	J	external joint number
3	Release Code	
4	θ_1	
5	θ_2	
6	θ_3	

Words 2 through 6 are repeated for each released support joint. The release code is given in Bits 12 through 17. (This is the same form used for member-release code.)

Force X release = 1 in Bit 17
Force Y release = 1 in Bit 16
Force Z release = 1 in Bit 15
Moment X release = 1 in Bit 14
Moment Y release = 1 in Bit 13
Moment Z release = 1 in Bit 12

KATR   77074* $\overline{A}^t$ matrix. It contains NJ data arrays referenced by internal joint number. Each data array contains the signed member numbers of all members incident to that joint.

1	NBRI	Number of incident branches
2	SIZE	Array size = $5\left(\dfrac{NBRI-1}{5}\right)+1+2$
3	M1	
4	M2	Member numbers incident on joint J: positive if J is start, negative if J is end of member.
NBRI+2	M_r	

5.4.3 Input member data

JPLS   77146   Incidence table, MEXTN long. It contains tains the external joint number of the

Name (continued)	Address (continued)	Description (continued)

5.4.3 (continued)

JPLS (continued)		joint at the start (plus end) of the member.
JMIN	77145	Similar to JPLS, but contains the joint number at the end (minus end) of the member.
MTYP	77144	First-level array, MEXTN long. Each word contains data concerning the member type and is made up as follows:

Bit Position	Content
S — 5	Member-constraint code
6 — 11	Member-release code for start of member
12 — 17	Member-release code for end of member
18 — 23	IMLOT
24 — 29	N = length of data array MEMB for this member
30 — 35	Not used

The bracket joining "Member-release code for start of member" and "Member-release code for end of member" is labeled: same as for KJREL

S	5 6		17 18	23 24	29 30	35
MEMB CONSTR CODE	MEMBER RELEASES		IMLOT	N		

IMLOT	N	Member Type
0	-	Member deleted
1	6	Prismatic (A_x, A_y, A_z, I_x, I_y, I_z stored in MEMB)
2	NSQ	STIFFNESS GIVEN
3	NSQ	FLEXIBILITY GIVEN
4	7*NS	VARIABLE N = NS*7. MEMB contains A_x, A_y, A_z, I_x, I_y, I_z, L for each segment. NS is the number of segments.
5	-	STEEL

MEMB	77142*	Second-level array, MEXTN data arrays. Each array is N words long, where N depends on the member type (see MTYP).
MTYP1	77100	First-level array, MEXTN long, similar to MTYP. The word is made up as follows:

Name (continued)	Address (continued)	Description (continued)

5.4.3 (continued)

MTYP1 (continued)

Bit Position	Content
S — 5	Not used
6 — 17	I = order number for data array KMEGA for this member. KMEGA(I) is the KMEGA array pertaining to this member.
18 — 35	Not used

5.4.4 Optional member input data

KPSI	77143	Member rotation table, MEXTN long. It contains the Beta angles for each member (zero length if no Beta angles are specified).
KYOUNG	77072	Table of Young's moduli for each member. MEXTN long if IYOUNG = 1; otherwise zero length.
KSHEAR	77071	Table of shear moduli for each member. MEXTN long if ISHEAR = 1; otherwise zero length.
KEXPAN	77070	Table of coefficients of thermal expansion for each member. MEXTN long if IEXPAN = 1; otherwise zero length.
KDENS	77067	Tables of densities or unit weights for each member. Length MEXTN if IDENS = 1; otherwise zero length.

5.4.5 Load input data

LINT	77073	Internal loading numbers. Loading conditions are assigned sequential numbers during input. To eliminate deleted and combination (that is, dependent) loading conditions, internal numbers are also assigned in PHAS1B, where the parameter NLDSI is also assigned. The relation between the two sequences of loading numbers is given by: $LINT(J)=JC$ where $1 \leq J \leq LEXTN$ and JC=1, NLDS
MLOAD	77076*	Member-load data arrays. MEXTN data arrays (see LOADS).
JLOAD	77075*	Joint-load data arrays. JEXTN data arrays (see LOADS).

Name (continued)	Address (continued)	Description (continued)

5.4.5 (continued)

LOADS   77141*
Load data-reference arrays. LEXTN data arrays containing headerwords, the loading name in BCD and cross-reference words to each joint and member load (in MLOAD and JLOAD) belonging to this loading condition.

The relation and makeup of the MLOAD, JLOAD, and LOADS arrays are now shown.

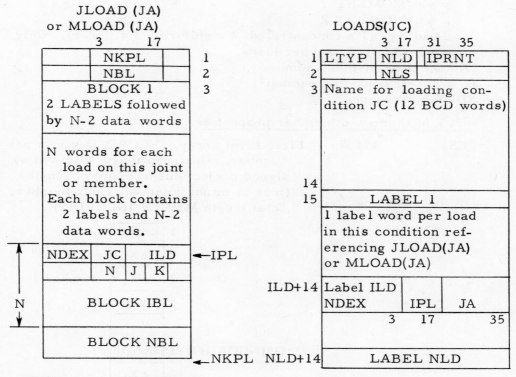

NDEX	N	Load Type
1	8	Joint load
3	14	Member-end load
4	8	Member distortion
2	6	Member load
5	8	Joint displacement

NLD = Number of loads in loading condition
NLS = Number of words available for labels
NBL = Actual number of loads. First word of load block zero if the load is deleted and is not counted.

Name	Address	Description
(continued)	(continued)	(continued)

5.4.5 (continued)

LTYP = Loading condition type (0 = deleted, 1 = independent, 2 = dependent)

N = Length of block including a two-word label

J = Load direction. Only applicable for member loads.

    J = 1   FORCE X

       = 2   FORCE Y

       = 3   FORCE Z

       = 4   MOMENT X

       = 5   MOMENT Y

       = 6   MOMENT Z

K = Load type (1 = concentrated, 2 = uniform, 3 = linear). Only applicable for member loads.

JA = Joint or member number

JC = External loading number

NDEX = load type

5.4.6 Computed member properties

KS	77137	First-level array, (NCORD+1) words per member. Unsigned length followed by signed projections of member length (plus to minus joint) for each member. Total length MEXTN*(NCORD+1).

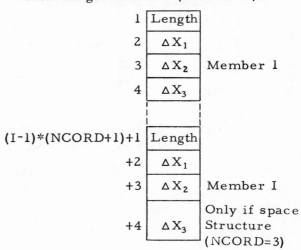

KMKST	77136*	Local-member stiffness table, containing each member stiffness (or flexibility) matrix in local-member coordinates, MEXTN data arrays, each NSQ long.
KSTDB	77135*	Member stiffness table, containing

Name	Address	Description
(continued)	(continued)	(continued)

5.4.6 (continued)

KSTDB (continued)   member stiffnesses in global coordinates. NB data arrays, each NSQ long.

KMEGA   77103*   Second-level array. NPR data arrays. NPR = total number of members with releases at the start. Auxiliary arrays (JF*JF) set up in MRELES-FIXM. Used in LOADPC-LSTOR.

5.4.7 Computed structure properties

KDIAG   77113*   Structure stiffness table, containing the diagonal stiffness submatrices. NJ data arrays, each JF×JF.

KOFDG   77112*   Structure stiffness table, containing the nonzero off-diagonal (JF*JF) submatrices in the lower triangle of ATKA, according to IOFDG. NB data arrays up to JRELES. Redefined NBB long in JRELES and SOLVER.

IOFDG   77111*   Address and position table for the <u>non-zero</u> submatrices of $\overline{A}^t K \overline{A}$. One data array per joint (or row of $\overline{A}^t K \overline{A}$). Each data array contains

         1st word: Address: current defined length of array = LDE
Decrement: current number of entries for existing submatrices in this row = LCU

         2nd word to (LCU+1): 1 word per existing submatrix in this row.

         Address: order position of submatrix itself in KOFDG table
Decrement: column order of submatrix

IOFC   77077*   Logical transpose of IOFDG. One data array per column of $\overline{A}^t K \overline{A}$ (only free-joint portion). One word per nonzero submatrix in that column stored the same way as words 2 to (LCU+1) in IOFDG. IOFC, however, does not have the leader word with LDE and LCU. Arrays are generated during forward

Name (continued)	Address (continued)	Description (continued)

5.4.7 (continued)

IOFC (continued) — sweep of SOLVER and used in backward sweep of SOLVER.

IFDT   77104*   Logical table for $\bar{A}^t K \bar{A}$. Each off-diagonal submatrix of the lower half of $\bar{A}^t K \bar{A}$ is represented by a single binary bit. Each data array of IFDT is 200 words long and can thus represent 200*36 = 7200 submatrices. Bits are stored from right to left by columns in each word representing $\bar{A}^t K \bar{A}$. A 1-bit representing an existing submatrix and a 0-bit a nonexisting matrix. For example, a 1 bit in the last position of the first word of IFDT(1) means that $\bar{A}^t K \bar{A}$ has a nonzero submatrix row 2, column 1. Thus the IFDT table can be used only to determine whether or not a certain submatrix is present or not. Information as to where that submatrix is stored is contained in IOFDG.

MEGAO   77107*   Auxiliary matrix if joint releases are present. Dimensions of flexibility. NDAT data arrays, each JF*NDAT*JF long. Only those data arrays are allocated that have at least one nonzero element. Each data array represents JF rows.

5.4.8 Computed load data

KPPLS   77133*   Member-end force matrix in global coordinates, containing the force vector in each member at the start (plus end). NLDS data arrays, each MEXTN*JF.

KPMNS   77132*   Similar to KPPLS, but for the end (minus end) of the members.

KUV   77131*   Member-distortion matrix in global coordinates, containing the distortions V for each member. NLDS data arrays, each MEXTN*JF.

KPPRI   77130   Effective joint-load matrix, containing the external loads P' applied to each free joint. After SOLVER it contains the solution displacements. NSTV*NLDS.

Name (continued)	Address (continued)	Description (continued)

5.4.8 (continued)

KR	77127*	Statics check matrix. Contains the applied loads at the free joints and the support-joint reactions as obtained from backsubstitution if a statics check is requested. NLDS second-level arrays, each NJ*JF long.
KPDBP	77102	Effective support-joint load matrix $\bar{\bar{P}}'$, containing $\bar{P}'$ after FOMOD, and $\bar{\bar{U}}'$ after DEFSUP (if there are support releases). Length JF*NDAT*NLDS.
		If there are no joint releases, it contains zeros after BAKSUB, with length JF*NDAT.

5.4.9 Miscellaneous

KSAVE	77126	Number of words used to store the initial data specifications after the first SOLVE THIS PART. (This is not a codeword referring to an array.)
KSRTCH	77125 to 77114	Scratch-array codewords

Chapter 6

DESCRIPTION OF SYSTEM COMPONENTS

In this chapter, a general description of the program components is given. The description serves primarily as a guide in studying the detailed flow charts and program listings in the Appendices A and B. For the convenience of persons intending to modify the system, the important coding conventions used in the system are also summarized in this chapter.

6.1 General Organization

Because of its size, the STRESS system has been partitioned both functionally and physically. The functional partition consists of the three phases of the processor:

1. Phase Ia — reading and decoding of input
2. Phase Ib — consistency checks and compilation
3. Phase II — execution

The physical partitioning of the system is accomplished by separating the system into program links, which are loaded, one at a time, in memory by the modified CHAIN subroutine described in Section 2.2. At the present, the processor is broken up into six links. The first link contains Phases Ia and Ib, and Links 2 through 6 make up the execution phase. If it were desired to make more core storage available for data, the system could conveniently be broken into ten program links. In addition to Phases Ia and Ib, the first link contains the procedures for the error returns from all execution links, as well as the normal return from the last (backsubstitution and output) link. Finally, this link also contains the programs for SELECTIVE OUTPUT and PRINT DATA procedures.

The general flow of the computation process is shown in the Over-All Flow Chart, Appendix A (page 93). The normal operational mode of the system treats the input in a two-level loop. The outer loop cycles on the independent problems in the input

58

batch, and is terminated when the input file is exhausted. Each
independent problem, or job, is initiated by a STRUCTURE statement and terminated when a SOLVE or FINISH statement is processed. The number of problems that can be handled in succession
is not limited. For each individual problem, the normal process
consists of an inner loop sequentially through Phases Ia, Ib, and
II until all modifications have been processed. A problem without
modifications obviously involves just one cycle through the inner
loop. Again, the number of modifications for a given problem is
not limited.

The normal process of the inner loop can be interrupted by one
of the following three occurrences:

1. A consistency error in Phase Ib, signifying that the problem (or the modification being processed) is not executable;
2. A fatal error in the execution phase; or
3. A memory overflow at any stage of the computation.

When any one of the three possibilities occurs, identifying
messages are printed out and control is transferred to the first
link. The system then is automatically placed in the scanning
mode. In this mode, any remaining statements pertaining to the
problem being processed are scanned only for possible input errors, but no further execution takes place. The scanning mode
is terminated either when the SOLVE or FINISH statements are
encountered, or when the STRUCTURE statement of a new problem is read.

The diagnostic information printed after an interruption identifies the type of interruption that has occurred, and, in the case
of a consistency or execution error, the problem statement can
be corrected and resubmitted. At the present, there is no automatic procedure available to further partition data arrays and/or
program segments in case of a memory overflow.

The general logic of the execution phase is shown in the Over-
all Execution Flow Chart, Appendix A (page 94). It can be seen
that, whenever a logical choice exists, the program link in core
memory determines the next link to be called.

In the succeeding three sections, the operation of the three
functional phases are described in greater detail.

6.2 Input Phase: Phase Ia

PHAS1A is called from the main program of Link 1 and controls all statement input. (A familiarity with Chapters 2, 3, and
5 will help the reader understand this section: Chapter 3 describes the structure of the input programs, while Chapter 5
lists the form of data storage.) Data storage is concerned with

the following scalars and arrays defined in Chapter 5:

Scalars	Arrays
ID	NAME
IMETH	MODN
NB	JTYP
NJ	KXYZ
NDAT	KJREL
NLDS	JPLS
	JMIN
	MTYP
	MEMB (second-level)
	KPSI
	KYOUNG
	KSHEAR
	LOADS ⎫
	MLOAD ⎬ (second-level)
	JLOAD ⎭

In addition numerous scalars are used for logical control during input and counters for consistency checking. Since input for initial specification and modification is handled by the same program, ICONT and IMOD are important branching scalars.

6.2.1 Array status during input. In order that the processor not be limited greatly by problem size, a dynamic memory-control system is used. It is desired then that during inputting only the arrays needed at any one point in the process be allocated and not released. For convenience, however, a few arrays are left in an allocated status during input. They are KXYZ, JPLS, JMIN, KPSI, JTYP, and the codeword arrays of MLOAD, JLOAD, and LOADS. All other arrays are allocated when needed and immediately released.

With an initial problem specification, these arrays cannot be allocated until the appropriate size descriptors, the NUMBER statements, are read. In addition to defining and allocating these arrays, subroutine SIZED (the subroutine called for by a NUMBER statement) defines KJREL, MTYP, and others. With a MODIFICATION OF LAST PART, these arrays are allocated as soon as the MODIFICATION statement is read. For MODIFICATION OF FIRST PART, the machine status prior to the release of these arrays is restored.

6.2.2 Error flagging. A single word, CHECK, is used to record the errors of commission and omission. Each bit in this word (only 21 bits are presently used) corresponds to an error, with the error indicated with the bit set to zero. Boolean statements are used to either "mask" out bits or "or" in bits. Upon reading the STRUCTURE statement, all commission-error bits are set to one.

If an error is detected during input, the inputting program terminates statement translation and branches to the appropriate point to mask out the corresponding bit in CHECK and print an identifying error message below the statement echo print (shown as "SEP" for "set error parameter" in the flow charts). Conversely, upon detection of a statement considered required for solution a bit is "or'd" into CHECK (shown as SCP for "set check parameter" in the flow charts) during translation. PHAS1B adds to CHECK during its consistency checking and analyzes the word for accumulated errors.

6.2.3 Data storage. ID is computed from the words in the type statement. The numbers used in the computation are the word positions in the dictionary as returned by function MATCH. Similarly IMETH is set after translating the METHOD statement. Subroutine SIZED may be entered after reading a NUMBER statement or during modification upon encountering a number, joint, or loading number higher than any previously encountered. When entering from a NUMBER statement during an initial problem specification, the number is stored and arrays defined and allocated. When so entering during modification, the only action performed is to store the number unless that number is greater than its corresponding maximum number counter (that is, MEXTN to NB), in which case redefinition of arrays is also performed and the maximum counter set to the number.

During joint, number, and loading data input, input counters are incremented during initial input and additions modification. JJC is incremented for free joints, JDC for support joints, JMIC for member incidences, JMPC for member properties, and NLDG for loading conditions. For change modification, the two counters that might change are JJC and JDC and then only if the joint type is changed. For deletions, the appropriate counter is decremented. For a member deletion, both JMIC and JMPC are reduced. At present the programs do not check for numbering commission errors before altering the counters. An example would be the specification of two joints with the same numbers. Program changes for this additional error detection are minor.

Storage of data in the arrays NAME, MODN, JTYP, KXYZ, JPLS, JMNS, and MTYP are performed in a straightforward manner. Part of MTYP is specified from the member properties and part from member releases. With no ordering restrictions desirable, the parts are treated separately by either packing parts separately or treating the word logically. KJREL is redefined five words longer for every new released joint. When joint releases are specified, the array is searched to determine if releases at that joint have already been specified. If not, a new five-word group is appended for an initial specification or

additions. The release code is then logically treated with the
appropriate word. The logical treatment is for initial specification or additions

$$B \quad U(IJ)=U(IJ)+A$$

where the B specifies a Boolean statement, U(IJ) is the word
where the code is stored, A is the newly specified code, and "+"
is the "OR" operator. For changes, the operation is

$$B \quad U(IJ)=(U(IJ)*Mask)+A$$

where "*" is the "AND" operator and the mask deletes all possible
releases. For deletions, the operation to eliminate the specified
components is

$$B \quad U(IJ)=-(-U(IJ)+A)$$

where "-" is the "COMPLEMENT" operator. In addition to joint
releases, member releases and tabulate codes in the LOADS arrays are so treated: the masking in the changes modification is
necessary to preserve the other parts of the storage word in these
latter cases.

Member properties and load data (including distortions and displacements) are stored independently of the structural type. All
of these data are read in subroutine READ and stored temporarily
in an array together with an array for label numbers. The label
numbers give the position in the subarray MEMB for the data
pieces for prismatic members. For variable-segment properties,
the label number gives the position in the group of words for the
segment, with seven words reserved for each segment.

Storage of load data is performed in a similar fashion in subroutine READ. The label numbers refer to positions for the data
in a load block appended on the subarray of MLOAD or JLOAD.
A cross-reference system is established with the subarray of
LOADS so that the loads can be referred to during modification
by loading conditions.

Modification of load, joint, or member data by changes super-
imposes specified quantities on the original data after locating
the data position in the case of loads. For deletion of loads, the
reference word in the array LOADS and the first word in the load
data block are set to zero, thereby erasing the load block. Additions operate as initial data input.

6.2.4 Selective output. The selective-output mode can be
used to print previously calculated results in an interpretive
manner. A solution upon which to operate must be available, so
that all print requests must follow a SOLVE THIS PART and precede a MODIFICATION OF FIRST PART if given.

The SELECTIVE OUTPUT statement initiates this mode by
setting IMOD=2, that is, CHANGES modification. This parameter

has been so set in order to use the program part under LOADING
to read the loading condition number upon which to operate without requiring additional programming. The program part used
for the branch on PRINT reads and retains an index for a word
designating an output type and calls subroutine SELOUT for each
joint or member number encountered.

Subroutine SELOUT sets parameters for the printing of results
performed by subroutine ANSOUT. SELOUT retains information
on the previous output request and determines what titles should
be printed. If the loading condition changes, the structure name,
the modification name, if any, and the loading title are printed.
The table heading is printed if the request type has changed, and
the joint-status label is printed if the joint type has changed.
These last two are printed in ANSOUT, but the logical parameters for the printing are set in SELOUT.

6.2.5 Termination. The input phase is exited for any of the
following reasons:

1. Reading a SOLVE or SOLVE THIS PART when not in the
   scanning mode.
2. Reading a SOLVE when in the scanning mode.
3. Reading a FINISH statement.

For the first case, processing continues with PHAS1B. After
the return from PHAS1B the process immediately returns to the
main program of Link 1. This main program checks whether to
proceed with execution, terminate the problem, or enter the
scanning mode, which returns to PHAS1A. The scanning mode
is not terminated with a SOLVE THIS PART, only a SOLVE or
FINISH. A problem is terminated after scanning by returning
to the main program, that is, ISCAN=2. The FINISH statement
simulates this form of termination by setting ISCAN=2 and returning to the main program.

A modification counter, ICONT, is incremented when SOLVE
THIS PART is read, or it is set to zero for SOLVE.

An interruption of processing may also occur as a result of a
memory overflow. This return, regardless of its logical location
in the processor, enters the start of the main program in Link 1.
If the overflow occurred during input (ISOLV=1) or after a SOLVE
THIS PART (ICONT≠0) additional input for the problem may
still remain to be processed. For this case the scanning mode
is also entered.

6.3 Checking Phase: Phase Ib

Subroutine PHAS1B controls and performs the input-consistency
checking and compilation function of the processor. These two

tasks are not, however, physically separated in the programs.

Joint processing involves three operations in one loop on all joint numbers, performed if the joint-input counters correspond to the NUMBER specification. The first operation is to prepare two tables of internal-external number correspondence. Internally, all support-joint numbers are greater than the number of free joints (NJ-NDAT). The internal numbers are assigned for support joints starting with NJ to NJ-NDAT+1 in the order they are encountered in the loop on sequential external numbers. Free-joint numbers are assigned from 1 to NJ-NDAT. For all existing joints that have loads applied to them, the load type for each load block is checked. For a free joint, joint displacements (INDEX= 5) specified at the joint constitute an input error. Conversely, joint loads (INDEX=1) specified at a support are not permissible unless the joint has been released in the direction of the load. In this latter case, a check is made of the array KJREL to see if the joint has been given some releases. If so, no error is flagged, but no check is made on the correspondence of the release and load directions. For joints deleted during the last input specification (JTYP=3), all loads on the joints are erased as well as their references in LOADS. A check is made to assure that no joint releases are given at a free joint.

Member processing primarily performs the implicit deletions. The incidences of each member are checked. If in error, the error is flagged. If one of the joints has been deleted, the member is deleted by zeroing its type and erasing its properties. For deleted members any existing loads are erased as well as the references in LOADS. The member counters are checked against the NUMBER specification after all implicit deletions have been performed.

Processing of the loading condition consists only of a check on the number given and the assignment of internal numbers, in a manner analogous to the internal joint numbers, for independent and dependent loading conditions. The bit pattern in CHECK is compared with a mask and all noncorrespondence causes an appropriate error message to be printed by subroutine PRERR as given in Section 5.1. If no errors were detected, the processing proceeds.

After the first specification with SOLVE THIS PART (ICONT=1) the input data are saved for later specification of MODIFICATION OF FIRST PART. This is done by creating a file with an image of the pertinent data in core at this time. All data from the location of TOP down to the next location available in the pool are saved. No information is lost if a memory reorganization has occurred during input since all file numbers of data on tape are less than the file number of the saved image and are referenced in the image. The restoration of this condition consists

of simply reading this file back into core and rewinding all scratch
tapes to make their position references consistent.

   Many arrays are defined at this point. For all but two of these
the sizes will not change during execution. The number of subarrays of KOFDG will be NB at the end of the generation of the
structural stiffness matrix $\overline{A}^t K \overline{A}$ and the codeword array is defined as NB long at this stage. The number of elements in the
subarrays of IOFDG will also grow during the solution of the equations, but the number needed up to the end of the generation of
$\overline{A}^t K \overline{A}$ is determined from the incidence. As the incidence tables
for the joints are compiled by members, the maxima of the inci-
dences are counted for each joint. The subarrays of IOFDG are
then defined to this length plus a small amount for expansion for
free joints.

6.4 Execution Phase: Phase II

   The execution phase consists of Links 2 through 6, and operates
as described in Sections 6.4.1 through 6.4.6.

6.4.1 Link 2

1. MEMBER. See also flow chart for MEMBER (page 151).
    Generates all member stiffnesses $K^* =$ KMKST in member
    coordinates. Loops on members 1 to MEXTN. Raw data
    (properties of cross sections, flexibility matrix, or stiff-
    ness matrix) are obtained from MEMB. If stiffness is given,
    KMKST is filled directly. If flexibility is given, it is di-
    rectly inverted and stored into KMKST. If cross-section
    properties are given, MEMFOD computes F^* and MEMBER
    computes $K^* = F^{*-1}$. Trusses are handled on a separate
    branch because their stiffness or flexibility in member co-
    ordinates has only one component and is therefore singular.
    If a singular flexibility matrix for a frame is encountered
    because not enough member properties have been given or
    because a singular flexibility matrix has been specified,
    the member is geometrically unstable (no stiffness matrix
    exists). This is considered a fatal error: a message is
    printed for each unstable member and processing of this
    problem is terminated upon return to the main program.
2. MRELES. See flow chart for MRELES (page 153). If there
    are any member releases (NMR≠0), MRELES modifies the
    stiffnesses KMKST corresponding to the released members
    and stores them back in KMKST.

       A loop on members is initiated and for each released
    member (Check MTYP) K^* is modified according to the
    procedure of Chapter 4: first for the releases at the end

of the member (if any), and then for those at the start. If
a member is so released that it becomes geometrically
unstable, a singular stiffness matrix K_{22} will result. The
illegal combination of releases is detected and an error
message is printed. Such errors are treated as fatal
(ISUCC set to 2), but all members are checked for illegal
releases before execution is terminated upon return to the
main program of Link 2. If a member is released at the
start, an auxiliary submatrix that appears during the release process is later needed in the load processor when
the fixed-end forces on the member are modified for the
prescribed releases. (See Section 4.2.) These matrices
are stored in a second-level array KMEGA. KMEGA matrices are stored only for members released at their start,
and a member-to-KMEGA correspondence table is kept in
MTYP1.

6.4.2 Link 3. LOADPC. See flow charts for LOADPC and
LOADPS (pages 166 and 168) and Section 4.3. This link processes the raw data for loads into contributions to member distortions, member forces, and joint loads in the kinematically
determinate system, taking into account member releases.
LOADPC is the monitor routine, with a loop on members (1 to
MEXTN) and one on joints (1 to JEXTN). For each nondeleted,
loaded member or joint, all corresponding load data in MLOAD
or JLOAD are allocated and control is transferred to LOADPS
which monitors the processing of one MEMBER or JOINT for all
loads in all load conditions. This method of processing is convenient because all load raw data are stored by joint and member.
However, it may require the repeated allocation and release of
the resulting internal arrays KUV, KPPLS, and KPMNS, because
these are partitioned by loading condition.

No fatal errors can result in LOADPC. Nonfatal errors are
detected in MEMBLD if member loads are prescribed on members for which the flexibility or stiffness matrix were given as
direct input. In such cases it is not possible to compute fixed-
end forces due to loads acting between the end joints of the member. A message indicating this incompatibility is printed and the
particular load is ignored.

6.4.3 Link 4

1. TRANS. See flow chart for TRANS (page 185). Subroutine
TRANS rotates all member stiffnesses K^* (reflecting member releases) from local-member coordinates into joint
coordinates with origin at the end joint of the member. A
single loop on members (1 to MEXTN) with a call to TRAMAT
and MATRIP inside the loop makes up the subroutine. The
rotated member stiffnesses K are stored in KSTDB.

2. ATKA. See flow chart for ATKA (page 186). Subroutine
    ATKA generates the logical tables for the structure stiff-
    ness matrix A^tKA (KDIAG, KODFG, IFDT, IOFDG) by
    bookkeeping from the incidence tables $\bar{A}$ (JPLS, JMIN) and
    the primitive stiffnesses K (KSTDB). If in the modification
    mode, the data arrays of IFDT, IOFDG, and KDIAG are
    first cleared. The tables representing the structural stiff-
    ness are generated by a loop on members (1 to MEXTN)
    and by adding to KDIAG, KOFDG, IFDT, and IOFDG the
    proper contributions in the rows and columns corresponding
    to the start and end joint of the (nondeleted) member. All
    tables are generated disregarding all effects of possible
    joint releases prescribed. Only the lower left half of A^tKA
    is generated and stored.

3. JRELES. See flow charts JRELES through STEP5 (page 187).
    This set of subroutines is called only if there are joint
    releases (NJR≠0). JRELES is the monitor routine to
    modify the tables generated in subroutine ATKA accord-
    ing to the theory of Chapter 4, to reflect the effect of
    joint releases. Subroutine FOMOD does the corresponding
    modification on the effective joint loads $\mathscr{P}'$. This pro-
    cedure eliminates all kinematic unknowns associated with
    support joints. (Joints are internally arranged so that
    support joints are numbered last.) In the case where
    all joints are support joints the auxiliary array stored
    in KPDBP by FOMOD represents the displacements of
    the support joints and is the solution to the governing
    equations.

    All formal matrix operations implied by the equations
of the procedure outlined in Chapter 4 are done by bookkeeping. Only those operations (multiplications of submatrices) are performed for which both submatrices are not
zero matrices. Also, all arrays, such as ΛR, $(\Lambda R)^t$ etc.,
are stored in logical tables. This results in an efficient
processing. The only operation which is done formally is
the inversion of the auxiliary matrix K_{33}. Subroutine STEP5
which performs the actual modification of A^tKA using the
auxiliary array MEGA0 is called only if there are any free
joints (NFJS≠0). It is noted that if no members connect
any two support joints the auxiliary arrays K_{33} and MEGA0
are hyperdiagonal and the necessary operations are greatly
simplified.

6.4.4 Link 5. SOLVER. See flow chart for SOLVER (page
200). Subroutine SOLVER solves the governing joint stiffness
equations for the free-joint displacements. The method used is
the Gauss elimination procedure for a symmetric system, operating on submatrices of order JF as elements, rather than on

scalars. Apart from certain subsections of the bookkeeping system (IOFDG, IFDT, IOFC) the maximum amount of data required in core at any one time consists of three submatrices of order JF, plus the total load vector $\mathscr{P}'$. This vector could also be partitioned into second-level arrays, preferably by joints. This would result in an improvement of speed and capacity for large problems, say NFJS larger than 200.

All the required operations are determined by the bookkeeping system, and only the required multiplications of submatrices are actually performed. The process consists of a forward sweep during which logically a triangular matrix is generated, and a backward sweep proceeding from the last free joint upward and solving successively for the displacements of the free joints. Since during the backward sweep the operations are on columns of the lower triangular matrix, the bookkeeping arrays IOFDG, representing rows, are completely released and corresponding column arrays IOFC are generated as rows of the system are eliminated during the forward sweep. Since new arrays in KOFDG, which were zero in the original matrix A^tKA, are generated during the solution process, the bookkeeping arrays IFDT, IOFDG, and the codeword array of KOFDG must constantly be updated as new arrays are generated. Also the codeword array must be redefined with larger length. This redefinition is done in increments depending on the size and the connectivity of the structure. Data arrays of IOFDG must also be redefined for greater length. These same problems of updating and redefining arrays are also present in JRELES, since the modification of $\bar{A}^tKA$ for joint releases can also imply new elements in KOFDG.

6.4.5 Link 6. BAKSUB. Subroutine BAKSUB is the monitor subroutine for the process of backsubstitution (solving for induced member distortions and member forces from the known joint displacements) and printing of answers in tabular form. The induced member distortions and member forces are directly added to the components of the kinematically determinate system, computed, and stored in LOADPC (see also Section 4.3). At the same time a statics check is performed by summing all member forces at each joint. A comparison of these sums (stored in KR) with the applied joint loads provides a useful check on the numerical accuracy (round-off errors) of the solution.

The whole process is accomplished by a first outer loop on external loading conditions (1 to LEXTN), and a second inner loop on members (1 to MEXTN). For each member and independent loading condition, the member distortions are obtained by vectorially subtracting the displacements of its end joints, then rotating them into member coordinates. The induced member-end forces are then obtained by premultiplying these distortions by the member stiffness K^*. These operations are done in sub-

routine AVECT. Note that the local stiffness table K* was computed in Link 2 and was not used after subroutine TRANS until
the time of backsubstitution. During the intervening processing,
the KMKST arrays are in a released status and could have been
dumped on tape if the memory space was needed for other information. The array is allocated again in BAKSUB and completely
released after BAKSUB. The statics check is performed by a
call to STATCK in the outer loop after completion of the loop on
members. If the loading condition is a combination, a call to
COMBLD is made instead of the entry to the inner loop. COMBLD
takes the proper linear combinations of all total member distortions, member forces, joint displacements, and reactions (statics
check KR) of the corresponding independent loading conditions.
All these independent loading conditions are already processed
at the time when the combination is treated. This is the reason
why combination loadings can be dependent only on previously
specified conditions. The last step inside the outer loop is a
series of calls to ANSOUT. Such a call is made for each group
of requested tabulated results for each loading condition. A maximum of four calls per loading condition in the case of a TABU-
LATE ALL request (member forces, member distortions, joint
displacements, reactions) is made.

6.5 General Coding Conventions

The following conventions adhered to throughout the system are presented for the convenience of persons intending to modify the system.

6.5.1 Clearing of arrays.
Since Phase Ia resets all codewords
to zero for every new problem but not for a modification, data
arrays can be assumed to contain zero only for new problems.
This makes it necessary in various places to clear certain data
arrays during execution phase if in the modification mode (for
example, IOFDG, KDIAG in ATKA). Clearing is done by a statement of the form CALL CLEAR (NCODWD), where NCODWD is
the codeword name of the data array to be cleared. No codeword
array should be cleared in this way, as this would destroy the
reference system for any future reorganization.
Second-level data arrays are cleared as follows:

        ICDWD2=ICDWD1+I
        CALL CLEAR (IU(ICDWD2))

The array name is ICDWD1. Subroutine CLEAR sets the array
elements to zero if the array is in core, or erases the file number if the array is on tape. The latter action makes the array
appear unused, and its elements will therefore be set to zero
upon allocation.

6.5.2 Packing and unpacking routines. For a number of purposes information about members, joints, loads, etc., is packed into various portions of the 36-bit machine word. Packing and unpacking functions have been written in FAP but can be called by FORTRAN to pack or unpack. Two such routines are used in the processor for storage according to various arrangements within the word. The two routines are PACKW, UPACW and PADP, UPADP.

1. PACKW, UPACW packs or unpacks the FORTRAN integers I, K, J, L, M into or from U(IA) according to the following figure:

PACKW (U(IA), I, K, J, L, M)

S12		17	23	29	35
I	K		J	L	M

2. PADP, UPADP packs or unpacks the FORTRAN integers I, J, K into or from U(IA)

PADP (U(IA), I, J, K)

S12		17		35
I	J		K	

See the listings of these routines for the format of the packed data.

6.5.3 Matrix multiplication routines. The execution phase uses four different service subroutines that perform matrix multiplication according to different conventions. These subroutines are all written in FORTRAN. Their efficiency could be increased by rewriting them in FAP.

MAMUL (Y, T, A, JS, JT, JJ) used by MRELES
MATRIP (K1, K2, NT) used by TRANS, ATKA, BAKSUB
MAPROD (N1, N2, N3, N4, IZ, JF, IND) used by JRELES
MAPRDT (N1, N2, N3, N4, IZ, JF, NP) used by SOLVER

6.5.4 Organization of the program links. Each link has a short main program that performs the calling of main subroutines, updating of the solution stage counter ISOLV, and testing for solution failure (ISUCC). Each link also contains the basic service subroutines, that is, START, PRER2, CLEAR, CHAIN, etc. The master subroutines in each link, that is, the first level of subroutine below the main program, generally perform the following functions:

1. Allocate needed arrays
2. Operation (generally, a series of loops on members and/or

joints), with various calls to lower level subroutines

3. Release arrays no longer needed in the next operation

Typical examples of such master subroutines are LOADPC, MRELES, MEMBER, etc.

Additional calls to ALOCAT and RELEAS occur at lower-level subroutines, especially with respect to second-level data arrays only temporarily needed, even within one link. (Example: KOFDG, IOFDG in SOLVER.)

6.5.5 Address computations in FORTRAN. Each subroutine contains the array names U and IU at the beginning of its common list (if any) and an equivalence statement for U and IU. This places both U and IU at the top of common starting at $(77461)_8$ and any data in the variable pool (or the first-data area) can be referenced as an element in the <u>one-dimensional array</u> U (for floating-point data) or IU (fixed point). This is possible because FORTRAN does not check whether the subscripts of an array exceed the size of the corresponding dimension statement. All data in the variable pool are therefore referenced as U (IADR) or IU (IADR) where IADR will usually be computed from codewords.

6.5.6 Allocation of arrays. Caution must be exercised when temporarily needed arrays are allocated. It must always be remembered that any call ALOCAT (or a redefinition of an array) could cause a memory reorganization to take place, depending on the size and makeup of the particular problem. Therefore any array subscript that is computed from a codeword, such as I=MTYP+JM, IA=IU(I), must be assumed to be destroyed by either an allocation or a redefinition of an array; that is, no such address computations can be "carried past" an allocation or redefinition. This frequently requires that an address be recomputed after such allocations. For example, if the second-level data array MEMB(I) is used in a loop on I, the codeword array will usually be allocated and released before and after the loop, respectively, but the allocation and release of the Ith data array must be done inside the loop. Because of this allocation inside the loop, it is not only necessary to compute the address of the data array anew each time but also the address of the codeword array itself, since the allocation of the Ith data array could relocate both the codeword array and the data arrays. Thus the following sequence must be used:

```
          CALL ALOCAT (MEMB, 0)
          DO 100 I=1, MEXTN
          CALL ALOCAT (MEMB, I)
          I1=MEMB+I
          I1=IU(I1)
  C       GET NTH WORD OF ITH ARRAY
```

```
        IADR=I1+N
        WORD=U(IADR)
 C      NECESSARY OPERATIONS ON WORD
 100    CALL RELEAS (MEMB, I)
        CALL RELEAS (MEMB, 0)
```

6.6 Bookkeeping System for the Solution of the Equations

6.6.1 The bookkeeping system. The linear structural analysis
problem for elastic lumped-parameter systems involves the solution of a set of simultaneous algebraic equations for the unknown
joint displacements ū' if the stiffness approach is used. These
equations are of the form:

$$\mathscr{K}U' = \bar{\mathscr{P}}' \qquad \mathscr{K} = \bar{A}^t K \bar{A} \tag{6.1}$$

Even though advantage is taken of symmetry and only the lower
half of $\bar{A}^t K \bar{A}$ is stored, for larger structures it is not possible to
store the complete matrices $\mathscr{K}$ and $\bar{\mathscr{P}}'$ in core at one time. These
matrices must be blocked into smaller arrays, operated on se-
quentially, and those parts that are currently not needed for computation stored in secondary bulk storage if necessary (for example, magnetic tape). The form and size of the matrix $\mathscr{K}$ depends on the connectivity of the structure and the number of kine-
matic degrees of freedom. For each free joint there exist JF
degrees of freedom (JF = 2, 3, or 6 depending on the structure
type). It is convenient to consider the submatrices of order JF
in $\mathscr{K}$ as basic elements, since there is one such diagonal stiffness
submatrix for every joint in the structure. Also one submatrix
exists in one half of $\mathscr{K}$ off the diagonal for every member in the
structure. These off-diagonal submatrices appear in $\mathscr{K}$ in the
rows and columns corresponding to the end joints of the member.
All submatrices of $\mathscr{K}$ corresponding to two joints that are not
connected by a member are zero. Most structures have a rela-
tively sparse stiffness matrix; that is, most of the off-diagonal
submatrices are zero. It is, therefore, efficient to store only
those elements of $\mathscr{K}$ that are known to be nonzero. Information
for status and referencing of the off-diagonal submatrices is
stored in three second-level arrays, IFDT, IOFC, and IOFDG;
the submatrices themselves are stored in two second-level arrays, KDIAG and KOFDG.

All diagonal submatrices are stored in a second-level array
KDIAG, ordered by internal joint number; that is, the Ith subarray of KDIAG contains the diagonal submatrix of $\mathscr{K}$ for the Ith
internal joint. A similar table, KOFDG, contains all the nonzero,
off-diagonal submatrices for the lower half of $\mathscr{K}$. This means
that initially there will be NB(=number of members) subarrays
in KOFDG. Since this table KOFDG does not directly tell where

in the matrix $\overline{\mathcal{K}}$ any particular submatrix is located, additional bookkeeping tables must be kept. Since it is convenient to keep as large a portion as possible of the logical makeup of $\overline{\mathcal{K}}$ in core, an array IFDT (see Section 5.4) is used in which each off-diagonal submatrix below the diagonal is represented by one binary bit. Thus, one 36-bit word can contain the information about 36 submatrices (of 4, 9, or 36 words each) whether or not the submatrix is nonzero. A zero submatrix is represented by a zero bit, a nonzero matrix, for which there is an array in KOFDG, by a one bit. This array, IFDT, is used during the solution of the equations to determine whether a matrix corresponding to a given pair of joints is present or not. If such a submatrix is found to exist, it still must be located within the table KOFDG. This information is stored in the second-level arrays IOFDG. One subarray exists for each hyperrow (a hyperrow consists of JF scalar rows corresponding to one joint), that is, for each joint of the structure. Each array contains one word per nonzero, off-diagonal submatrix below the diagonal in that hyperrow. This word contains in the decrement the hypercolumn number, where the submatrix is located, and in the address part the position of the submatrix within the KOFDG table. Thus, the off-diagonal submatrix corresponding to row J and column K (J > K) is addressed at several stages as follows:

1. Find out whether the submatrix exists: M=2.

      CALL ADRESS (J, K, NAD, M)

ADRESS will look up the bit picture in IFDT. If the bit corresponding to column K, row J is 1, ADRESS returns the position of the corresponding array in KOFDG in NAD by using IOFDG; if not, NAD=0 is returned. The fourth argument is an operation code for ADRESS. If the array exists, it can then be referenced by

      KF=KOFDG+NAD

so that

      N=IU(KF)
      U(N+1) is the element (1, 1) of the subarray.

2. Add a submatrix: M=0.
ADRESS will place a bit in the bit picture corresponding to column K and row J.
3. Search a column for the next nonzero array: M=1 and 4.
ADRESS will scan the bit picture of the column K from any position in KOFDG corresponding to the first nonzero subarray. Row J will give the hyperrow number at which the nonzero subarray is found. None exists if NAD=0. For M=1, the column is scanned to J=NFJS; for M=4, to J=NJ.

88

4. Determine the number of subarrays in a column: M=3.
ADRESS will count the number of 1 bits in the hypercolumn K
and return the result in NAD.

A simple structure is given in **Figure 6.2**, and Figure 6.1
shows the makeup of the various arrays before the reductions
of the equations for the simple structure.

JOINT	1	2	3	4	5
1	KDIAG(1)				
2	KOFDG(2)	KDIAG(2)			
3	0	KOFDG(3)	KDIAG(3)		
4	KOFDG(1)	0	0	KDIAG(4)	
5	0	KOFDG(4)	0	0	KDIAG(5)

← JF →

Figure 6.1. Schematic for matrix K for structure of Figure 6.2.

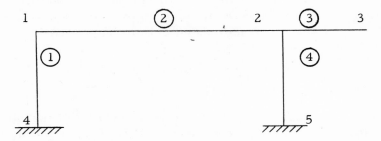

Figure 6.2. Sample structure.

The array IFDT consists of only one second-level array 200
words long. Since the structure is very small, its stiffness
matrix image has only one nonzero word, which is made up as
follows (see Section 5.5):

S 1 2   26   30   35

```
          0  0  0  1  0  1  0  1 0 1
          4  3    2        1      hypercolumn nr
```

There are 5 arrays IOFDG, 1 each for hyperrows 1, 2, 3, 4,
and 5; IOFDG(1) is of zero length.

IOFDG(2)	S	17	35 S		17	35
	Used length =1	Defined length			1	2

Leader word   Word for KOFDG(2)

	S		35	
IOFDG(3)	Leader word		2	3

Word for KOFDG(3)

IOFDG(4)	Leader word		1	1
IOFDG(5)	Leader word		2	4

The submatrix number in KOFDG is the same as the member number as a result of the order of operations in forming $\mathscr{K}$.

6.6.2 Elimination of support joints.

If any support-joint releases are given, the displacements associated with the released components are eliminated from the equations in the joint-release subroutines (JRELES) according to the method of Section 4.2.3, so that during the actual solution (subroutine SOLVER) only the free joints must be considered. The elimination of the unknowns at the released supports involves the modification of all off-diagonal submatrices that correspond to those members running between free joints connected to the same released support joint

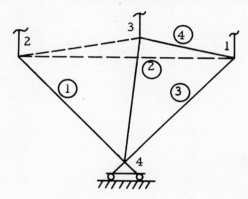

Figure 6.3. Modification of the structural stiffness due to joint releases.

and those diagonal matrices that correspond to the same free joints. If no member exists between such joints, the elimination procedure will result in a new submatrix corresponding to a fictitious member. For example, the elimination of the release joint 4 in Figure 6.3 would change the diagonal stiffness for joints 1, 2, and 3, and the off-diagonal stiffness of member 4; a new stiffness matrix corresponding to a fictitious member between 1 and 2, and 2 and 3 would be created.

If, during the solution, new off-diagonal matrices are created, KOFDG and the bookkeeping arrays IFDT and IOFDG must be updated and expanded for the new array. The codeword array of KOFDG and the data arrays of IOFDG change their length as the

elimination proceeds. For efficiency these length changes are
by blocks. Each time a new element is needed, the defined length
of the array must be checked against the required length and, if
necessary, the array must be redefined with a greater length.

   6.6.3 Solution for the free-joint displacements. The solution
of the linear algebraic equations for the free-joint displacements
u'

$$A^t \mathcal{K} A \, u' = \mathcal{P}' \tag{6.2}$$

is done in subroutine SOLVER. The matrices $\mathcal{P}'$ and $A^t \mathcal{K} A$ already reflect the effect of joint and member releases, if any. A
modified Gauss-elimination procedure for symmetrical systems,
consisting of a forward and a backward sweep, is used. (See
flow chart for SOLVER in Appendix A, page 200.) Instead of
eliminating the scalar rows one row at a time by operating on
scalar quantities, SOLVER operates on matrices, thus eliminating
a number of rows at a time. At present all operations are done
on joint submatrices of order JF; that is, elimination during the
forward sweep is done on sequential joint members (JF scalar
rows at a time). Thus, if a joint stiffness submatrix is designated by a_{ij} and the load subvectors by c_i, during elimination of
joint k, proper multiples of the hyperrow k are subtracted from
all hyperrows $n > k$, according to the following formulas:

$$a_{ni} \rightarrow a_{ni} - a_{nk} a_{kk}^{-1} a_{ki} \tag{6.3}$$

$$c_n \rightarrow c_n - a_{nk} a_{kk}^{-1} c_k \tag{6.4}$$

for all integers n, i

$$k < n \leq NFJS \tag{6.5}$$
$$k < i \leq n \tag{6.6}$$

   This procedure leads to implicit zero off-diagonal submatrices
in the complete hypercolumn k. The forward sweep consists of
three nested loops. The outermost loop is on k, the hyperrow
being eliminated. A call to ADRESS searches in IFDT to find the
next existing submatrix in column k, starting at row $n + 1$ (the
first time $n = k$). If such a submatrix is found at a new row n,
a search is made in column k ($k < i < n$), which is equivalent to
a search in row k between k and n. For each nonzero matrix k, i
found (via calls to ADRESS), Formula 6.3 is applied to change the
corresponding submatrix a_{ni} in row n. During the process of
altering row n, IOFDG(n) is used to find the position of a_{ni} in
KOFDG. Thus during the forward sweep, data arrays of IOFDG
must be allocated and released in the second of the three nested
loops. However, during the backward sweep they would have to
be allocated in the innermost loop, because the operations for the

backward sweep proceed row by row from the diagonal to the right (or therefore by column from the diagonal down). For large problems this would cause a prohibitive number of memory reorganizations. For this reason the arrays IOFDG are totally released successively during the forward sweep, and corresponding arrays IOFC are created for the backward sweep (see Section 5.5 for IOFC). These arrays contain essentially the information of IOFDG in transposed form; that is, they are arranged by column below the diagonal, rather than by row.

During the forward sweep an upper triangular matrix is formed. The backward sweep is therefore only a successive computation of unknown subvectors of u' for one joint at a time, starting with the last joint, NFJS, and proceeding upwards. As the displacements are computed, they are stored in the array KPPRI in place of the previously stored joint-load vectors. The operations to be performed to compute the displacements of joint n are as follows:

$$\sum_{i=n}^{NFJS} a'_{ni} u'_i = c'_n \qquad\qquad (6.7)$$

where a'_{ni} and c'_n are the reduced submatrix n, i and load subvector n at the end of the forward sweep. Since during the backward sweep all u'_i are already known for $n < i \leq NFJS$, the nth subvector becomes

$$u'_n = a'_{nn}\left[c'_n - \sum_{i=n+1}^{NFJS} a'_{ni} u'_i \right] \qquad\qquad (6.8)$$

At the time of the backward sweep, all nonzero submatrices a'_{ni} are listed sequentially in the bookkeeping array IOFC(n). Thus, only one such array must be allocated (and released) per free joint during the backward sweep.

Chapter 7

EXTENSION OF THE SYSTEM

Extension and modification of the STRESS processor will generally involve three considerations: translation and data input;
storage of input and generated data; and the position in, and control of, the process. Each of these aspects has been treated separately in the preceding chapters. The following examples are
given to illustrate the procedure used in extending the system.
The reader should note how the existing process is analyzed and
followed to the point of modification, how the modification is in-
serted, and how the modification interacts with the memory-usage
procedure.

7.1 Member Geometry Output

This first example illustrates the ease of insertion of additional
output requests. It is assumed that we required an additional
statement of the form

> PRINT GEOMETRY

or

> TABULATE GEOMETRY

which is to generate an output table of the form

MEMBER NUMBER	START JOINT	END JOINT	LENGTH	PROJEC- TIONS	TANGENTS (RISES)

The required output can be provided directly in Phase Ia, provided that at the time the new statement is encountered, the number of joint-coordinates input (JJC + JDC) equals the number of
joints specified (NJ), and the number of member-incidences input
(JMIC) equals the number of members specified (NB). If the above
two cases do not hold, the request cannot be processed.

Examination of the listing of subroutine PHAS1A shows that in
processing both the TABULATE statements (program statement 400
et seq.) and the PRINT statements (program statement 1700 et
seq.) we use LIST2 for further input decoding. LIST2 contains

eight words at the present (DISPLA, DISTOR, REACTI, FORCES, DATA, ALL, MEMBER, and JOINT — see listing of LISTS). We therefore add a ninth word, GEOMET, and change LIST2 accordingly:

```
LIST2 DEC   9
      BCI   9,...GEOMET
```

With this change, a return from MATCH with K = 9 will occur whenever the word GEOMETRY is encountered.

In order to accommodate the new alternative, the following two changes must be made:

a. Replace statement 402 by

```
402   IF(K-5)420,1720,4021
4021  IF(K-9)430,1712,1712
```

b. Replace statement 1701 by

```
1701  IF(K-5)1702,1720,1711
1711  IF(K-9)1700,1712,1712
```

The processing of the request can then be continued with a statement 1712. The processing starts by checking for the scan mode and for whether the request can be satisfied. If it can be, a CALL is made to a new subroutine GEOPR and the output is advanced to a new page through a CALL to subroutine PRERR. If the tests fail, an identifying message is printed. In either case, control returns to Statement 10 to read a new input statement. The process therefore can be specified as follows:

```
1712  GO TO (1713,10) ISCAN
1713  IF (JMIC-NB) 1716,1714,1716
1714  IF(JJC+JDC-NJ) 1716,1715,1716
1715  CALL GEOPR
      CALL PRERR(1)
      GO TO 10
1716  PRINT 1717
1717  FORMAT (44 DATA INCOMPLETE. REQUEST CANNOT
      1BE PROCESSED)
      GO TO 10
```

The new subroutine GEOPR can now be completed. The subroutine will have to deal with four arrays: the coordinate table KXYZ, the incidence tables JPLS and JMIN, and the member-type table MTYP. The latter is needed to check for members that may have been deleted (MTYP=0). Since the subroutine is called from PHAS1A, all arrays needed except MTYP are already allocated. In a subroutine called during the execution phase, it may not be assumed that this condition holds, and any arrays needed must be allocated at the beginning of the subroutine and released just prior to returning to the calling subroutine.

In order to avoid overflow when a member is nearly perpendicular to one of the coordinate axes, the convention will be made that if $\cot \theta_i < 1'$, the output for $\tan \theta_i$ will be the constant 9999.9999 with the sign of $\tan \theta_i$.

The following listing of the program should be studied carefully, especially with regard to the use and significance of the subscript notation.

```
      SUBROUTINE GEOPR
      DIMENSION D(3), T(3)
      DIMENSION U(2),...
      COMMON U, IU,..., NB, NCORD,..., KXYZ, JPLS,
             JMIN, MTYP,...
```

The latter two statements should be duplicated from PHAS1A.

```
      CALL ALOCAT (MTYP)
  10  IF(NCORD-2) 11, 11, 12
  11  PRINT 13
      GO TO 20
  12  PRINT 14
  13  FORMAT (title for planar structures)
  14  FORMAT (title for space structures)
  20  DO 100 M=1, MEXTN
      NTYP=MTYP+M
      NPLS=JPLS+M
      NMIN=JMIN+M
```

The addresses of the Mth entry in the corresponding arrays are NTYP, NPLS, and NMIN. A check for nonexisting members can now be made.

```
      IF(IU(NTYP))30, 100, 30
  30  IPLS=IU(NPLS)
      IMIN=IU(NMIN)
```

The joint numbers of the two joints incident of member M are IPLS and IMIN.

```
      KPLS=KXYZ+(IPLS-1)*3
      KMIN=KXYZ+(IMIN-1)*3
```

The starting (base) addresses in the coordinate table of the coordinates of the two joints are KPLS and KMIN plus one.

```
      SPAN=0.0
  40  DO 50 I=1, NCORD
      KPI=KPLS+I
      KMI=KMIN+I
      D(I)=U(KMI)-U(KPI)
  50  SPAN=SPAN+D(I)*D(I)
```

The addresses of the Ith coordinates of the two joints are KPI and

KMI; D(I) is therefore the projection of the member length on the Ith axis.

```
        SPAN = SQRTF(SPAN)
        TEST=0.0002909*SPAN
  60    DO 70 I=1,NCORD
        IF(ABSF(D(I))- TEST)61,62,62
  61    T(I)=SIGNF(9999.9999,D(I))
        GO TO 70
  62    T(I)=D(I)/SPAN
  70    CONTINUE
  80    PRINT 81,M,IPLS,IMIN,SPAN,(D(I),I=1,NCORD),
        1(T(I),I=1,NCORD)
  81    FORMAT (3II,7F15.4)
 100    CONTINUE
        CALL RELEASE(MTYP)
        RETURN
        END
```

This completes the modification of the system.

7.2 Joint Coordinate Computation

In many applications, it is convenient to define some joint co-ordinates as a function of previously defined coordinates. As an example, the programming additions will be described to include a joint defined by the intersection of two lines defined by two pairs of joints. The input specifications for this addition might be the new statement

        JOINT J (ON) INTERSECTION J1 J2 J3 J4

where the word ON may be included or omitted. The two lines are defined by J1 to J2 and J3 to J4. Any additional descriptive words may be added before the four additional joint numbers. The tabular form would have the header JOINT INTERSECTIONS. The defining joints must all have been previously given. The joint J is assumed to be free unless the word S or SUPPORT follows the last joint number. Two possible errors would have to be detected: either one of the joints has not been given, or the two lines do not intersect.

The word JOINT is translated by PHAS1A, which then trans-fers to JTDAT for further translation. All of the following changes apply to subroutine JTDAT only and should be studied in conjunction with the listing of that subroutine. JTDAT starts by reading an integer or a word given in LIST3. The words ON and INTERSECTION need to be added to LIST3, with the action on detecting ON being to read the next word and the action on detecting INTERSECTION, to transfer to statement 2180.

In LISTS the statements for LIST3 become

```
LIST3  DEC    8
       BCI    6,...
       BCI    1,0000ON
       BCI    1,INTERS
```

The statement 202 is used to skip superfluous words, and should be changed to read

202 GO TO (203,203,203,203,203,200,200,203),I2

Statement 209 transfers on the branch number for reading the rest of the statement. This is changed to read

209 GO TO (210,220,230,230,250,96,96,2180),K2

When statement 209 is reached, J has been read and stored. With a tabular header, K2 has been set when the header is translated, and J read in PHAS1A. Otherwise, J is read in JTDAT.

Two existing arrays will be needed in this program addition, JTYP and KXYZ. The reader should refer to Section 5.3 for their meaning and storage form. The element in JTYP corresponding to a joint number must be nonzero if the joint has been defined. Four integers must follow in the translation process. The reading of these numbers and checking for nonexisting joints is accomplished by

```
2180   DO 2182 IJ=1,4
       I3=MATCH(0,K,2)
       IF(I3-2)94,2181,94
2181   L=JTYP+K
       IF(IU(L))2182,47,2182
2182   LABL(IJ)=K
```

The joint numbers are stored in LABL for processing.

Numbering the points in the order of input and defining the direction cosines of line 1-2 as l_1, m_1, and n_1, and of line 3-4 as l_2, m_2, and n_2, the condition that an intersection exists[1] is

$$(m_1n_2 - m_2n_1)(x_3 - x_1) + (n_1l_2 - n_2l_1)(y_3 - y_1)$$
$$+ (l_1m_2 - l_2m_1)(z_3 - z_1) = 0$$

The distance from point 1 to the intersection is given by

$$r = \left(\frac{(x_3 - x_1)^2 + (y_3 - y_1)^2 + (z_3 - z_1)^2 - \{l_2(x_3 - x_1) + m_2(y_3 - y_1) + n_2(z_3 - z_1)\}^2}{(m_1n_2 - m_2n_1)^2 + (n_1l_2 - n_2l_1)^2 + (l_1m_2 - l_2m_1)^2} \right)^{\frac{1}{2}}$$

---

[1]D. F. Gregory, A Treatise on the Application of Analysis to Solid Geometry, Whittaker, Simpkin, and Bell, London, 1852.

If the denominator equals zero, the lines are parallel. The direction cosines can be computed by

```
          DIMENSION DRCS(3,2)
          DO 2185 I=1,2
          DIST=0.
          DO 2183 II=1,3
          IK=KXYZ+(LABL(2*I)-1)*3+II
          IL=KXYZ+(LABL(2*I-1)-1)*3+II
          DRCS(II,I)=U(IK)-U(IL)
   2183   DIST=DIST+DRCS(II,I)**2
          DIST=SQRTF(DIST)
          DO 2184 II=1,3
   2184   DRCS(II,I)=DRCS(II,I)/DIST
   2185   CONTINUE
```

Coordinate differences used in the formula are computed by

```
          DIMENSION DIFF(3)
          DO 2186 I=1,3
          IK=KXYZ+(LABL(3)-1)*3+I
          IL=KXYZ+(LABL(1)-1)*3+I
   2186   DIFF(I)=U(IK)-U(IL)
```

The intersection test and the parallel test can be calculated together:

```
          SUM=0.
          DIN=0.
          DIST=0.
          DO 2187 I=1,3
          IK=XMINF(I,6-2*I)+1
          IL=(3*I*I-13*I+16)/2
          TEMP=DRCS(IK,1)*DRCS(IL,2)
         1-DRCS(IK,2)*DRCS(IL,1)
          SUM=SUM+DIFF(I)*DIFF(I)
          DIN=DIN+TEMP*TEMP
   2187   DIST=DIFF(I)*TEMP
          IF(ABSF(DIST)-1.E-5)2188,2188,46
   2188   IF(ABSF(DIN)-1.E-5)46,46,2189
   2189   R=SQRTF((SUM-(DRCS(1,2)*DIFF(1)+DRCS(2,2)
         1*DIFF(2)+DRCS(3,2)*DIFF(3))**2)/DIN)
```

Computing and storing the coordinates is done by

```
          IK=KXYZ+(LABL(1)-1)*3
          IL=KXYZ+(J-1)*3
          DO 2190 I=1,3
          II=IK+I
          IJ=IL+I
```

```
2196   U(IJ)=U(II)+R*DRCS(I,1)
       GO TO 210
```

The existing joint-coordinate translation portion can be used to read the joint type and increment the counters since no decimal data follow the last joint number, and no data are to be stored. The last transfer then is to statement 210.

Additional statements for error messages are needed. An error message specifying that an expected integer is not found is included in PRERR with flagging of CHECK and printing of the message already programmed in PHAS1A.

The error-return code from JTDAT need only be extended in PHAS1A to include this error with NE=11. Additions to JTDAT at the end might be

```
94   NE=11
     GO TO 10
47   PRINT 9
B    CHECK=CHECK*777777773777
     GO TO 10
 9   FORMAT (53H THE USE OF DATA FOR AN
     1UNDEFINED JOINT IS ATTEMPTED)
46   PRINT 8
B    CHECK=CHECK*777777767777
     GO TO 10
 8   FORMAT (42H LINES CONNECTING JOINTS DO
     1NOT INTERSECT)
```

The masking of CHECK shown is for unacceptable statements and for incorrect structural data. The GO TO statement in PHAS1A following statement 200 should read

```
GO TO (10,48,90,91,92,93,98,99,96,97,94),NE
```

7.3 Member Loads in Global Coordinates

7.3.1 Input requirements for the new statement.

This example illustrates the programming steps required to implement in the STRESS processor programs a new command that specifies member loads in the global-, rather than a member-, coordinate system.

Such a command is convenient in cases where member loads are acting in a coordinate system that is not parallel to the member-coordinate system of that member, such as vertical loads on an inclined roof beam shown in Figure 7.1.

To implement this capability it is only necessary to generalize the already available statement for member loads. The present statement is of the form

$$M \left\{ \begin{array}{l} \text{FORCE} \quad \begin{array}{l} \text{X} \\ \text{Y} \\ \text{Z} \end{array} \\ \text{MOMENT} \begin{array}{l} \text{X} \\ \text{Y} \\ \text{Z} \end{array} \end{array} \right\} \left\{ \begin{array}{l} \text{CONCENTRATED} \\ \text{UNIFORM} \\ \text{LINEAR} \end{array} \right\} W_A \ W_B \ L_A \ L_B$$

where it is assumed that the direction label X, Y, or Z refers to the member-coordinate system of member M.

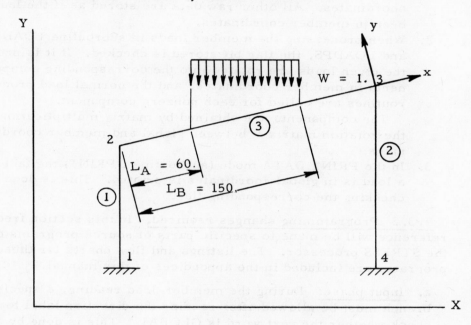

Figure 7.1. Member load in global correspondence.

To implement the desired altered statement we make the following convention: if (and only if) the direction label refers to the global coordinate system, an additional word "GLOBAL" is given after the direction label.   Thus the altered command becomes

$$M \left\{ \begin{array}{l} \text{FORCE} \quad \begin{array}{l} \text{X} \\ \text{Y} \\ \text{Z} \end{array} \\ \text{MOMENT} \begin{array}{l} \text{X} \\ \text{Y} \\ \text{Z} \end{array} \end{array} \right\} \text{GLOBAL} \left\{ \begin{array}{l} \text{CONCENTRATED} \\ \text{UNIFORM} \\ \text{LINEAR} \end{array} \right\} W_A \ W_B \ L_A \ L_B$$

For the example shown in the figure, the command to specify the vertical load would then be

3 FORCE Y GLOBAL UNIFORM -1.0 60. 150.

It is noted that the distances (L_A = 60.) and (L_B = 150.) still refer to the distance along the member axis. This feature, however, is convenient for uniform loads extending over the entire
member (L_A = L_B = 0.).

7.3.2 **Implementation.** The following procedure of implementation will be used:

1. In input phase a flag is set if a member load refers to global
    coordinates. All other raw data are stored as if the load had
    been in member coordinates.
2. When processing the member loads in subroutine LOADPC
    and LOADPS, the flag bit stored is checked. If it is present,
    the given loads are rotated into the corresponding compo-
    nents in member coordinates, and the normal load processor
    routines are called for each nonzero component.

    The components are obtained by matrix multiplication with
    the rotation matrix R between global and member coordi-
    nates.
3. In the PRINT DATA mode (subroutine DPRINT) the fact that
    a load is in global coordinates is printed. This is done by
    checking the corresponding flag.

7.3.3 **Programming changes required.** In this section frequent
reference will be made to specific parts of source programs of
the STRESS processor. The listings and flow charts for these
programs are included in the appendices of this manual.

a. **Input phase.** During the member-load reading, a special
branch must be allowed after reading the direction label to
check whether the next word is GLOBAL. This is done by altering subroutine READ as follows: a 38th branch is added to
statement 105 going to a new statement 2000 (word 38 in LIST5
is now "GLOBAL").

```
      105  GO TO(106,...,2000),K
      2000 NGLOB=1
           GO TO 100
```

The packing statement before statement 150 is used to store
the flag NGLOB into an unused portion of the load descriptor
word

```
      CALL PACKW(IU(IJ),NGLOB,N,LABL(1),LABL(2),0)
      NGLOB=0
```

This stores the flag into the prefix of the second headerword
of the load block for the raw data belonging to this member
load. (See Section 5.4.5.)

In addition subroutine LIST5 must be changed to read as follows:

```
        LIST5  DEC    38
               BCI    9,00...
               BCI    9,...
               BCI    8,...
               BCI    8,...
               BCI    3,...00IZDISTOR
               BCI    1,GLOBAL
               END
```

b. Execution phase. The only changes required are in the
member-load processor subroutines:

Additional statements are required in subroutine LOADPS
to unpack and check the flag NGLOB set in READ. If the member load is in global coordinates (NGLOB=1), a special newly
written subroutine GLOMLD is called to rotate the given loads
into member coordinates and call MEMBLD for each component.

The required changes in subroutine LOADPS are (a) statement after statement 4, change to

CALL UPACW(U(NLS),NGLOB,N,J,K,B)

and (b) before statement 206, replace "GO TO 30" by

```
        IF(NGLOB)2000,30,2000
  2000  CALL GLOMLD
        GO TO 149
```

Subroutine GLOMLD must be newly written and monitors the processing of the member load under consideration on member JM.

7.3.4 Subroutine GLOMLD. The following special cases are
to be considered.

Trusses: Since member forces on trusses must be in the
member x direction, such forces can only be
given in global coordinates if the member is
parallel to one of the global axes. If such a
load is specified so that force components normal to the member axis result, only the component in the member-axis direction will be
considered. Moment components will be neg-
lected.

Plane frame: Moment components of prescribed member
loads can only act around the Z (or z) axis.

Plane grid: Force components of prescribed member loads
can only act along the Z (or z) axis.

The flow chart for subroutine GLOBLD is shown in Figure 7.2.
The following conventions are used in the subroutine:

J = load direction in member coordinates
   (J=1,2,3 means force x,y,z, respectively)

(J=4, 5, 6 means moment x, y, z, respectively)
F = applied force in global coordinates

A listing of the subroutine follows.

```
C        FORTRAN PROGRAM FOR GLOMLD
         SUBROUTINE GLOMLD
C        COMMON STATEMENT MUST AGREE WITH
            MEMBLD
         COMMON U, IU, Y, T...
         DIMENSION...
         EQUIVALENCE...
      1  CALL TRAMAT (JM, 2)
C        SAVE RAW DATA (INTENSITIES) IN TEMP
         TEMP1=U(NLS+1)
         TEMP2=U(NLS+2)
         JTE=J
         JKS=1
         J=1
         GO TO (10, 20, 30, 10, 40)ID
C        TRUSSES
     10  ICN=1
         GO TO 900
C        PLANE FRAME
     20  IF(J-6)21, 22, 22
C        IF J=6 MZ IS GIVEN, GLOBAL=LOCAL
     22  ICN=1
         GO TO 900
     21  ICN=2
         GO TO 900
C        PLANE GRID
     30  IF(J-3)31, 32, 31
     32  ICN=1
         GO TO 900
     31  ICN=2
         J=4
         JTE=J-3
C        SHIFT COMPONENTS BY 3 SINCE MOMENT
C        IS GIVEN
         GO TO 900
C        SPACE FRAME
     40  ICN=3
         IF(J-3)900, 900, 42
     42  J=4
    900  DO 950 I=1, ICN
         U(NLS+1)=TEMP1*T(JKS, JTE)
         IF(K-2)54, 54, 53
     53  U(NLS+2)=TEMP2*T(JKS, JTE)
```

```
C          K=3 MEANS LINEAR LOAD, 2 INTENSITIES
C          GIVEN
C          NLS IS A POINTER TO RAW DATA UNDER
C          CONSIDERATION
   54  CALL MEMBLD(U(JC+1))
       NDEX=2
       CALL LSTOR
       J=J+1
  950  JKS=JKS+1
C          RESTORE RAW DATA IN MLOAD
       U(NLS+1)=TEMP1
       U(NLS+2)=TEMP2
       RETURN
       END
```

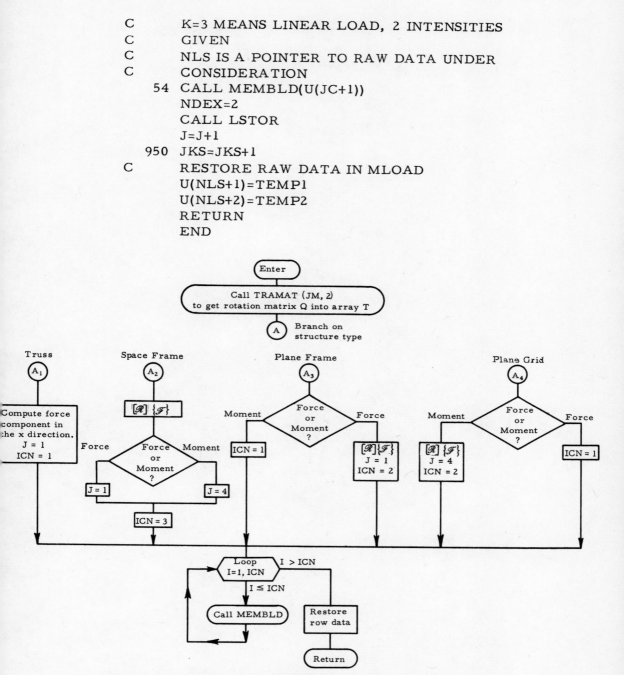

Figure 7.2.   Flow chart for new subroutine GLOMLD.

We shall not discuss here the changes in subroutine DPRINT that are necessary to differentiate between global and member coordinates for the member loads on the output. They would merely consist of one check of the corresponding bit NGLOB, and a corresponding message or label on the output.

Appendix A

FLOW CHARTS

OVER-ALL LOGICAL FLOW CHART

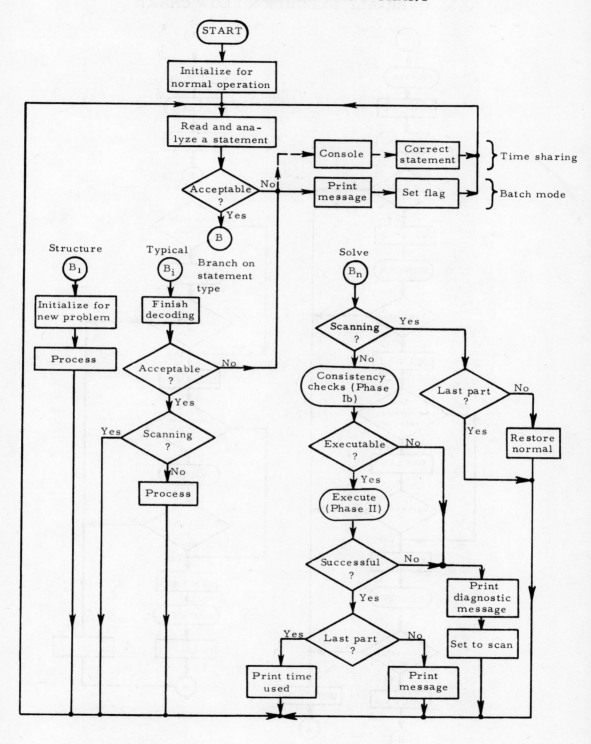

94

OVER-ALL EXECUTION FLOW CHART

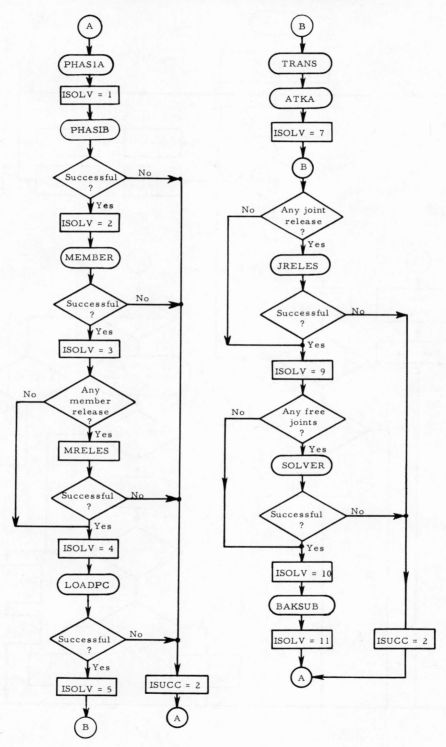

MAIN LINK 1

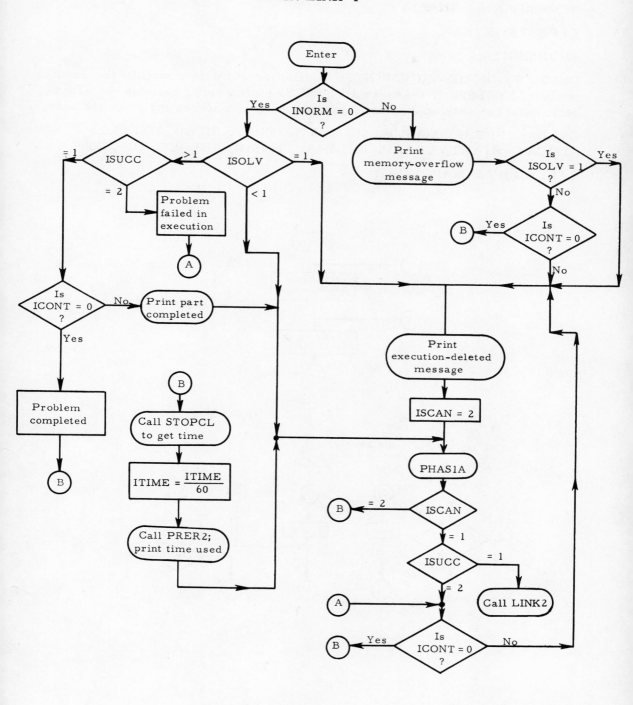

SUBROUTINE: PHAS1A

TYPE: FORTRAN

ARGUMENTS: None

DESCRIPTION OF ARGUMENTS: Subroutine PHAS1A controls the input phase of STRESS. It reads in all STRESS statements, decodes the alphabetic data and sets appropriate parameters, and stores the numeric data.

CALLS: PRERR, MATCH, MEMDAT, JTDAT, SIZED, DEFINE, ALOCAT, START, UPADP, RELEAS, PHAS1B, SELOUT, DPRINT

CALLED BY: MAIN LINK1

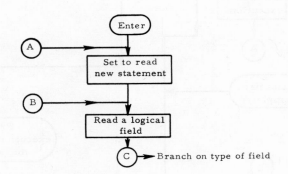

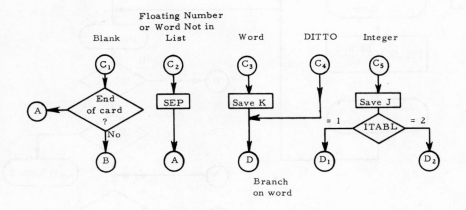

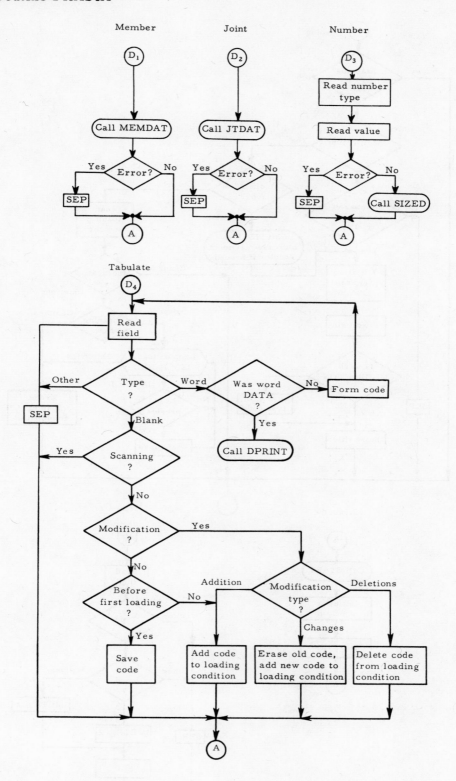

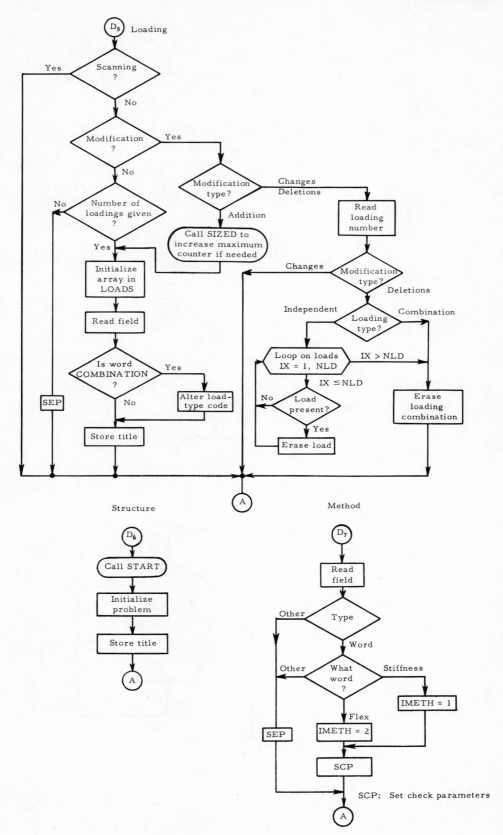

SCP: Set check parameters

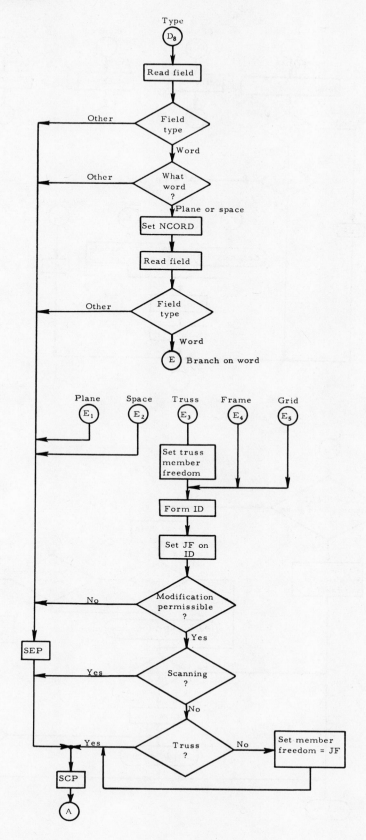

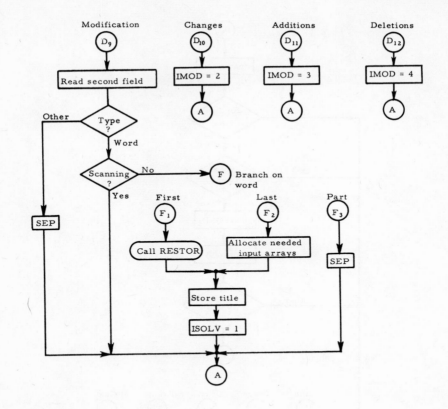

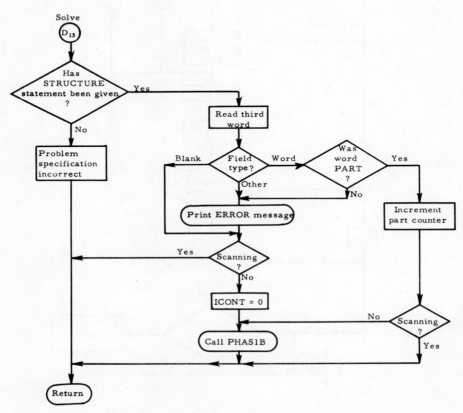

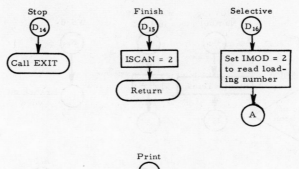

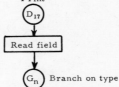

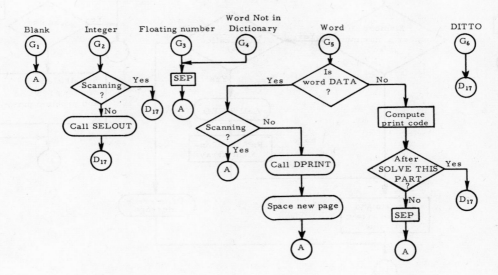

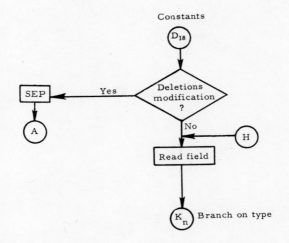

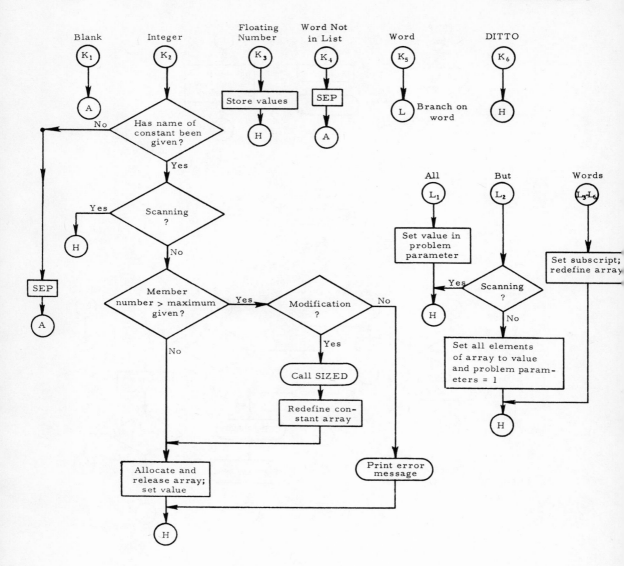

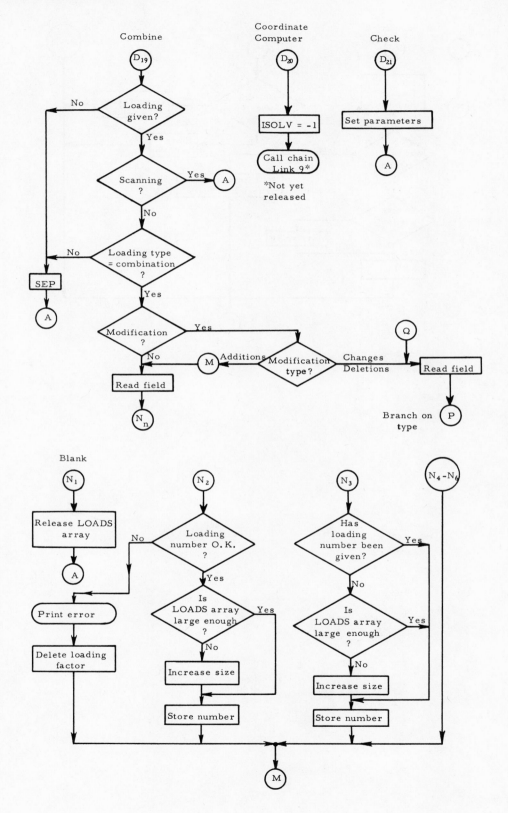

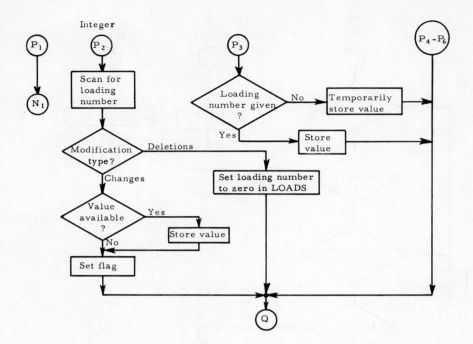

SUBROUTINE: LISTS

TYPE: FAP

ARGUMENTS: None

DESCRIPTION OF PROGRAM: Subroutine LISTS consists of ten lists of words (each list available under its own entry point) that are used as a dictionary by subroutine MATCH. MATCH compares an inputted word with a specified list (entry points: LIST 1, LIST 2, etc.). The first six letters only are used.

CALLED BY: All translating routines

ENTRY POINT	WORDS
LIST 1	MEMBER, JOINT, NUMBER, TABULA, LOADIN, STRUCT, METHOD, TYPE, MODIFI, CHANGE, ADDITI, DELETI, SOLVE, STOP, FINISH, SELECT, PRINT, CONSTA, COMBIN, COORDI, CHECK
LIST 2	DISPLA, DISTOR, REACTI, FORCE, DATA, ALL, MEMBER, JOINT
LIST 3	COORDI, RELEAS, LOADS, LOAD, DISPLA, NUMBER
LIST 4	JOINT, MEMBER, SUPPOR, LOADIN
LIST 5	X, Y, Z, FORCE, MOMENT, DISPLA, ROTATI, A, I, END, START, CONCEN, UNIFOR, LINEAR, P, W, WA, WB, LA, LB, S, F, SUPPOR, FREE, AREA, INERTI, L, LENGTH, STEEL, BETA, AX, AY, AZ, IX, IY, IZ, DISTOR
LIST 6	FROM, GOES, NUMBER, INCIDE, PROPER, RELEAS, CONSTR, DISTOR, LOAD, LOADS, PRISMA, STIFFN, FLEXIB, VARIAB, STEEL, END
LIST 7	PRISMA, STIFFN, FLEXIB, STEEL, VARIAB
LIST 8	ALL, BUT, E, G, CTE, DENSIT
LIST 10	FIRST, LAST, PART
LIST 11	PLANE, SPACE, TRUSS, FRAME, GRID

SUBROUTINE: MEMDAT

TYPE: FORTRAN

ARGUMENTS: None

DESCRIPTION OF PROGRAM: Subroutine MEMDAT controls the reading and storage of input data that deal with members.

CALLS: MATCH, READ, ICNT, DEFINE

CALLED BY: PHAS1A

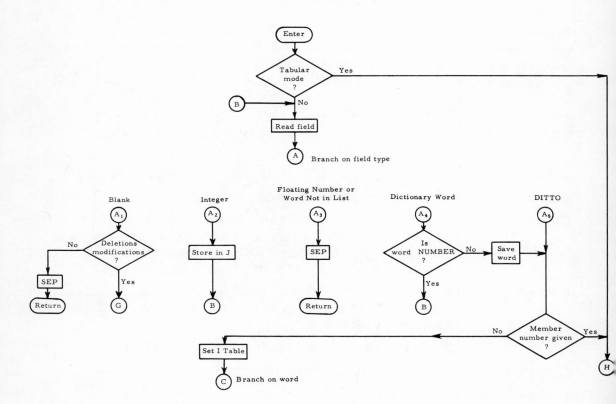

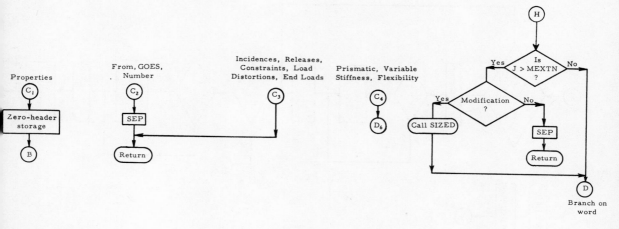

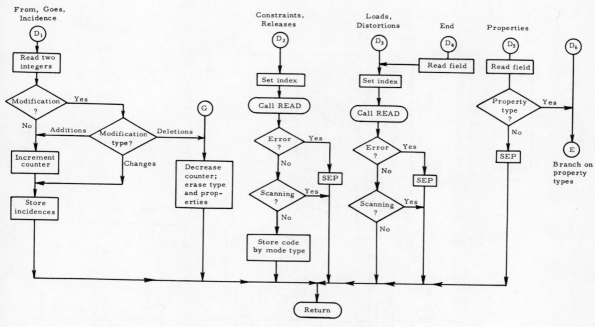

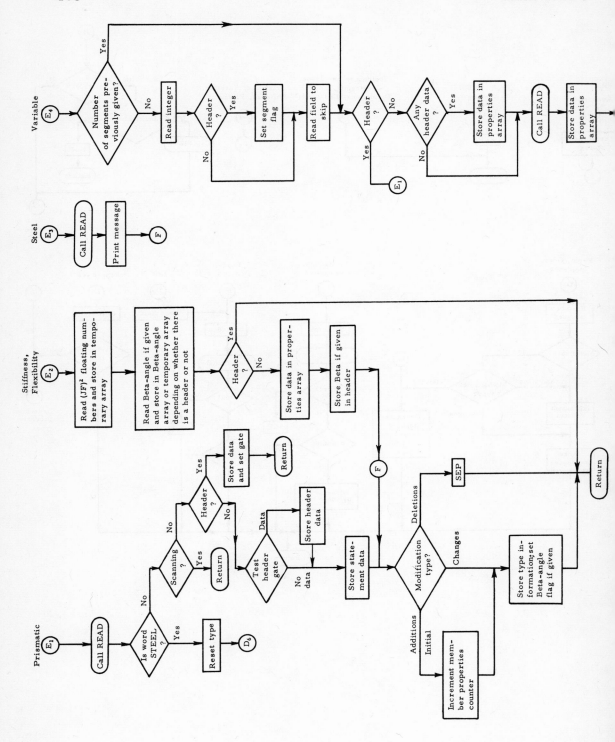

SUBROUTINE: SIZED

TYPE: FORTRAN

ARGUMENTS: J, K, L

DESCRIPTION OF ARGUMENTS:

Argument L gives the type of input statement being processed:

L = 1: Indicates that there is a NUMBER statement during the initial
data input or during the CHANGES mode of MODIFICATION;

L = 2: Indicates that during MODIFICATION a joint member or loading is greater than JEXTN, MEXTN, or LEXTN.

J	L	Calling SIZED from statement	K
1	1	NUMBER OF JOINTS	Number of joints
	2	JOINT COORDINATES	Number of a joint
2	1	NUMBER OF MEMBERS	Number of members
	2	MEMBER INCIDENCES	Number of a member
	2	MEMBER PROPERTIES	Number of a member
3	1	NUMBER OF SUPPORTS	Number of supports
4	1	NUMBER OF LOADINGS	Number of loadings
	2	LOADING HEADER	Highest loading number

DESCRIPTION OF PROGRAM: Subroutine SIZED defines and allocates,
or redefines during MODIFICATION, the various arrays (whose sizes are
functions of size statements) used to store input data; it also sets the corresponding size parameters.

CALLS: DEFINE, ALOCAT

CALLED BY: PHAS1A, MEMDAT, JTDAT

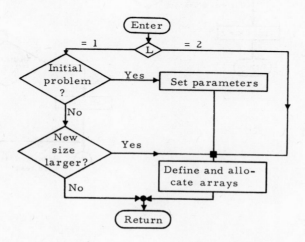

SUBROUTINE: JTDAT

TYPE: FORTRAN

ARGUMENTS: None

DESCRIPTION OF PROGRAM: Subroutine JTDAT controls the reading
and storage of input data that deal with joints.

CALLS: MATCH, READ, ICNT, DEFINE, SIZED

CALLED BY: PHAS1A

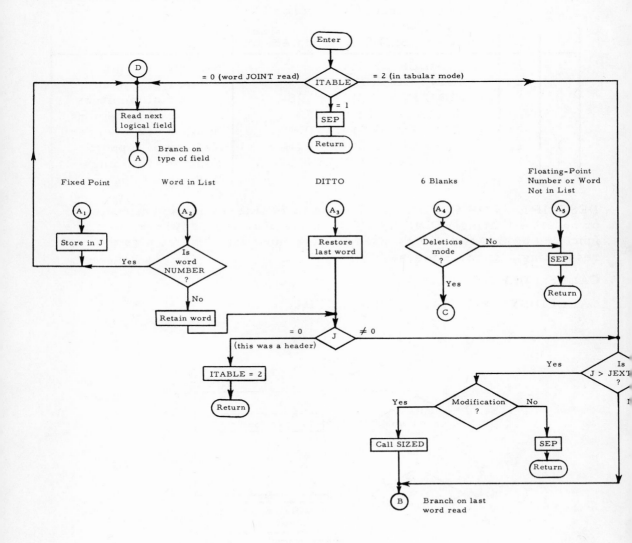

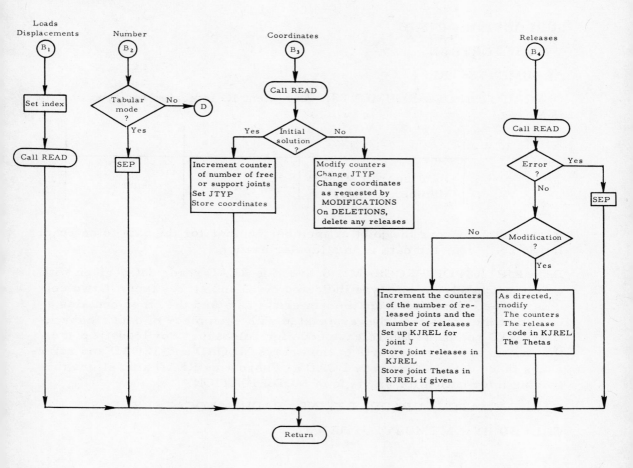

SUBROUTINE: READ

TYPE: FORTRAN

ARGUMENTS: IG, J

DESCRIPTION OF ARGUMENTS: Argument IG indicates the type of data
to be read.

IG =	1	2	3
	Joint coordinates Member properties	Releases	Loads

Argument J conveys the joint or member number for the data to be read.
If J is not given, the data is in a tabular header.

DESCRIPTION OF PROGRAM: Subroutine READ reads data, given with
or without labels, of the type indicated by IG and stores them in two columns: one column contains the numerical data and the other contains a
code that indicates the label associated with each piece of data. Subroutine READ compacts the release code for releases. For loads it places
the data in appropriate parts of the arrays MLOAD and JLOAD and references these data in the array LOADS. Subroutine READ also alters this
loading information for any type of MODIFICATION.

CALLS: MATCH, UPADP, DEFINE, PADP, ALOCAT, RELEAS

CALLED BY: MEMDAT, JTDAT

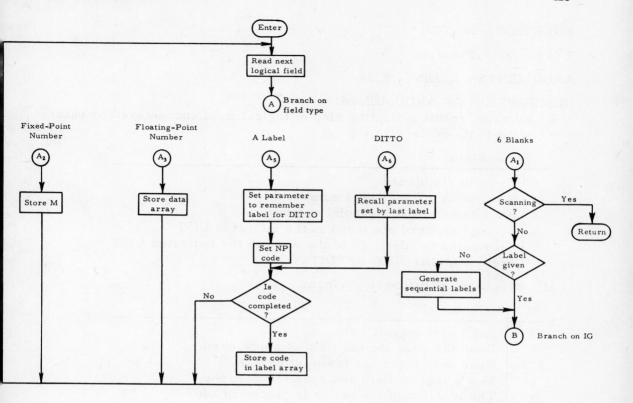

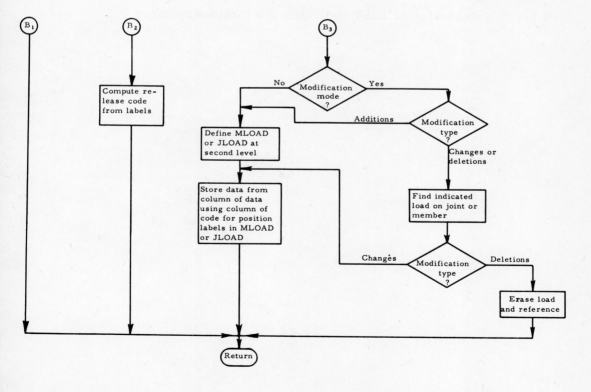

FUNCTION: MATCH

TYPE: FAP Function

ARGUMENTS: I, LIST, K, M

DESCRIPTION OF ARGUMENTS:
  I: Function result indicating kind of logical field encountered or nature
     of contents of K

I	Field
1	Field is six blanks
2	K contains a fixed-point number
3	K contains a floating-point number
4	K contains word not found in the indicated LIST
5	K contains the position of the word in the indicated LIST
6	Field contains "DO" or "DITTO"

  M: Specifies operation to perform

M	Operation
0	Read next logical field
1	Read the first logical field of a new card
2, 3	Read only a number (skip letters)
4	Skip a logical field and read the next one
5	The location of the buffer is placed in LIST

  LIST: Unless M = 5, LIST indicates the dictionary name

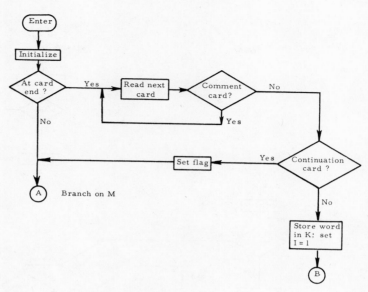

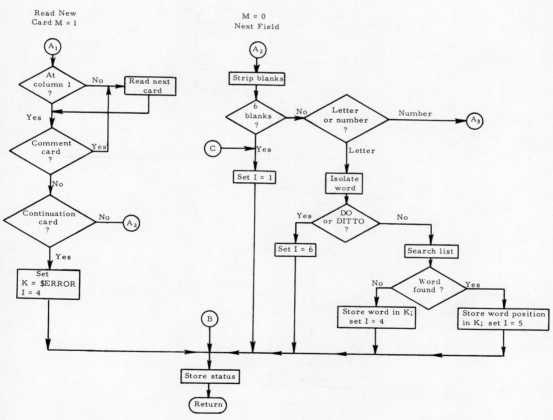

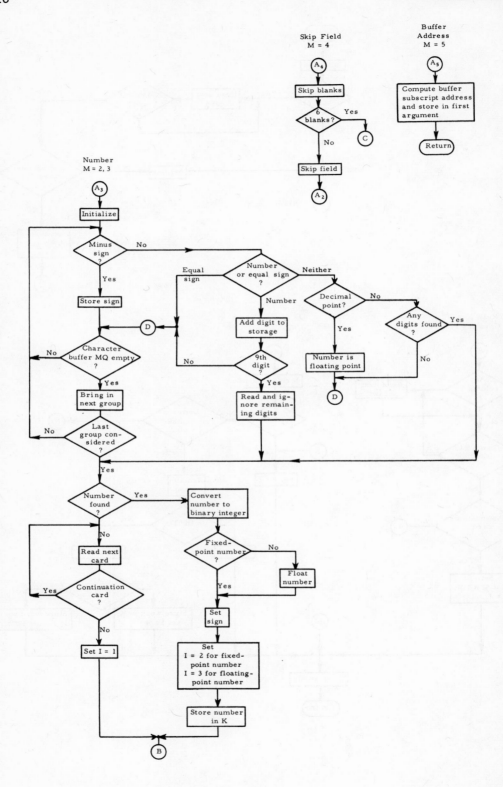

SUBROUTINE: DPRINT

TYPE: FORTRAN

ARGUMENTS: None

DESCRIPTION OF PROGRAM: Subroutine DPRINT prints out in tabular form the structural and loading data associated with the state of the problem at the time DPRINT is called.

CALLS: ALOCAT, UPACW, RELEAS

CALLED BY: PHAS1A

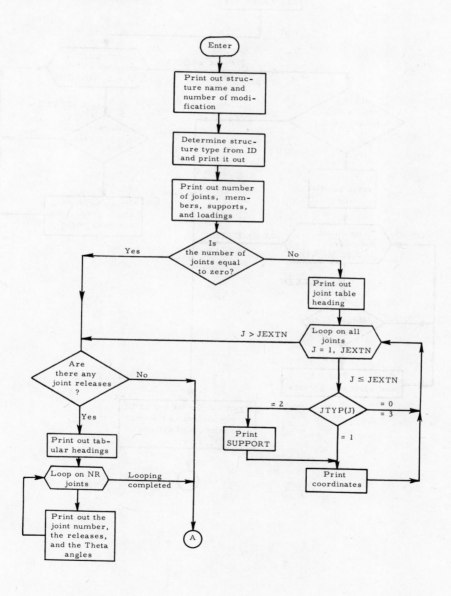

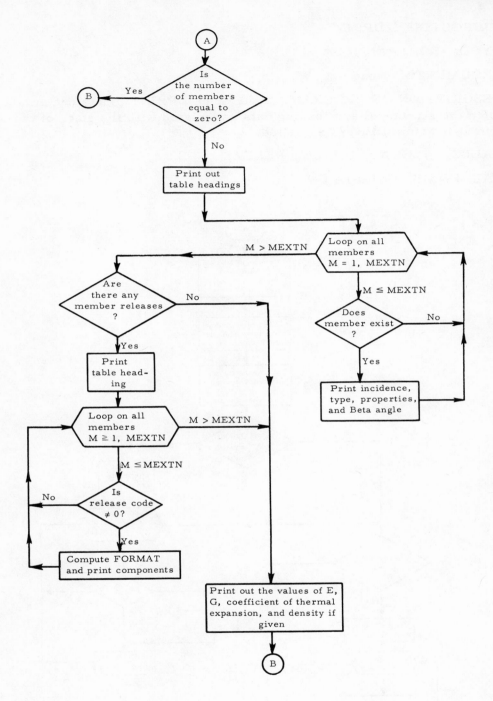

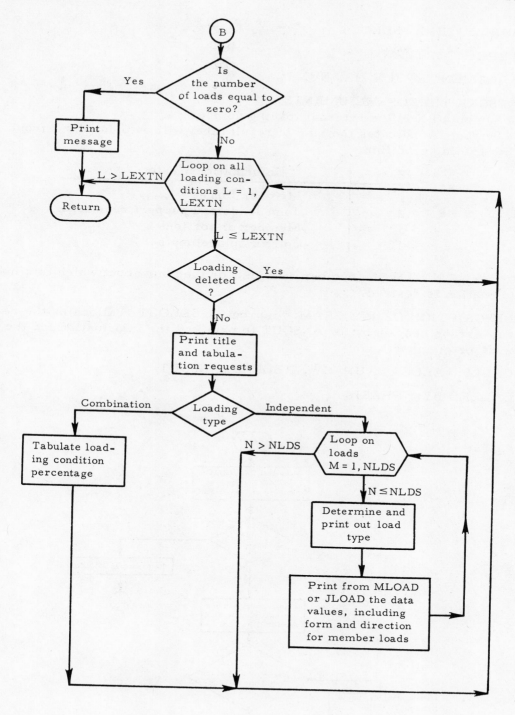

SUBROUTINE: SELOUT

TYPE: FORTRAN

ARGUMENTS: JLX, IZ, NUM

DESCRIPTION OF ARGUMENTS:
   Symbol JLX is the external loading number
   Symbol IZ indicates the kind of results that will be printed out from
this loading condition.

IZ	Information Wanted
1	Member forces
2	Joint loads and support reactions
3	Member distortions
4	Joint displacements

   Symbol NUM is the number of a joint or member about which this information is desired.

DESCRIPTION OF PROGRAM: Subroutine SELOUT reallocates the result arrays necessary for ANSOUT to print out the information for the
joint or member.

CALLS: ALOCAT, UPACW, RELEAS, ANSOUT

CALLED BY: PHAS1A

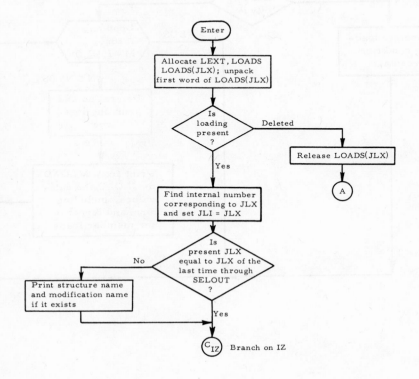

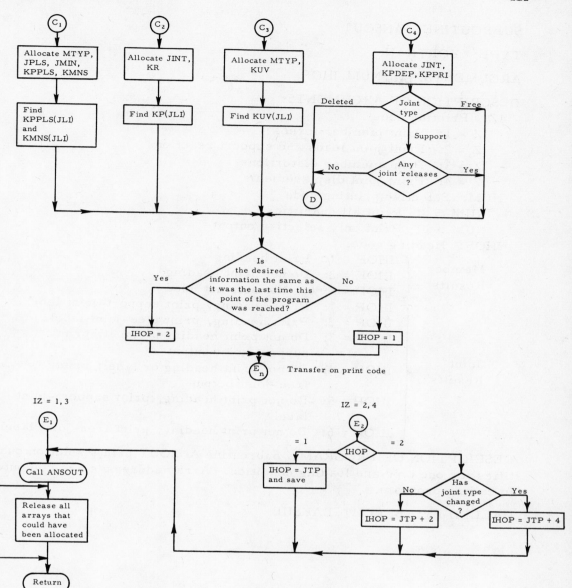

JTP = 3 − Joint type $\begin{cases} 1: \text{support} \\ 2: \text{free} \end{cases}$

SUBROUTINE: ANSOUT

TYPE: FORTRAN

ARGUMENTS: IZ, NUM, IHOP

DESCRIPTION OF ARGUMENTS:

IZ: Printing code

   IZ = 1: Print member forces
   IZ = 2: Print joint loads and support reactions
   IZ = 3: Print member distortions
   IZ = 4: Print joint displacements

NUM: Selective printing code

   NUM ≤ 0: Print all quantities
   NUM > 0: Print only selective output

IHOP: Heading code.

Member Results
$\left\{\begin{array}{l}\text{IHOP = 1: Print heading}\\ \text{IHOP = 2: Do not print heading}\\ \text{IHOP = 3}\rightarrow\text{6: Not set, not used}\end{array}\right.$

Joint Results
$\left\{\begin{array}{l}\text{IHOP = 1: Print heading, print support-joint label}\\ \text{IHOP = 2: Print heading, print free-joint label}\\ \text{IHOP = 3: Do not print heading or label; print support-joint output}\\ \text{IHOP = 4: Do not print heading or label; print free-joint output}\\ \text{IHOP = 5: Do not print heading, print support-joint label}\\ \text{IHOP = 6: Do not print heading, print free-joint label}\end{array}\right.$

DESCRIPTION OF PROGRAM: Subroutine ANSOUT prints solution quantities for one type and loading condition. Array addresses are computed in calling program.

CALLED BY: SELOUT, BAKSUB

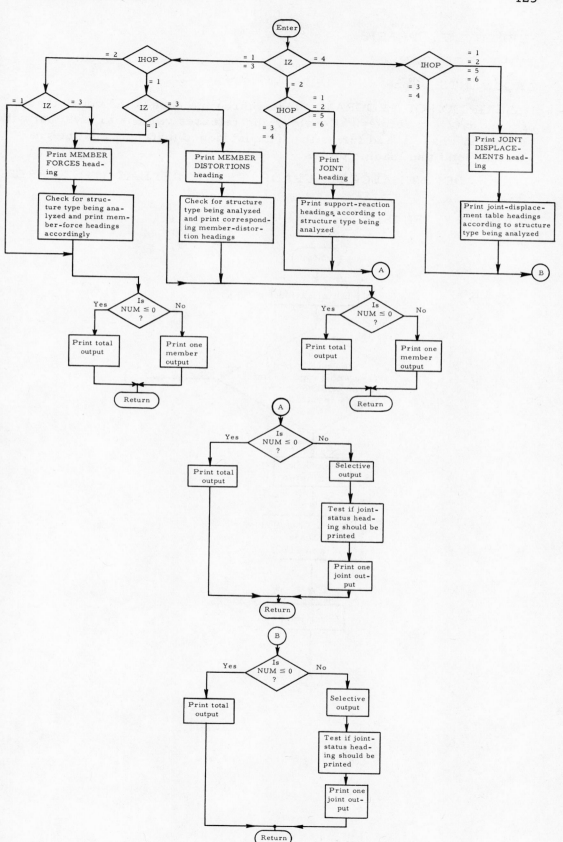

SUBROUTINE: PHAS1B

TYPE: FORTRAN

ARGUMENTS: None

DESCRIPTION OF PROGRAM: This subroutine checks the consistency
of input data. It defines, allocates, and releases various arrays that will
be used in Phase II and forms the joint incidence matrices and internal-
external joint and loading tables.

CALLS: DEFINE, ALOCAT, UPADP, PRER2, RELEAS, ITEST, PRERR,
SAVE

CALLED BY: PHAS1A

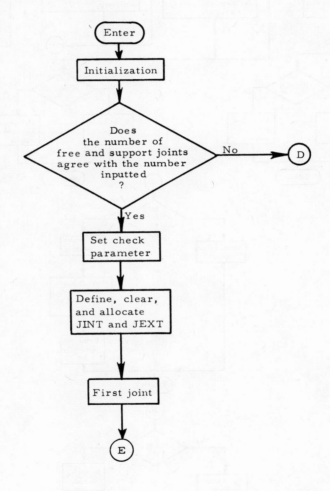

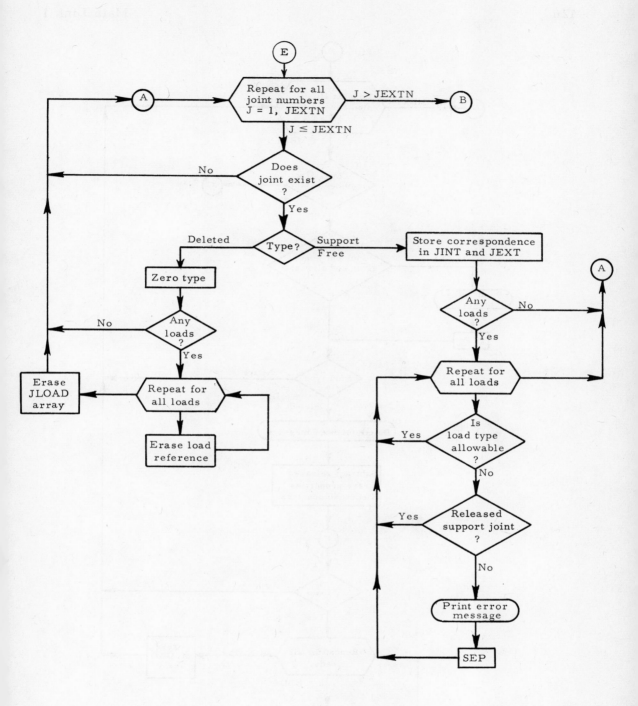

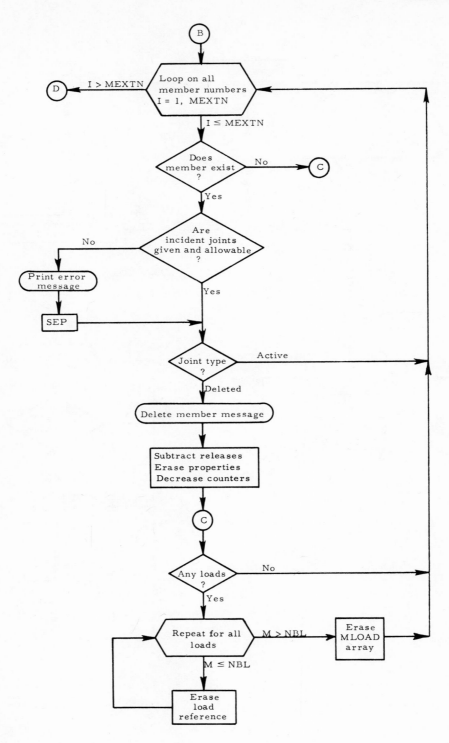

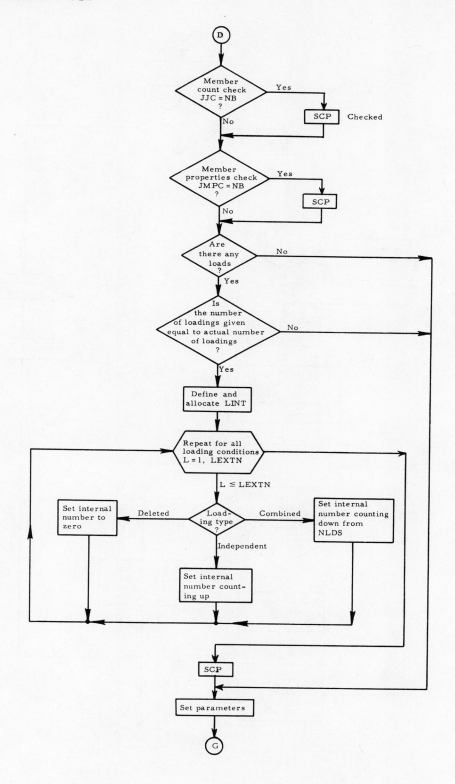

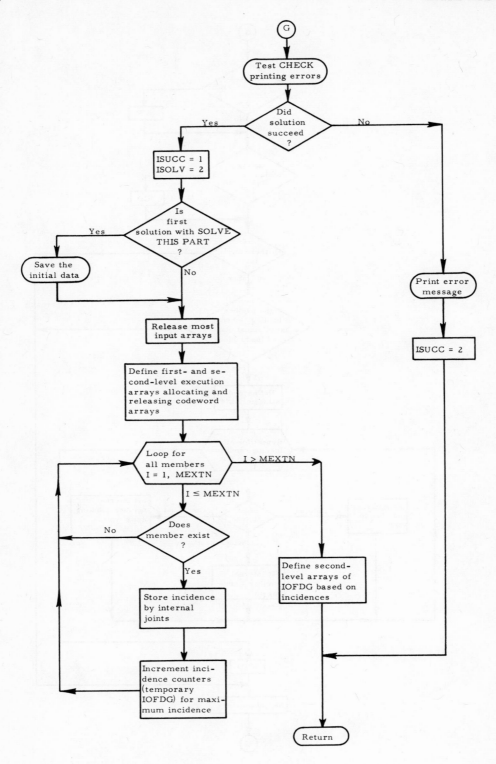

SUBROUTINE: RESTOR

TYPE: FORTRAN

ARGUMENTS: None

DESCRIPTION OF PROGRAM: Subroutine RESTOR returns to core the
input data of the initial specification of the problem for a MODIFICATION
OF FIRST PART and rewinds the scratch tapes.

CALL: FILES

CALLED BY: PHAS1A

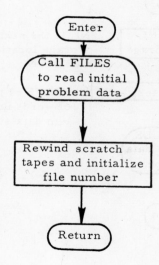

SUBROUTINE: SAVE

TYPE: FORTRAN

ARGUMENTS: None

DESCRIPTION OF PROGRAM: Subroutine SAVE stores on tape all of the input data of the problem when called.

CALLS: UPADP, FILES

CALLED BY: PHAS1B

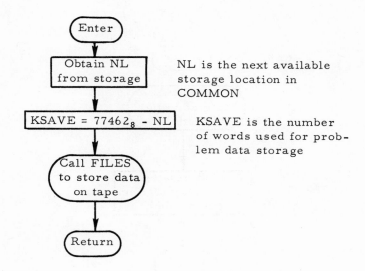

Enter

Obtain NL from storage → NL is the next available storage location in COMMON

$KSAVE = 77462_8 - NL$ → KSAVE is the number of words used for problem data storage

Call FILES to store data on tape

Return

SUBROUTINE: ITEST

TYPE: FAP

ARGUMENTS: A, B, L, I

DESCRIPTION OF ARGUMENTS:
   A: logical word to be tested
   B: Mask used in testing
   L: If 0, no diagnostic printing
   I: Set by ITEST
       I = 0 if A matches B
       I = 1 if A ≠ B

DESCRIPTION OF PROGRAM: Argument A is logically matched against
B. If all bits of A match B, I is set to 0. If any bits do not match, PRERR
is used to print $(L + K)^{th}$ error message, where K = bit position of A in
error (K is counted from the right of A).

CALLS: PRERR

CALLED BY: PHAS1B

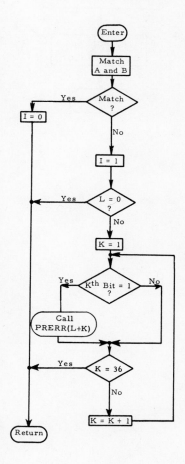

FUNCTION: ICNT

TYPE: FAP

ARGUMENT: I

DESCRIPTION OF ARGUMENTS: The word I contains the release codes.

DESCRIPTION OF PROGRAM: Function ICNT counts the number of joint
or member releases on a single support joint or member.

CALLED BY: PHAS1B

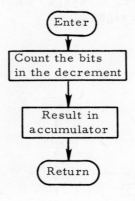

SUBROUTINE: CLEAR(NAM)

TYPE: FAP

ARGUMENT: NAM

DESCRIPTION OF ARGUMENT: A codeword

DESCRIPTION OF PROGRAM: Subroutine CLEAR is called if an array must be set to zero before the next use.

CALLED BY: Many programs

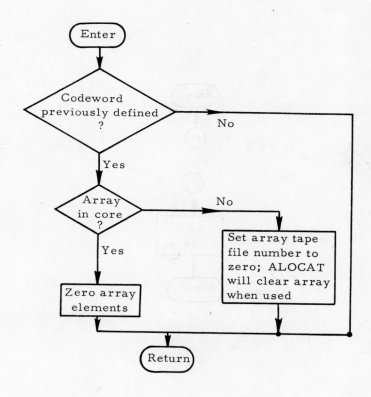

SUBROUTINE: PRERR

TYPE: FORTRAN

ARGUMENTS: J

DESCRIPTION OF ARGUMENTS: Argument J determines which one of the 36 error messages contained in PRERR is to be printed out.

DESCRIPTION OF PROGRAM: Subroutine PRERR prints one message with no numerical information.

CALLED BY: PHAS1A, ITEST

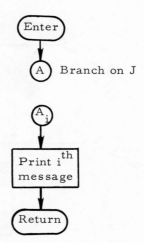

SUBROUTINE: PRER2

TYPE: FORTRAN

ARGUMENTS: J, N, A

DESCRIPTION OF ARGUMENTS:
  J: Indicates which one of the 18 error messages is to be printed
  N and A: Parameters printed in the message

CALLED BY: PHAS1A, READ, PHAS1B, etc.

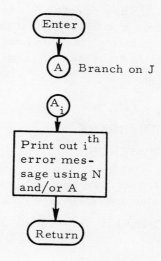

Allocator: Pertaining to Subroutines START, DEFINE,
ALOCAT, RELEAS, REORG, FILES, DUMP, DUMPER,
SSLAD, STER.

The dynamic memory allocation used in this process is accomplished
with a small series of subroutines. The majority of these routines are
coded in FAP and are grouped into a single subroutine with multiple entries, which are ALOCAT, DEFINE, RELEAS, REORG, START, and
DUMP. Routine REORG is called from other parts of the allocator (that
is, in the FAP subroutine) or from the modified library routine CHAIN
if the next program link would overlay data. Other portions of this FAP
routine are called by more than one of the entries and therefore are shown
in separate flow charts. They are SSLAD and STER.

Two FORTRAN routines are also part of the allocator: FILES for tape
handling and DUMP for printing intermediate dumps.

The allocator routines may be called from any program.

SUBROUTINE: START

TYPE: FAP

ARGUMENTS: NT, NCW, NMAX

DESCRIPTION OF ARGUMENTS:
  NT: Not used
  NCW: Size of the fixed arrays and codewords
  NMAX: Size of the space available for arrays in Link 1

DESCRIPTION OF PROGRAM: Subroutine START initializes storage of
arrays.

CALL: FILES

CALLED BY: PHAS1A

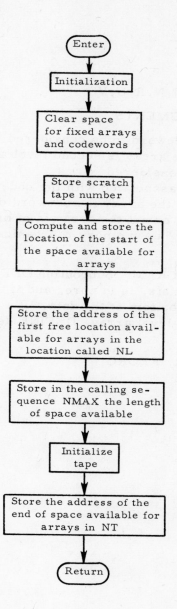

SUBROUTINE: DEFINE

TYPE: FAP

ARGUMENTS: NAME, J, N, R, U, V

DESCRIPTION OF ARGUMENTS:

NAME: Indicates codeword to be considered
J: Present only if the array is second level and indicates the secondary codewords to be considered
N: Size of the array associated with the codeword
R: Indicates whether the array is codeword or data
U: Indicates whether or not the array is of ordinary preference
V: Indicates whether in a memory reorganization the array must be erased or saved

DESCRIPTION OF PROGRAM: Subroutine DEFINE sets up codewords for arrays. If the array exists, is in core, and if the size is altered, DEFINE moves the array to the bottom of the data. If a second-level array is being redefined as a first-level array, DEFINE destroys the data-array back-references.

CALLS: SSLAD, STER, REORG

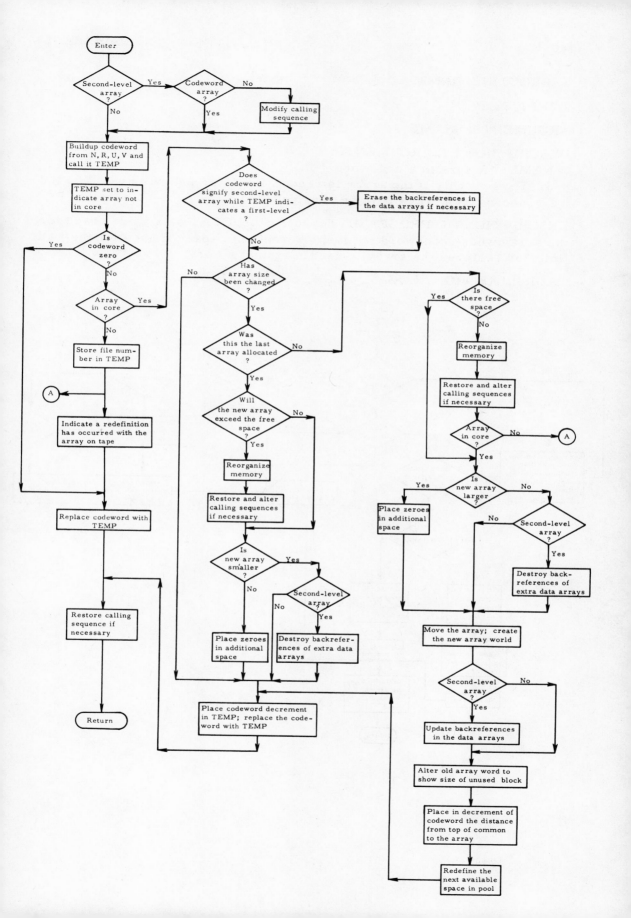

SUBROUTINE: ALOCAT

TYPE: FAP

ARGUMENTS: NAME, J

DESCRIPTION OF ARGUMENTS:

  NAME: Refers to a codeword
  J: Present only if the array is second level and indicates which of the
     secondary arrays is to be considered

DESCRIPTION OF PROGRAM: Subroutine ALOCAT sets aside space for
the array whose codeword is given, zeroes this space if used for the first
time, or restores the array to needed status in core.

CALLS: REORG, ALOCAT, STER, SSLAD, FILES

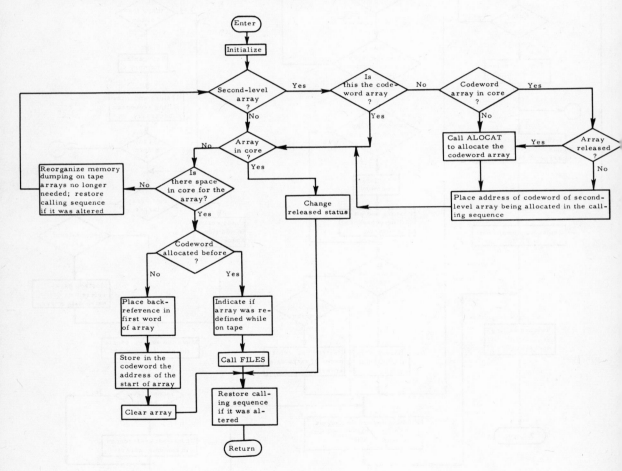

SUBROUTINE: RELEAS

TYPE: FAP

ARGUMENTS: NAME, J, u, v

DESCRIPTION OF ARGUMENTS:

  NAME: The codeword

  J: Present only if the array is second level and indicates which of the secondary arrays is to be considered

  u: Indicates whether or not the array is of ordinary preference

  v: Indicates whether the array is to be erased or saved in a memory reorganization

DESCRIPTION OF PROGRAM: Subroutine RELEAS alters the codeword if u and v are given. It also modifies the array word to show that the array is not needed.

CALLS: SSLAD, STER

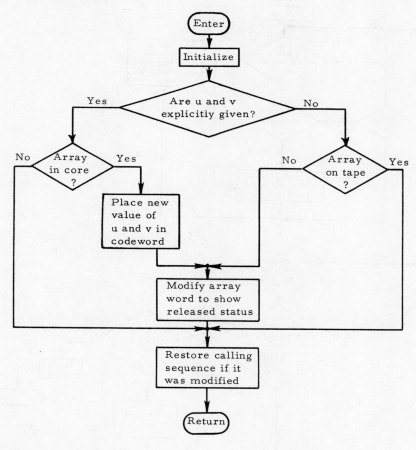

SUBROUTINE: REORG

TYPE: FAP

ARGUMENTS: None

DESCRIPTION OF PROGRAM: Subroutine REORG compacts data in core memory toward the top, erasing arrays that are no longer needed and placing on tape those not presently needed.

CALLS: FILES, CHAIN

CALLED BY: CHAIN, ALOCAT, DEFINE

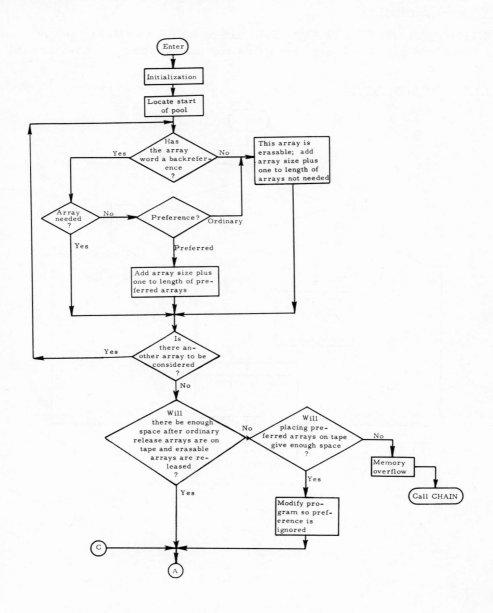

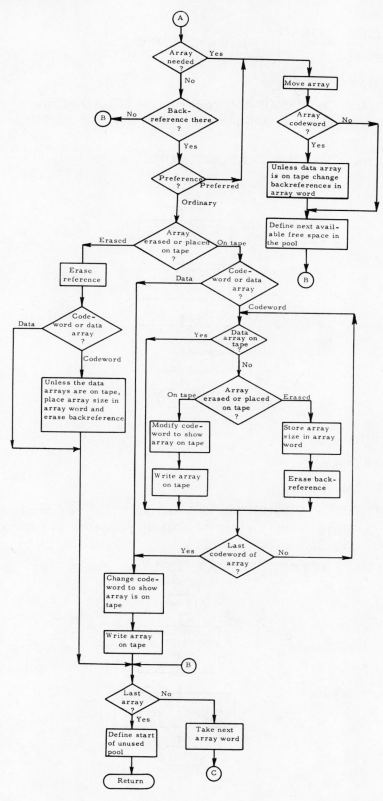

SUBROUTINE: SSLAD

TYPE: FAP

ARGUMENTS: None

DESCRIPTION OF PROGRAM: Subroutine SSLAD computes the address of a second-level array and places it in the calling sequence of the calling program.

CALLED BY: DEFINE, ALOCAT, RELEAS

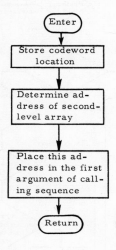

SUBROUTINE: STER

TYPE: FAP

ARGUMENTS: None

DESCRIPTION OF PROGRAM: Subroutine STER restores codeword address in calling sequence when the calling sequence has been altered by SSLAD.

CALLED BY: DEFINE, ALOCAT, RELEAS

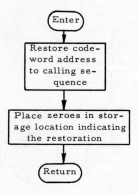

SUBROUTINE: DUMP

TYPE: FAP

ARGUMENTS: NAME, J

DESCRIPTION OF ARGUMENTS:
  NAME: The codeword
  J: Present only if the array is second level and indicates which of the
      secondary arrays is to be considered

DESCRIPTION OF PROGRAM: Subroutine DUMP prints the array which
has the codeword NAME, J.

CALLS: ALOCAT, RELEAS, DUMPER

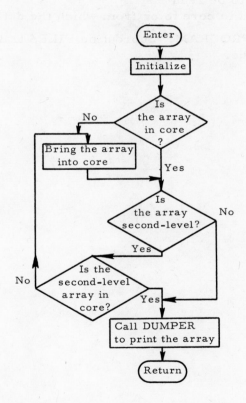

SUBROUTINE: FILES

TYPE: FORTRAN

ARGUMENTS: NOP, NT, NFILE, NCOUNT, ARRAY

DESCRIPTION OF ARGUMENTS:
  NOP: Indicates which operation is to be performed

NOP	Operation
1	Initialize tape
2	Write an array on tape
3	Read an array from tape
4	Read an array that has been redefined
5	Empty buffer

  NT: Tape number
  NFILE: File number
  NCOUNT: Size of the array to be written or the expected size of the
          array to be read
  ARRAY: Location in core to or from which the data is taken

DESCRIPTION OF PROGRAM: Subroutine FILES transmits information
between core and tape.

CALLS: CHAIN

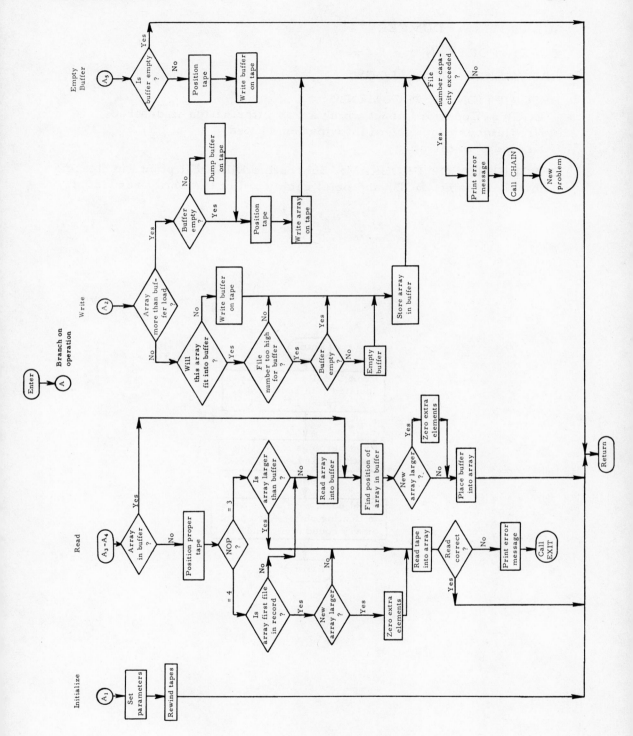

SUBROUTINE: DUMPER

TYPE: FORTRAN

ARGUMENTS: K, N, LOCW

DESCRIPTION OF ARGUMENTS:
  LOCW: Codeword about which array information is desired
  N: Number of pieces of information wanted
  K: Index to the array

DESCRIPTION OF PROGRAM: Subroutine DUMPER prints in floating point, fixed point, BCD, and octal each word of the array requested.

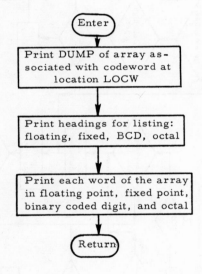

SUBROUTINE: PACKW

The following packing and unpacking programs are combined into one subroutine.

1.   ENTRY: PACKW

    TYPE: FAP

    ARGUMENTS: A, I, J, K, L, M

    DESCRIPTION OF PROGRAM:

> Word A is set to zero
> Then Bits 15, 16, 17 of I → Bits S, 1, 2 of A
>   Bits 3 to 17 of J → Bits 3 to 17 of A
>   Bits 12 to 17 of K → Bits 18 to 23 of A
>   Bits 12 to 17 of L → Bits 24 to 29 of A
>   Bits 12 to 17 of M → Bits 30 to 35 of A

2.   ENTRY: UPACW

    TYPE: FAP

    ARGUMENTS: A, I, J, K, L, M

    DESCRIPTION OF PROGRAM:

> Bits S, 1, 2 of A → Decrement of I
> Decrement of A → Decrement of J
> Bits 18 to 23 of A → Decrement of K
> Bits 24 to 29 of A → Decrement of L
> Bits 30 to 35 of A → Decrement of M

3.   ENTRY: PADP

    TYPE: FAP

    ARGUMENTS: A, I, J, K

    DESCRIPTION OF PROGRAM:

> Word A is set to zero
> Then Bits 15, 16, 17 of I → S, 1, 2 of A
>   Decrement of J → Decrement of A
>   Decrement of K → Address of A

4.   ENTRY: UPADP

    TYPE: FAP

    ARGUMENTS: A, I, J, K

    DESCRIPTION OF PROGRAM:

> Decrement of A → Decrement of J
> Address of A → Decrement of K
> S, 1, 2 of A → Decrement of I

150

MAIN LINK 2

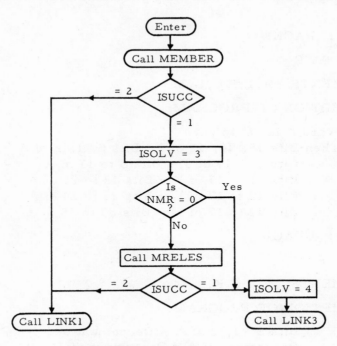

Member: Stiffness matrices at minus end of members are computed
         in member coordinates

NMR: Number of member releases

MRELES: Modifies stiffness matrices of released members

ISUCC = 1: Successfully processed

SUBROUTINE: MEMBER

TYPE: FORTRAN

ARGUMENTS: None

DESCRIPTION OF PROGRAM: For every member, this subroutine computes its stiffness matrix considering it cantilevered and using local-member coordinates at the minus end. The stiffness matrix is either taken directly from input or is obtained from the inversion of the flexibility matrix, which is either set up by subroutine MEMFOD or taken from input directly.

CALLS: ALOCAT, PRER2, MEMFOD, XSIMEQF

CALLED BY: MAIN LINK2

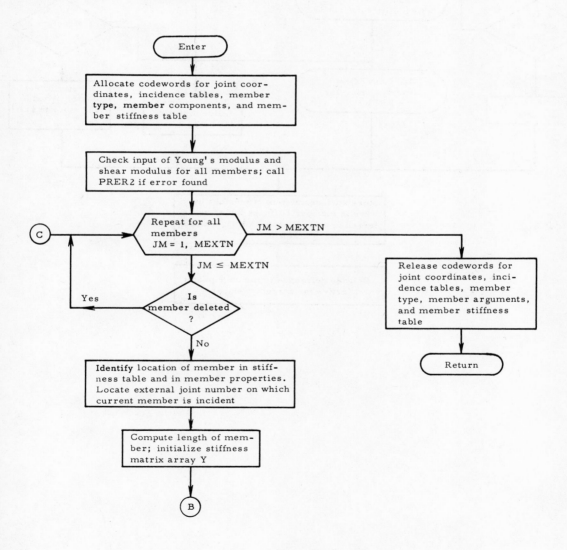

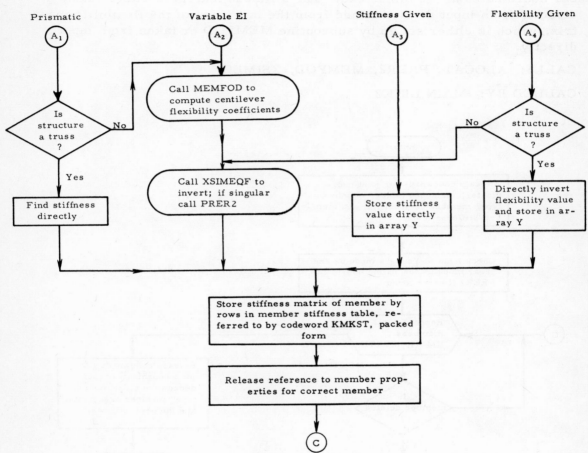

SUBROUTINE: MRELES

TYPE: FORTRAN

DESCRIPTION OF PROGRAM: This subroutine modifies the stiffness matrix of each member at minus end (in local-member coordinates) into an effective stiffness matrix to account for releases at one or both ends of the member. Also FIXM is called, for a positive start release, to transform fixed-end forces to right end of member.

CALLS: ALOCAT, DEFINE, COPY, CARRY, MAMUL, FIXM, PERMUT, PRER2, BUGER

CALLED BY: MAIN LINK2

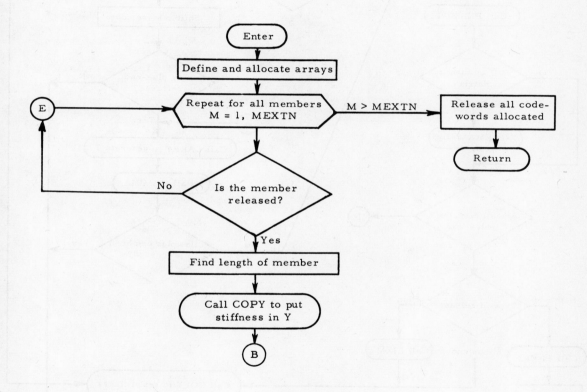

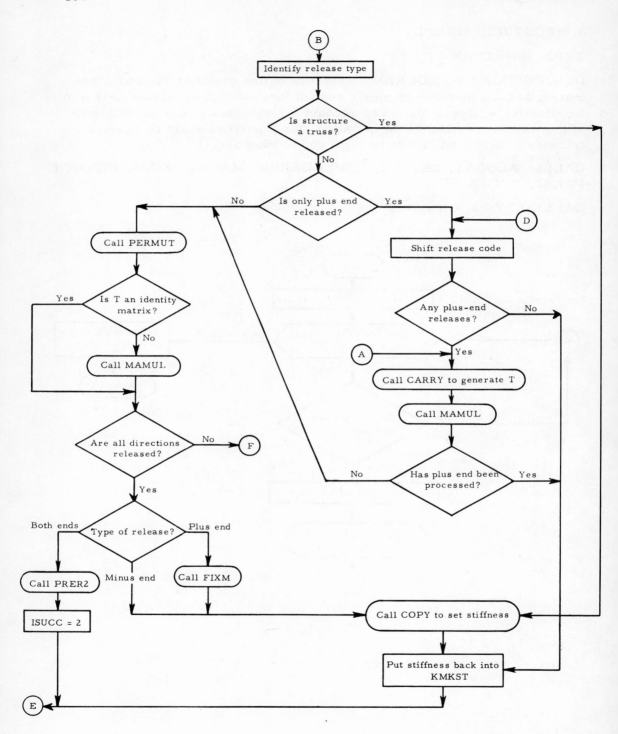

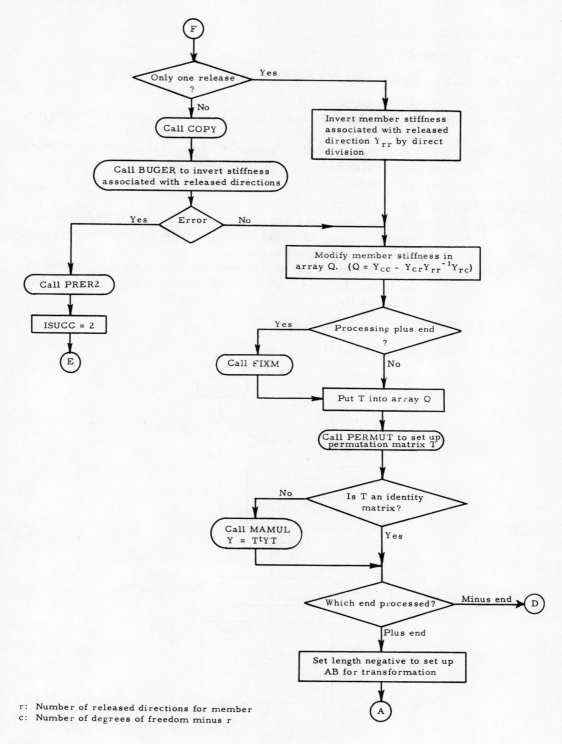

r: Number of released directions for member
c: Number of degrees of freedom minus r

SUBROUTINE: MEMFOD

TYPE: FORTRAN

ARGUMENTS: STOP, SP, IFOD

DESCRIPTION OF ARGUMENTS:
  STOP: Distance of applied load from left end of member or length of
         member
    SP: Array containing section properties for all segments of member
         being considered
  IFOD: Operation code

DESCRIPTION OF PROGRAM:
  IFOD = 1: Program computes cantilever flexibility coefficients of a
             nonprismatic straight member as well as of a prismatic
             member (from MEMBER).
  IFOD = 2: Program computes cantilever deflections at right end of a
             member because of applied concentrated load (from
             MEMBLD).

CALLS: PRER2

CALLED BY: MEMBER, MEMBLD

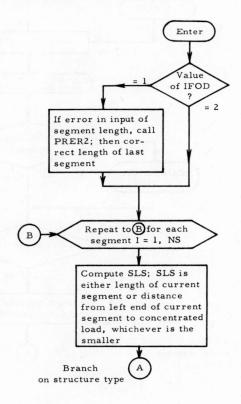

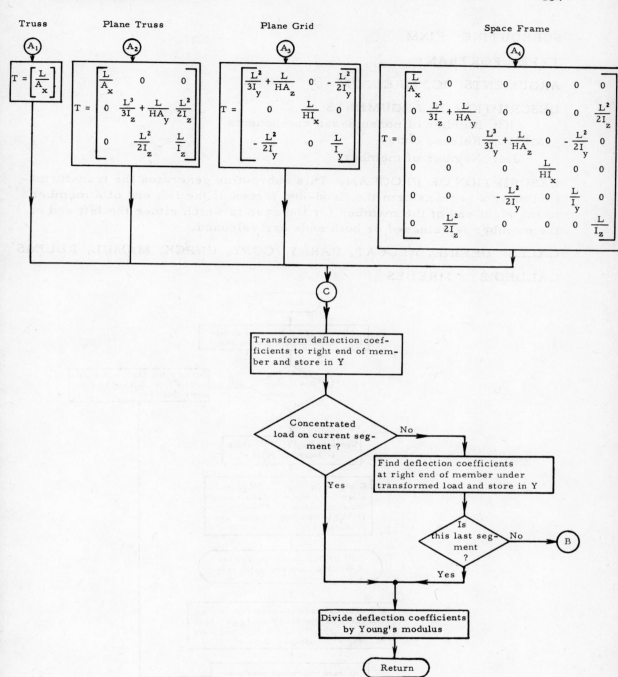

SUBROUTINE: FIXM

TYPE: FORTRAN

ARGUMENTS: IC, MREL, JM

DESCRIPTION OF ARGUMENTS:
    IC: Number of nonreleased components
 MREL: Release code
   JM: Number of member

DESCRIPTION OF PROGRAM: This subroutine generates the transformation matrix to transform the fixed-end forces at the left end of a member to the right end of the member for the case in which either the left end of the member is released or both ends are released.

CALLS: DEFINE, ALOCAT, CARRY, COPY, UNPCK, MAMUL, RELEAS

CALLED BY: MRELES

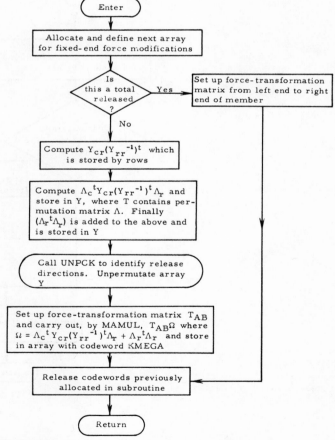

r: Number of released directions for member
c: Number of degrees of freedom minus r

SUBROUTINE: COPY

TYPE: FORTRAN

ARGUMENTS: A, B, JF, IX, ICM

DESCRIPTION OF ARGUMENTS:
  A, B: Square matrices that are transferred or copied
   JF: Size of square matrices A and B
   IX: Integer that controls purpose of subroutine
  ICM: Either codeword that locates storage of copied array or integer
        that indicates array to be copied

DESCRIPTION OF PROGRAM:
  IX = -2: Array A is filled with unit matrix.
  IX = -1: Array A is added columnwise to one-dimensional array U
            such that $U(ICM + 1) = A(1, 1) + U(ICM + 1)$, etc.
  IX = 0:   Array A is filled with zeroes.
  IX = 1:   If ICM > 0, the one-dimensional array U is transferred by
            columns into array A such that $A(1, 1) = U(ICM + 1)$, etc.
            If ICM = 0, the array B is transferred into corresponding
            locations in the array A.
  IX = 2:   Array A is transferred by columns to one-dimensional array
            U such that $U(ICM + 1) = A(1, 1)$, etc.

CALLED BY: MRELES, MAMUL, PERMUT

SUBROUTINE: CARRY

TYPE: FORTRAN

ARGUMENTS: ID, JF, SS

DESCRIPTION OF ARGUMENTS:
   ID: Identification of structure type
   JF: Size of force-transformation matrix (for example, 6 if space
        frame, etc.)
   SS: Length of member

DESCRIPTION OF PROGRAM: This program generates the force-transformation matrix T, which transforms the generalized force vector at the
minus end of a member to a statically equivalent force vector at the plus
end of the member.

CALLED BY: MRELES, LSTOR

This subroutine generates the following types of transformation matrix:

Plane Truss

$$T = \begin{bmatrix} 1 & 0 \\ 0 & 1 \end{bmatrix}$$

Space Truss

$$T = \begin{bmatrix} 1 & 0 & 0 \\ 0 & 1 & 0 \\ 0 & 0 & 1 \end{bmatrix}$$

Plane Frame

$$T = \begin{bmatrix} 1 & 0 & 0 \\ 0 & 1 & 0 \\ 0 & SS & 1 \end{bmatrix}$$

Plane Grid

$$T = \begin{bmatrix} 1 & 0 & 0 \\ 0 & 1 & 0 \\ -SS & 0 & 1 \end{bmatrix}$$

Space Frame

$$T = \begin{bmatrix} 1 & 0 & 0 & 0 & 0 & 0 \\ 0 & 1 & 0 & 0 & 0 & 0 \\ 0 & 0 & 1 & 0 & 0 & 0 \\ 0 & 0 & 0 & 1 & 0 & 0 \\ 0 & 0 & -SS & 0 & 1 & 0 \\ 0 & SS & 0 & 0 & 0 & 1 \end{bmatrix}$$

SUBROUTINE: PERMUT

TYPE: FORTRAN

ARGUMENTS: MREL, JF, ID

DESCRIPTION OF ARGUMENTS:
  MREL: Release code
  JF: Size of permutation matrix T
  ID: Identification of structure type, which is changed to 0 if T is
        not identity matrix

DESCRIPTION OF PROGRAM: This program sets up a permutation matrix T that, when it premultiplies a matrix A, will upon descending from top row by row shift the rows corresponding to released directions to the bottom of the matrix, followed upwards by subsequently released rows.

CALLS: COPY

CALLED BY: MRELES

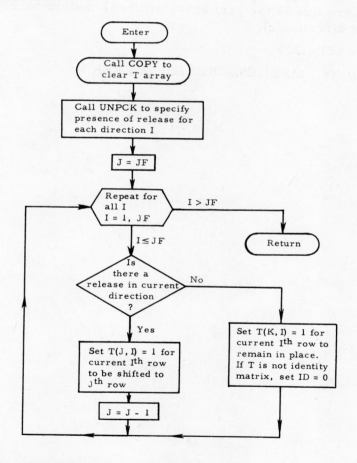

K: Current row I minus number of released rows
J: Final row JF minus number of previously released rows

SUBROUTINE: BUGER

TYPE: FAP

ARGUMENTS: N1, JQ1, JQ2, NTP, N2, N3, NV

DESCRIPTION OF ARGUMENTS:
          N1: Maximum size of array to be inverted
  JQ1 = JQ2: Number of rows in square array to be inverted
        NTP: Codeword for matrix to be inverted
          N2: Codeword for unit matrix or relative address in Common
                for unit matrix
          N3: Scale factor by which determinant of NTP is multiplied
          NV: One-dimensional erasable array, at least JQ1 elements
                in length

DESCRIPTION OF PROGRAM: Subroutine BUGER sets up calling sequence
for XSIMEQF, which gives the inversion of the stiffness matrix (that is,
S_{rr}^{-1} where this is the part of the stiffness matrix corresponding to the
released direction r).

CALLS: XSIMEQF

CALLED BY: MRELES, JRELES, SOLVER

SUBROUTINE: MAMUL

TYPE: FORTRAN

ARGUMENTS: Y, T, A, JS, JT, JJ

DESCRIPTION OF ARGUMENTS:

Y, T, A: Program does matrix multiplication $T \cdot Y \cdot A$ where sizes of
arrays are: $T: JT \times JS$
$Y: JS \times JS$
$A: JX \times JT$

JJ: Operation code

DESCRIPTION OF PROGRAM:

JJ = -1: Matrix multiplication TY

JJ = 0: Matrix multiplication TYA^T

JJ = 1: Matrix multiplication T^TYA

Result of matrix multiplication is stored in array Q. Subroutine COPY
is called to copy contents of A into array Y.

CALLS: COPY

CALLED BY: MRELES

Main Link 2

SUBROUTINE: UNPCK

TYPE: FAP

ARGUMENTS: MM, I, ID

DESCRIPTION OF ARGUMENTS:

MM: Release code for member being considered
I: 1 to JF, specifies direction of release in member word (JF: number of degrees of freedom)
ID: Structure type

DESCRIPTION OF PROGRAM: This program considers bits 12-17 of the release code MM. It returns MM ≠ 0 if release is in the Ith direction. Bit b of word MM corresponds to direction I, where I = 18 - b.

Type of Release	Truss I	Truss b	Grid I	Grid b	Plane Frame I	Plane Frame b	Space Frame I	Space Frame b
F_x	1	17			1	17	1	17
F_y					2	16	2	16
F_z			1	17			3	15
M_x			2	16			4	14
M_y			3	15			5	13
M_z					3	15	6	12

CALLED BY: PERMUT, FIXM

MAIN LINK 3

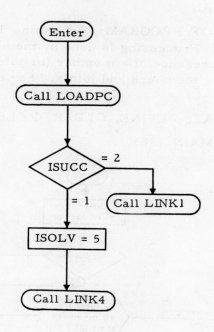

SUBROUTINE: LOADPC

TYPE: FORTRAN

DESCRIPTION OF PROGRAM: Subroutine LOADPC monitors the processing of loads. Processing is done by members and by joints. LOADPC checks for the presence of a member (or joint) and for the presence of loads on existing members and joints. For each existing load LOADPS is called.

CALLS: ALOCAT, DEFINE, CLEAR, RELEAS, LOADPS

CALLED BY: MAIN LINK3

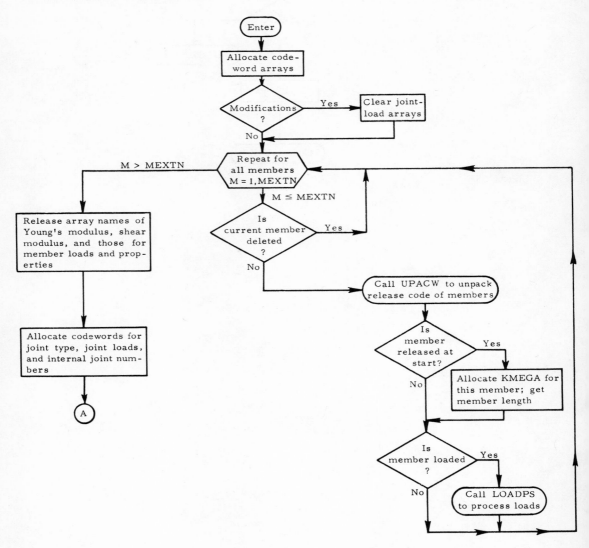

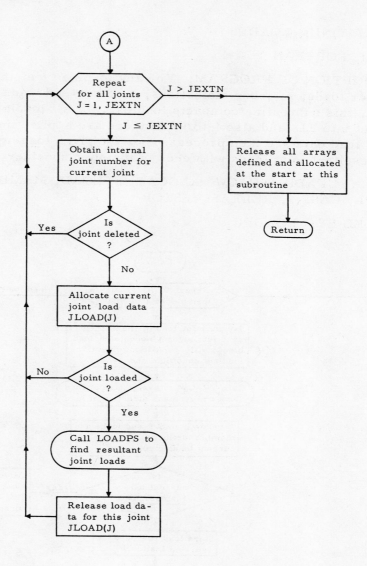

SUBROUTINE: LOADPS

TYPE: FORTRAN

DESCRIPTION OF PROGRAM: For any loading type, that is, joint loads, member loads, member distortions, member-end loads or joint displacements, this subroutine compacts load data for all loadings on a member or joint into PL (and also into PR if there are applied member-end loads). These load data are then processed for the given loading type, to finally find resultant joint loads with ends of member considered completely fixed.

CALLS: UPADP, UPACW, ALOCAT, JTLOAD, MEMBLD, MDISTN, JDISPL, CASE 2, RELEAS, LISTOR

CALLED BY: LOADPC

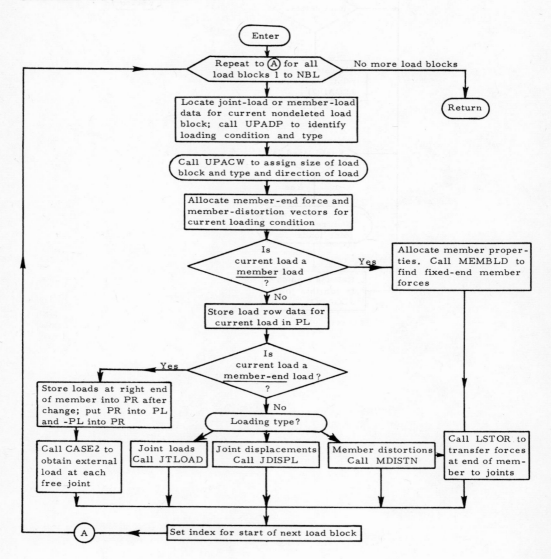

SUBROUTINE: MDISTN

TYPE: FORTRAN

DESCRIPTION OF PROGRAM: If member distortions are given for member JM, subroutine MDISTN adds these to KUV for member JM, then multiplies given distortion by member stiffness $\overline{K}^*$ to get contribution to KPPLS, KPMNS, and KPPRI (or KPDBP). Product of $\overline{K}^*$ and distortion is fixed-end force at minus end and is stored in PR (for CASE 2). Quantity -PR is translated over member to get plus-end force vector, temporarily stored in PL.

CALLED BY: LOADPS

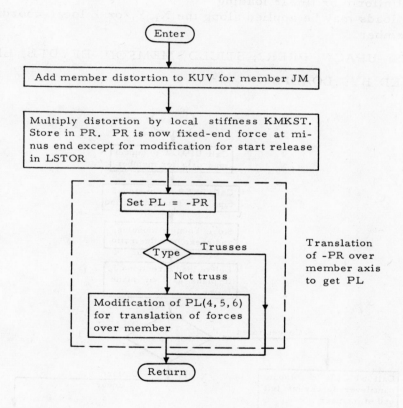

SUBROUTINE: MEMBLD

TYPE: FORTRAN

ARGUMENTS: SP

DESCRIPTION OF ARGUMENTS: The array SP contains section properties for all segments of member being considered.

DESCRIPTION OF PROGRAM: For a typical member, treating it as a cantilever fixed at the left end, this program computes cantilever forces at the left end of the member and cantilever deflections at the right end of the member under two types of loading:
1. Concentrated load
2. Uniform or linear loading

These loads may be applied along the X, Y, or Z local-coordinate axes of the member.

CALLS: UPACW, PRER2; STICLD, MEMFOD, EFVDTL, LINEAR

CALLED BY: LOADPS

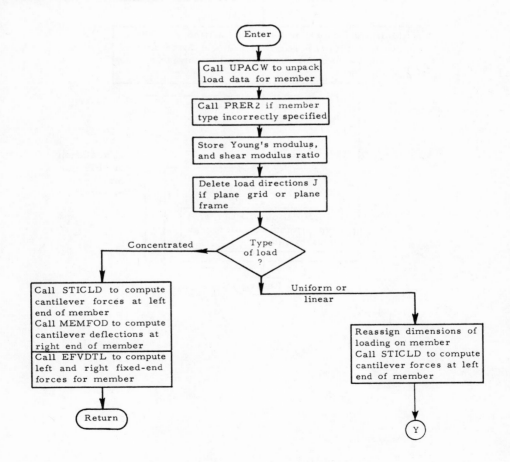

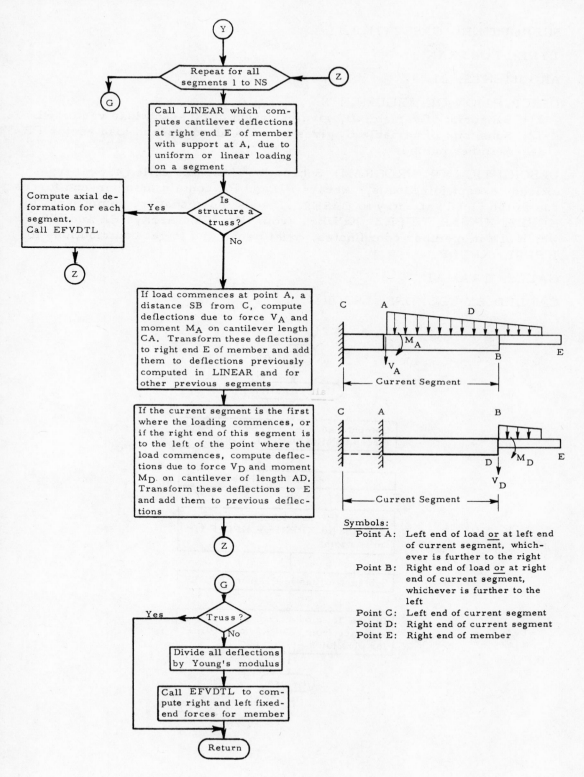

Symbols:

Point A: Left end of load <u>or</u> at left end
         of current segment, which-
         ever is further to the right

Point B: Right end of load <u>or</u> at right
         end of current segment,
         whichever is further to the
         left

Point C: Left end of current segment

Point D: Right end of current segment

Point E: Right end of member

SUBROUTINE: CASE2 (J1, J2)

TYPE: FORTRAN

ARGUMENTS: J1, J2

DESCRIPTION OF ARGUMENTS:
   J1: Subscript of variable U, gives location of plus-end load vector PL
   J2: Subscript of variable U, gives location of minus-end load vector PR
   JM: Member number

DESCRIPTION OF PROGRAM: Subroutine CASE2 is called for all types of loads except joint loads. Arrays PL and PR contain member-end forces due to the "load" at entry to CASE2. It adds the necessary components to KPPLS, KPMNS, KPPRZ, KPDBP, from PL, PR. Arrays PL and PR, which are in member coordinates, must be rotated to get contributions to KPPRI, KPDBP.

CALLS: TRAMAT

CALLED BY: LOADPS, LSTOR

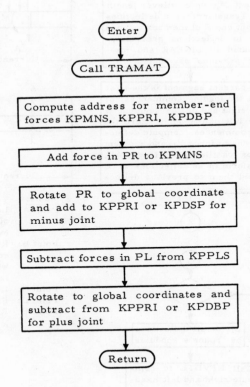

SUBROUTINE: JDISPL

TYPE: FORTRAN

DESCRIPTION OF PROGRAM: This routine processes prescribed joint displacements (for support joints, in nonreleased directions). By using the transpose of the incidence table (array KATR), subroutine JDISPL computes contributions to the member distortions KUV for all members incident to this joint. Using the stiffness matrix for each member, this routine obtains the corresponding fixed-end forces KPPLS and KPMNS (reflecting given member releases).

CALLS: ALOCAT, TRAMAT, LSTOR, RELEAS

CALLED BY: LOADPS

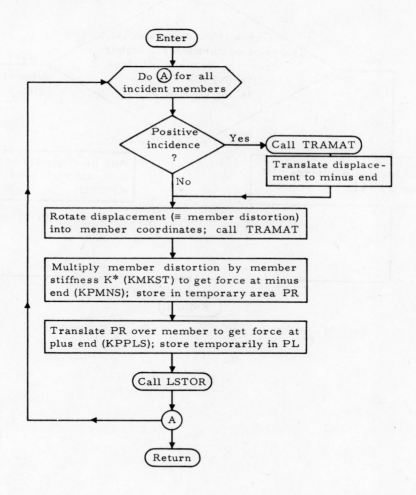

SUBROUTINE: JTLOAD

TYPE: FORTRAN

DESCRIPTION OF PROGRAM: This routine stores contributions to the effective joint-load vectors P' and $\bar{\bar{P}}'$ (arrays KPPRI and KPDBP) for free and (released) support joints, respectively.

CALLED BY: LOADPS

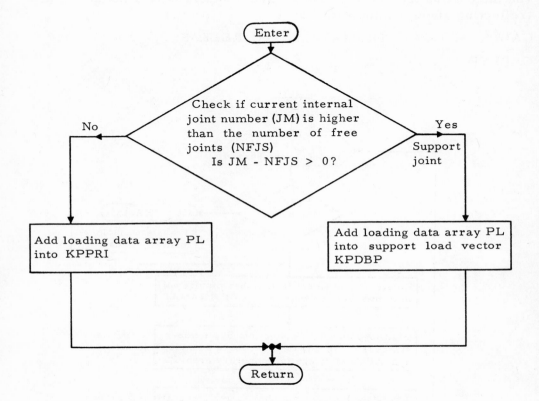

SUBROUTINE: LSTOR

TYPE: FORTRAN

ARGUMENTS: None

DESCRIPTION OF PROGRAM: This subroutine modifies the fixed-end
force vectors for members with releases prescribed at the plus end
(start), using the auxiliary matrix KMEGA saved by MRELES (see Section 4.2). The fixed-end force vectors are temporarily stored in V, W.

CALLS: CASE2

CALLED BY: LOADPS, JDISPL

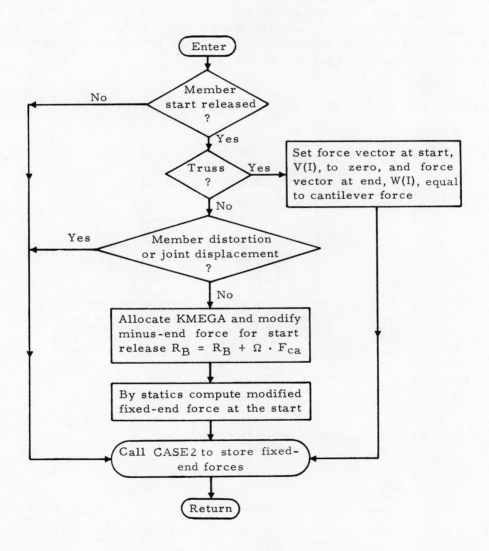

SUBROUTINE: LINEAR

TYPE: FORTRAN

ARGUMENTS: SL, CC

DESCRIPTION OF ARGUMENTS:
  SL: Length from left end of member to right end of segment being
      considered
  CC: Length from left end of member to left end of uniform or linear
      load on segment being considered

DESCRIPTION OF PROGRAM: This subroutine computes deflections
and rotations at right end of a segment under uniform or linear loading
and transforms to the right end of the member. The values computed
are cantilever deflections, and the fixed support is taken either at the
left end of the uniform or linear load or at the junction of the previous
segment, whichever is further to the right.

CALLS: PRER2

CALLED BY: MEMBLD

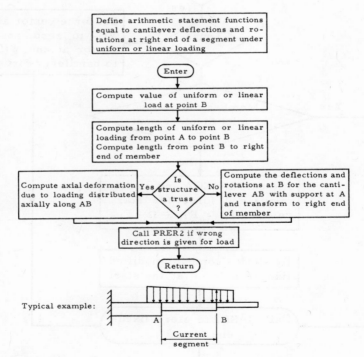

Symbols:

Point A: Left end of load or at left end of current segment, whichever is
         further to the right

Point B: Right end of load or at right end of current segment, whichever
         is further to the left

SUBROUTINE: EFVDTL

TYPE: FORTRAN

DESCRIPTION OF PROGRAM: This program computes right-end force vectors (assumed as acting on member) for a member by multiplying the member stiffness matrix by the cantilever deflections vector. Finally, left-end force vectors are calculated by statics.

CALLED BY: MEMBLD

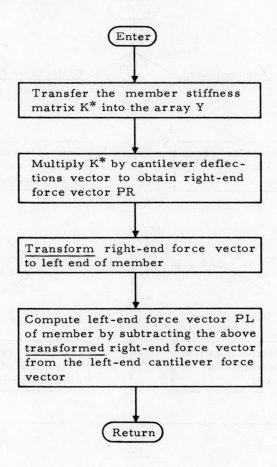

SUBROUTINE: STICLD

TYPE: FORTRAN

DESCRIPTION OF PROGRAM: For a cantilever fixed at the left end, this program computes by statics the cantilever forces at the left end of the member resulting from the applied loading, which may be concentrated, uniform, or linear.

CALLED BY: MEMBLD

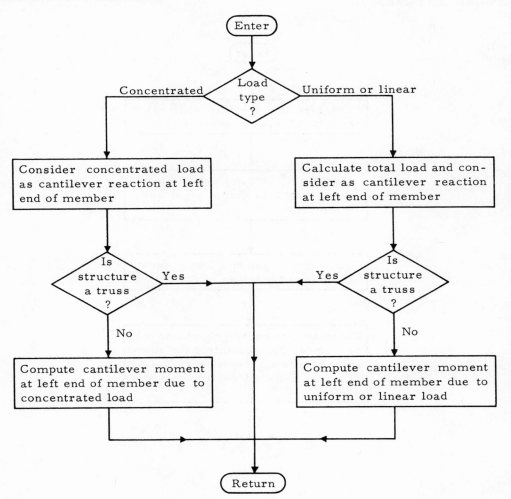

SUBROUTINE: TRAMAT

TYPE: FORTRAN

ARGUMENTS: JM, JT

DESCRIPTION OF ARGUMENTS:
      JM: Member number
      JT: Operation code
 JT = 1: Form translation matrix only, minus node to plus node
 JT = 2: Form rotation matrix only, global to local

DESCRIPTION OF PROGRAM: TRAMAT computes the force translation
or force rotation matrix for member JM. The matrix is formed using
the array KS (member length and projections of member axes on global
axes ΔX, ΔY, ΔZ). The matrix is stored into T(I, J) located at 77415.

The vector rotation matrix from global X, Y, Z coordinates to member
local x, y, z coordinates may be thought of as the product of two rotation
matrices, R' and R. The R matrix rotates the global axes first into an
intermediate X', Y', Z' system with $\beta \equiv 0$, and the R' matrix rotates the
intermediate system (through an angle β) into the local coordinates.

The rotation matrix from global to intermediate coordinates will be de-
veloped first. Consider a directed member of length L from plus to minus

$$\Delta X = X^{(-)} - X^{(+)}$$

$$\Delta Y = Y^{(-)} - Y^{(+)}$$

$$\Delta Z = Z^{(-)} - Z^{(+)}$$

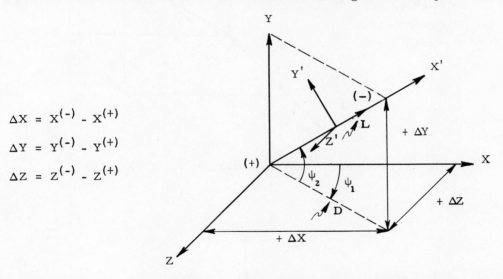

joints with projections of ΔX, ΔY, ΔZ in the X, Y, Z global frame. These
are shown positive as are the two angles ψ_1 and ψ_2. The X' intermediate
axis is along the member from plus to minus joints; the Y' axis lies in
the plane of X' - Y, directed such that its projection on the Y axis is positive. All systems are orthogonal. The unsigned projected length in the
X - Z plane is

$$D = +\sqrt{(\Delta X)^2 + (\Delta Z)^2}$$

The angle ψ_1 may have any value between 0 and 360 degrees; the angle ψ_2 has limits:

$$-90° < \psi_2 < + 90°$$

If the member is perpendicular to the X - Z plane (D = 0), a special case arises that will be considered later.

With these definitions, the member is considered to rotate about the Y axis in the general case. For plane structures, ψ_1 is identically zero and the member rotates about the Z axis.

The elements of R are direction cosines. Any element R_{ij} is the cosine of the angle between the ith intermediate axis and the jth global axis.

For example, the first column of R may be thought of as the projections of a unit vector in the X direction on the X', Y', Z' axes:

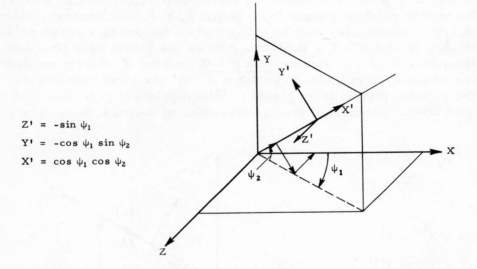

$$Z' = -\sin \psi_1$$
$$Y' = -\cos \psi_1 \sin \psi_2$$
$$X' = \cos \psi_1 \cos \psi_2$$

Projection of unit vector in Y direction on X', Y', Z' axes:

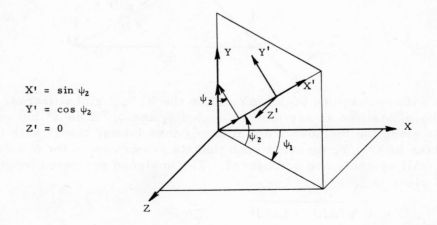

$$X' = \sin \psi_2$$
$$Y' = \cos \psi_2$$
$$Z' = 0$$

Projection of unit vector in Z direction on X', Y', Z' axes:

$$Z' = \cos \psi_1$$
$$Y' = -\sin \psi_1 \sin \psi_2$$
$$X' = \sin \psi_1 \cos \psi_2$$

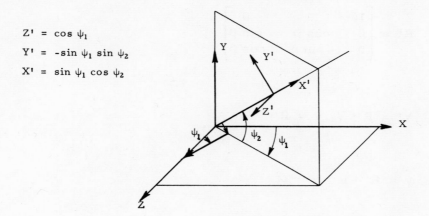

$$V' = R \cdot V$$

where

$$V' = \begin{Bmatrix} X' \\ Y' \\ Z' \end{Bmatrix}, \quad V = \begin{Bmatrix} X \\ Y \\ Z \end{Bmatrix}$$

and

$$R = \begin{bmatrix} \cos \psi_1 \cos \psi_2 & \sin \psi_2 & \sin \psi_1 \cos \psi_2 \\ -\cos \psi_1 \sin \psi_2 & \cos \psi_2 & -\sin \psi_1 \sin \psi_2 \\ -\sin \psi_1 & 0 & \cos \psi_1 \end{bmatrix}$$

The rotation from intermediate to local coordinates is merely a plane rotation of the Y', Z' axes through the angle β:

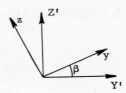

$$v = R' \cdot V'$$

where

$$v = \begin{Bmatrix} x \\ y \\ z \end{Bmatrix}, \quad V' = \begin{Bmatrix} X' \\ Y' \\ Z' \end{Bmatrix}$$

and

$$R' = \begin{bmatrix} 1 & 0 & 0 \\ 0 & \cos\beta & \sin\beta \\ 0 & -\sin\beta & \cos\beta \end{bmatrix}$$

Since

$$V' = R \cdot V, \quad v = R' \cdot V'$$

then

$$v = R' \cdot R \cdot V = \mathscr{R} \cdot V$$

and

$$\mathscr{R} = \begin{bmatrix} \cos\psi_1\cos\psi_2 & \sin\psi_2 & \sin\psi_1\cos\psi_2 \\ (-\cos\psi_1\sin\psi_2\cos\beta - \sin\psi_1\sin\beta) & \cos\psi_2\cos\beta & (-\sin\psi_1\sin\psi_2\cos\beta + \cos\psi_1\sin\beta) \\ (\cos\psi_1\sin\psi_2\sin\beta - \sin\psi_1\cos\beta) & (-\cos\psi_2\sin\beta) & (\sin\psi_1\sin\psi_2\sin\beta + \cos\psi_1\cos\beta) \end{bmatrix}$$

In general (consistent with the limits on ψ_1 and ψ_2)

$$\cos\psi_1 = \frac{\Delta X}{D} \text{ (sign of } \Delta X)$$

$$\sin\psi_1 = \frac{\Delta Z}{D} \text{ (sign of } \Delta Z)$$

$$\cos\psi_2 = \frac{D}{L} \text{ (always positive)}$$

$$\sin\psi_2 = \frac{\Delta Y}{L} \text{ (sign of } \Delta Y)$$

For the special case $D = 0$ (member perpendicular to the X - Z plane), $\psi_1 \equiv 0$, $\psi_2 = \pm 90$, therefore

$$\cos\psi_1 = +1, \quad \sin\psi_1 = 0, \quad \cos\psi_2 = 0, \quad \sin\psi_2 = \frac{\Delta Y}{L}$$

For plane structures, $\psi_1 \equiv 0$, $\beta = n \cdot 90°$, $n = 0, \pm1, \pm2, \pm3$.

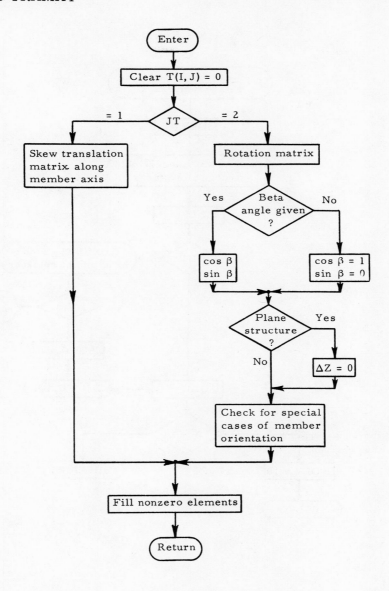

184

MAIN LINK 4

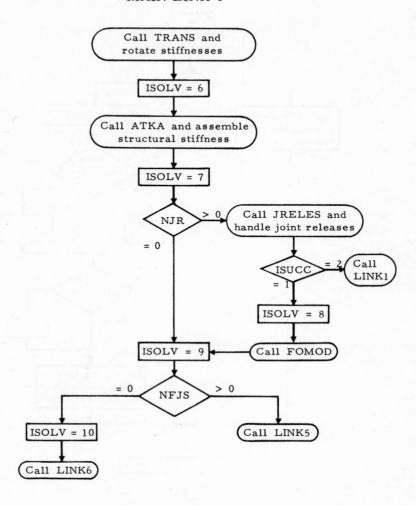

SUBROUTINE: TRANS

TYPE: FORTRAN

ARGUMENTS: None

DESCRIPTION OF PROGRAM: This subroutine rotates the local member stiffness matrix into global (joint)coordinates. It obtains the rotation matrix from TRAMAT and uses subroutine CLEAR to initialize the data arrays of KSTDB where the member stiffness in global coordinates is stored.

CALLS: CLEAR, TRAMAT, MATRIP, RELEAS

CALLED BY: MAIN LINK2

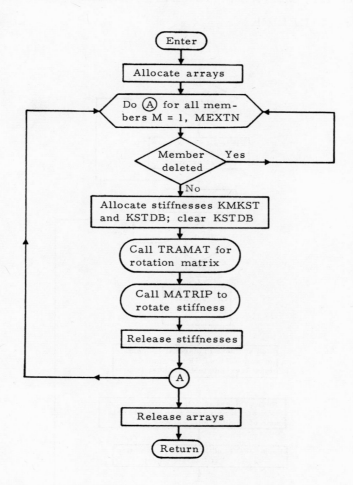

SUBROUTINE: ATKA

TYPE: FORTRAN

ARGUMENTS: None

DESCRIPTION OF PROGRAM: Subroutine ATKA generates the lower right half of the structural stiffness matrix $\overline{A}^t K \overline{A}$ from the member stiffnesses in (end) joint coordinates and the member-incidence tables. Using a loop on member stiffnesses the diagonal stiffnesses are successively added to KDIAG, and off-diagonal stiffnesses are stored in KOFDG. The symbolic connections by matrix arrays IFDT and IOFDG are generated.

CALLS: ALOCAT, MATRIP, TRAMAT, RELEAS

CALLED BY: MAIN LINK2

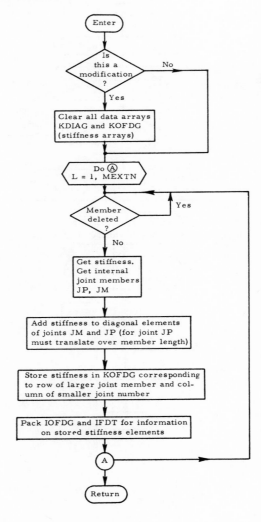

SUBROUTINE: JRELES

TYPE: FORTRAN

ARGUMENTS: None

DESCRIPTION OF PROGRAM: This routine controls the modification of stiffness matrix $\overline{A}^T K \overline{A}$ for the effect of support-joint releases.

CALLS: ALOCAT, DEFINE, TTHETA, UNPCK, STEP2, BUGER, PRER2, RELEAS, STEP5.

CALLED BY: MAIN LINK3

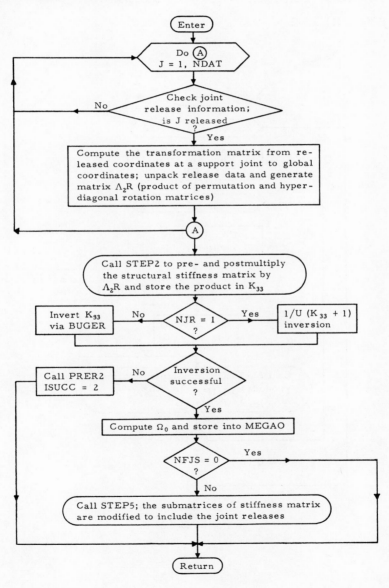

SUBROUTINE: STEP2

TYPE: FORTRAN

ARGUMENTS: None

DESCRIPTION OF PROGRAM: Subroutine STEP2 performs the matrix triple product $(\Lambda_2 R)\bar{\bar{A}}^t K \bar{\bar{A}} (\Lambda_2 R)^t = K_{33}$ by bookkeeping. The structural stiffness matrix $\Lambda_2 R$ is stored in KV in packed form by JRELES. Its product K_{33} is the structural stiffness matrix associated with released support-joint components.

CALLS: MAPROD, DEFINE, ALOCAT, RELEAS

CALLED BY: JRELES

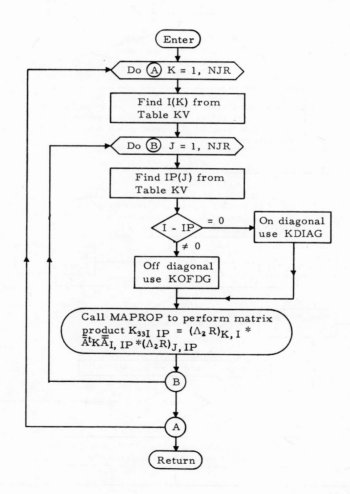

SUBROUTINE: MATRIP

TYPE: FORTRAN

ARGUMENTS: K1, K2, NT

DESCRIPTION OF ARGUMENTS:
   K1: Subscript for pool-variable U or IU; position of first element of the matrix to be multiplied by T, with respect to U(0)
   K2: Subscript for pool-variable U or IU; position of first element of resultant product with respect to U(0)
   NT: Operation code
      NT = 1: Transpose premultiplier
      NT = 2: Transpose postmultiplier

DESCRIPTION OF PROGRAM: Subroutine MATRIP performs the triple matrix product of order JF as follows (JF = 2, 3 or 6):
   NT = 1: $U(K2) = T^t \cdot U(K1) \cdot T$
   NT = 2: $U(K2) = T \cdot U(K1) \cdot T^t$
The matrix T is always located at U(37) and is dimensional T(6, 6), not packed. However, U(K1) and U(K2) are packed to JF $\times$ JF.

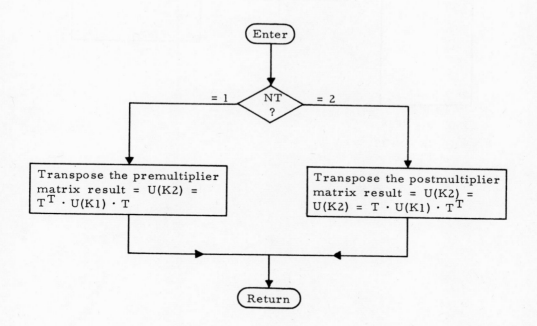

SUBROUTINE: STEP5

TYPE: FORTRAN

ARGUMENTS: None

DESCRIPTION OF PROGRAM: It modifies the free-joint structural stiffness A^tKA as affected by joint releases, does the product $A^tK\bar{\bar{A}}\ (\Omega_0)\ \bar{\bar{A}}^tKA$, and subtracts the result from A^tKA.

CALLS: ADRESS, ALOCAT, DEFINE, UPADP, PADP, RELEAS

CALLED BY: JRELES if there are free joints.

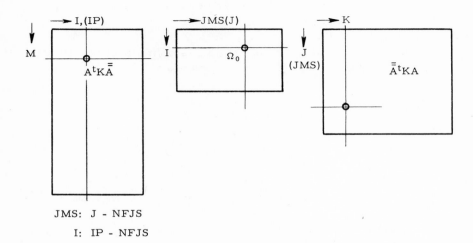

JMS: J - NFJS

I: IP - NFJS

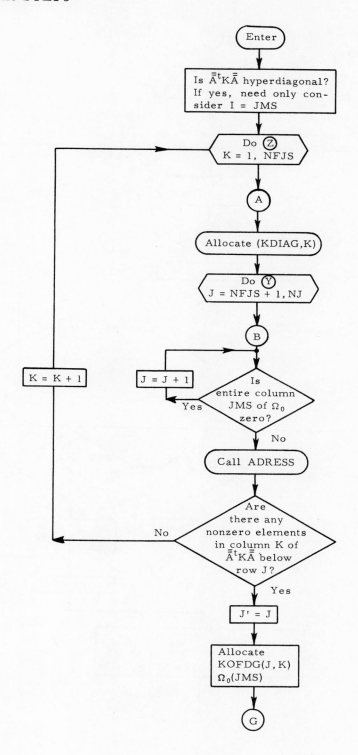

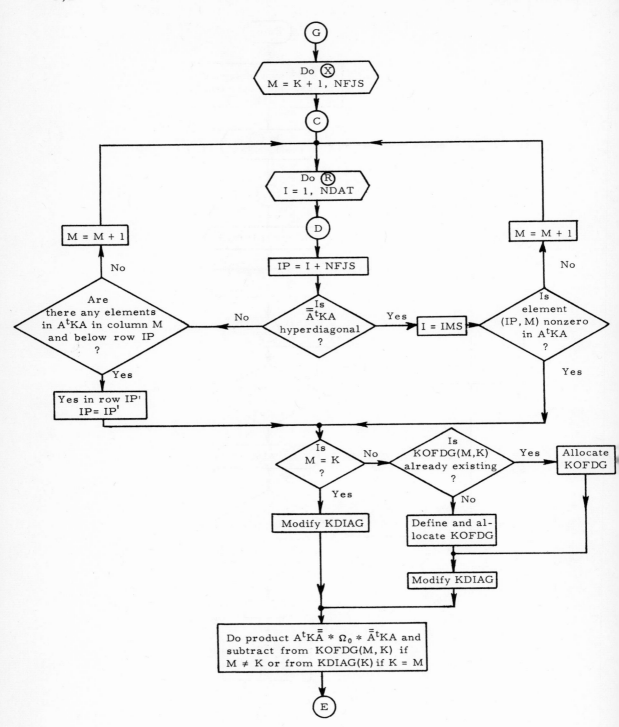

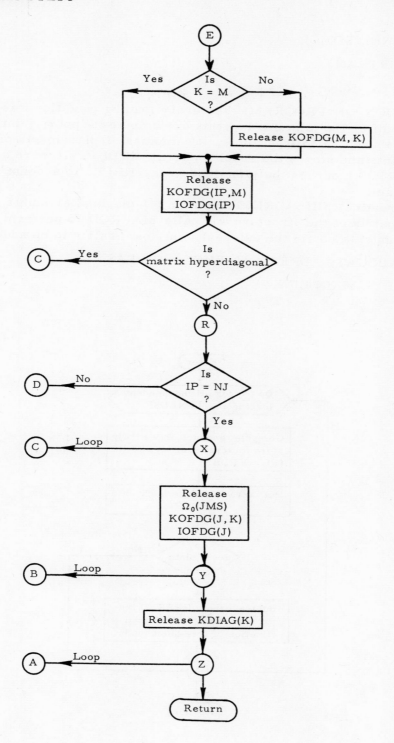

SUBROUTINE: FOMOD

TYPE: FORTRAN

ARGUMENTS: None

DESCRIPTION OF PROGRAM: For every loading condition this subroutine modifies the joint-load matrix or computes support-joint displacements. If there are no free joints, it computes displacements of released-
support joints and stores the result in array KPDBP. If there are free
joints, FOMOD modifies the joint-load array KPPRI to account for the
joint releases.

The subroutine calls ADRESS to obtain off-diagonal elements of the
stiffness matrix of the structure and calls MAPROD to perform the double
matrix product to obtain the external joint-load matrix to be added to KPPRI.

CALLS: ALOCAT, ADRESS, MAPROD, RELEAS

CALLED BY: MAIN LINK4

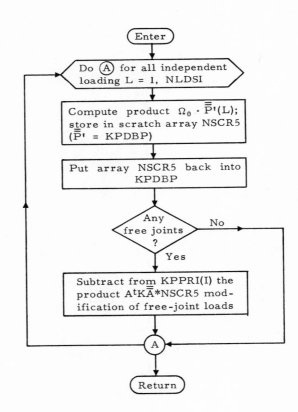

SUBROUTINE: ADRESS

TYPE: FAP

ARGUMENTS: N, I, NP, IT

DESCRIPTION OF ARGUMENTS:
   N: Number of the row under consideration in matrix $\overline{A}^t K \overline{A}$
   I: Number of the column under consideration in matrix $\overline{A}^t K \overline{A}$
  NP: Subscript obtained in routine ADRESS which is used to find code-
       word for off-diagonal stiffness matrix of member joining N-I
   IT: Operation code, determining the operations carried out by routine
       ADRESS. Each operation is defined in the description of pro-
       gram.

DESCRIPTION OF PROGRAM: Subroutine ADRESS performs five different functions depending on IT:
   1. IT = 0: Place a bit into IFDT corresponding to submatrix $\overline{A}^t K \overline{A}(N,I)$
      $N > I \geq 1$. This function does not allocate IOFDG.
   2. IT = 1: Determine which submatrix [$A^t KA(K,I)$, NFJS $\geq$ K $\geq$ N >
      I $\geq$ 1], in the Ith column of $A^t KA$ starting at the Nth row,
      is the first nonzero submatrix encountered at or below the
      N^{th} row. The routine returns the row number K (in N) and
      its position in KOFDG (in NP). If there are no nonzero ele-
      ments in that column below row N, NP = 0. Routine allo-
      cates IOFDG(K). The absence or presence of a nonzero
      submatrix is determined by the absence or presence, re-
      spectively, of a bit in the corresponding bit picture IFDT
      of the joint stiffness matrix. This bit picture is arranged
      by columns and is stored in words of IFDT, which is a
      second-level array, each data array being 200 words in
      length. Subroutine ADRESS will allocate and release the
      data arrays of IFDT as needed but does not do anything to
      the codeword array.
         Array IOFDG(K) represents the structure of the row K
      of $\overline{A}^T K \overline{A}$ below the diagonal, indicating the position of non-
      zero subarrays. Since the data arrays of IOFDG are de-
      fined in blocks, the first word in each data array contains
      two pieces of information:
         a. The decrement contains the current number of non-
            zero submatrices in row K below the diagonal.
         b. The address part contains the defined length of
            IOFDG(K).
               Succeeding words also contain two pieces of infor-
         mation:
         a. The decrement contains the column position of the
            submatrix in $\overline{A}^t K \overline{A}$.
         b. The address contains the position of the submatrix in
            ROFDG.

3. IT = 2: Determine whether a bit is present in position N, I. If not, return NP = 0. If it is, call ALOCAT IOFDG(N), RETURN, NP=NUMBER OF SUBMATRIX N, I IN KOFDG. Do not touch N or I.

4. IT = 3: Determine how many bits are present in column I. Return the number in NP. There is no allocation of IOFDG.

5. IT = 4: Same operation as IT = 1 except search goes to row NJ, not NFJS.

CALLS: ALOCAT, RELEAS

CALLED BY: DEFSUP, FOMOD, STEP5, SOLVER

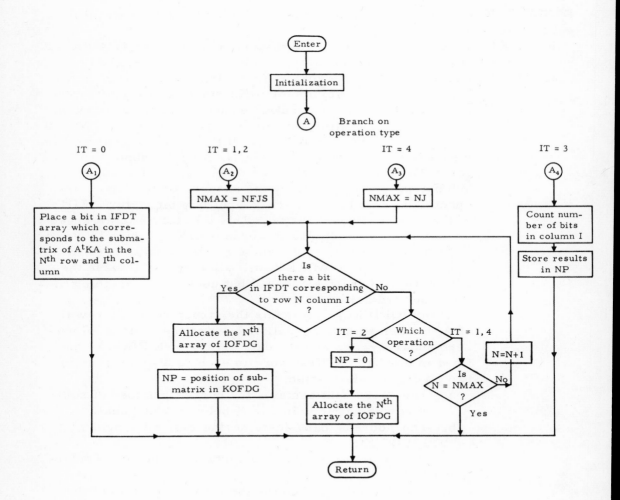

SUBROUTINE: TTHETA

TYPE: FORTRAN

ARGUMENTS: KJND, ID, JF

DESCRIPTION OF ARGUMENTS:
  KJND: Subscript used for finding joint-release information. First
        word of array contains number of joints with releases.
  ID: Structural-type identification number
  JD: Number of degrees of freedom

DESCRIPTION OF PROGRAM: Subroutine TTHETA computes the rotation matrix from released coordinates to global-coordinate axes for a released support. Subroutine COPY is used to clear Q and the transformation matrix is stored in Q(6, 6) which starts in U(73). The transformation matrix is obtained for any structural type.

CALLS: COPY

CALLED BY: JRELES

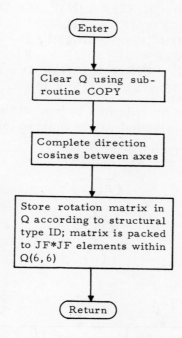

SUBROUTINE: MAPROD

TYPE: FORTRAN

ARGUMENTS: N1, N2, N3, N4, IS, JF, IND

DESCRIPTION OF ARGUMENTS:
    U(N1): First element of premultiplier A
    U(N2): First element of second matrix in triple product B (B is square
            matrix)
    U(N3): First element of postmultiplier C
    U(N4): First element of resultant matrix D
       IS: Number of rows in A equal number of columns in C
       JF: Number of rows in A equal size of B
      IND: Operation code

DESCRIPTION OF PROGRAM: Subroutine MAPROD performs matrix
double or triple product, depending on IND:
    IND = 0: $U(N4) = U(N4) + U(N1) \cdot U(N2) \cdot U(N3)$; $D = D + A \cdot B \cdot C$
             $[IS \times IS] = [IS \times JF] \times [JF \times JF] \times [JF \times IS]$
    IND = 1: $U(N4) = U(N4) + U(N1) \times U(N2)^t \cdot U(N3)$; $D = D + AB^tC$
             $[IS \times IS] \times [IS \times JF] \times [JF \times JF] \times [JF \times IS]$
    IND = 2: $U(N4) = U(N4) + U(N2) \times U(N3)$; $D = D + B \cdot C$
             $[JF \times IS] = [JF \times JF] \times [JF \times IS]$
    IND = 3: $U(N4) = U(N4) + U(N2) \times U(N3)^t$; $D = D + B \cdot C^t$
             $[JF \times JF] = [JF \times JF] \times [JF \times JF]$
    IND = 4: $U(N4) = U(N4) + U(N2)^t \times U(N3)$; $D = D + B^t \cdot C$
             $[JF \times IS] = [JF \times JF] \times [JF \times IS]$

All matrices are stored by columns in one-dimensional arrays. The
resultant matrix product is added to the matrix located at U(N4).

CALLED BY: STEP2, FOMOD

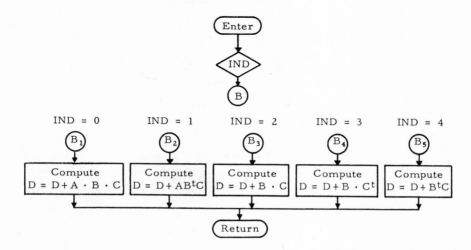

MAIN LINK 5

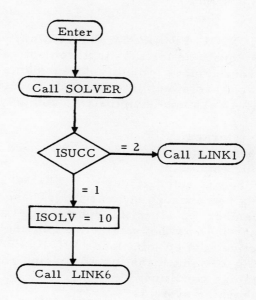

SUBROUTINE: SOLVER

TYPE: FORTRAN

ARGUMENTS: None

DESCRIPTION OF PROGRAM: Subroutine SOLVER solves the joint-equilibrium equations for the displacements of all free joints.

This subroutine uses a modified Gaussian elimination procedure (forward and backward sweep) operating on submatrices as elements. The matrix $A^t KA$ is assumed symmetric, and only the lower diagonal matrix is treated.

Method of storage: $KU' = P'$

1. Diagonal submatrices K: KDIAG, second-level arrays (one submatrix for each joint.)

2. Off-diagonal submatrices of K: KOFDG, second-level arrays (one matrix per <u>nonzero</u> submatrix). These submatrices are ordered by member for all initial submatrices and consecutively for all submatrices as they are defined during elimination.

3. Effective applied joint loads P': KPPRI, first-level arrays including all joint loads for all loading conditions. They are ordered by loading condition.

4. Bookkeeping systems:

   a. IOFDG: Coded representation of the off-diagonal elements, second-level arrays (one datum per row of $A^t KA$.) Only nonzero elements are stored. Decrement contains column order. Address contains position in KOFDG of the corresponding submatrix.

   b. IOFC: Used in the backward sweep of SOLVER. It contains the same information (except first word of each array) as IOFDG, but the information is by columns. IOFDG is released during forward sweep.

   c. IFDT: A second-level array (200 words per array) with representation of complete left lower half of $A^t \overline{KA}$, one binary bit per submatrix. Bit is 1 if submatrix (I, K) exists; bit is 0 if it does not exist. The storage order is by column, column 1 first. In each word bits are ordered from right to left.

5. Displacements U' are stored in KPPRI after SOLVER.

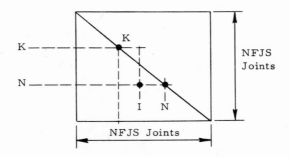

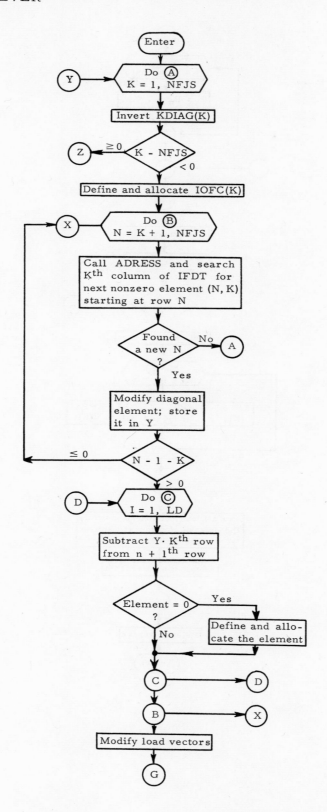

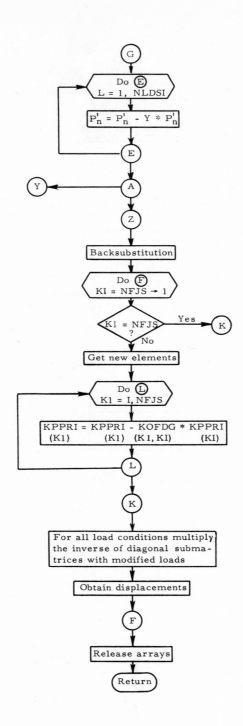

SUBROUTINE: MAPRDT

TYPE: FORTRAN

ARGUMENTS: N1, N2, N3, N4, IZ, KF, INP

DESCRIPTION OF ARGUMENTS:
  $U(N1+1)$: First element of premultiplier
  $U(N2+1)$: First element of matrix B (square)
  $U(N3+1)$: First element of postmultiplier C
  $U(N4+1)$: First element of resultant matrix D
  IS = $|IZ|$ = number of rows in A = number of columns in C
  KF: Number of columns in A = size of B
  IND = $|INP|$ = operation code.

DESCRIPTION OF PROGRAM: Subroutine MAPRDT performs triple or double matrix multiplication of arrays whose U subscripts are N1, N2, and N3. Whether this multiplication is triple or double is determined by the operation codes.

  IND = 1 $D_0 = A \cdot B \cdot C$
  IND = 2 $D_0 = A \cdot B^t \cdot C$
  IND = 3 $D_0 = B \cdot C$
  IND = 4 $D_0 = B \cdot C^t$
  IND = 5 $D_0 = B^t \cdot C$

The matrix D_0 is first stored in a working area. Whether D_0 is either directly stored in D, added to D, or subtracted from D is determined by the signs of IZ and INP.

  1. IZ < 0   INP > 0 $D = D + D_0$
  2. IZ < 0   INP ≤ 0 $D = D - D_0$
  3. IZ ≥ 0   INP ≥ 0 $D = -D_0$
  4. IZ ≥ 0   INP > 0 $D = D_0$

CALLED BY: SOLVER

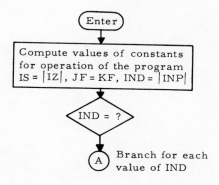

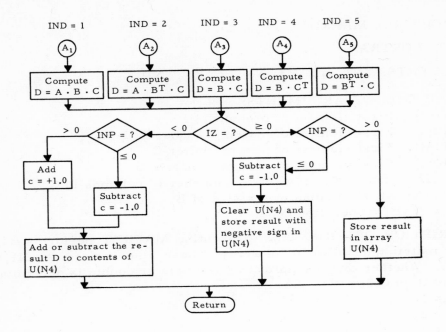

MAIN LINK 6

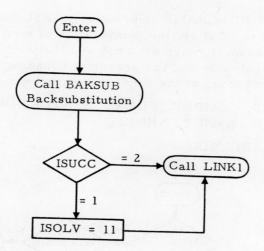

SUBROUTINE: BAKSUB

TYPE: FORTRAN

DESCRIPTION OF PROGRAM: Subroutine BAKSUB monitors the back-substitution process, that is, the computation of member distortions, member forces, support reactions, and applied joint loads from the induced joint displacements and the prescribed loading components. It also controls the printing of the requested output via subroutine ANSOUT.

CALLS: ALOCAT, COMBLD, CLEAR, STATCK, DEFSUP, DEFINE, NEWADR, UPACW, AVECT, ANSOUT

CALLED BY: MAIN LINK6

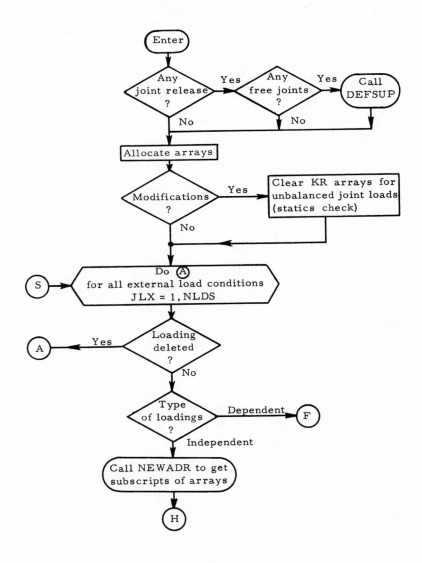

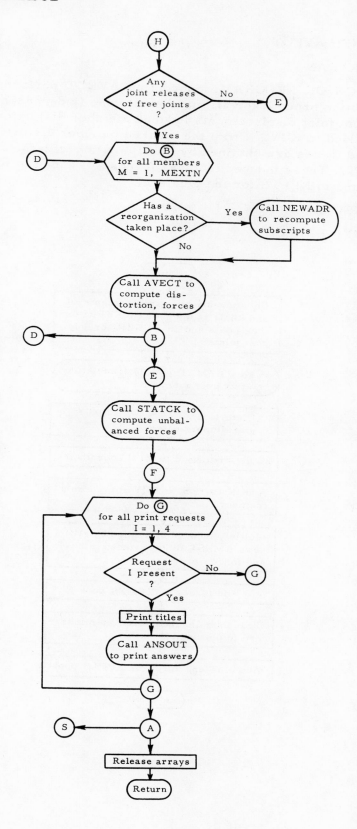

SUBROUTINE: AVECT

TYPE: FORTRAN

DESCRIPTION OF PROGRAM: Subroutine AVECT performs the actual backsubstitution process for all members of one (independant) loading condition. From joint displacements induced member distortions are computed and added to KUV. From the induced member distortions the induced member-end forces are obtained by multiplication with the member stiffness KMKST. These induced member forces are added to KPMNS for the minus end and to KPPLS for the plus end.

CALLS: TRAMAT

CALLED BY: BAKSUB

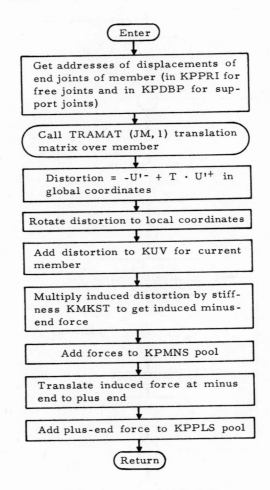

Enter

Get addresses of displacements of end joints of member (in KPPRI for free joints and in KPDBP for support joints)

Call TRAMAT (JM, 1) translation matrix over member

Distortion = -U'⁻ + T · U'⁺ in global coordinates

Rotate distortion to local coordinates

Add distortion to KUV for current member

Multiply induced distortion by stiffness KMKST to get induced minus-end force

Add forces to KPMNS pool

Translate induced force at minus end to plus end

Add plus-end force to KPPLS pool

Return

SUBROUTINE: COMBLD

TYPE: FORTRAN

ARGUMENTS: None

DESCRIPTION OF PROGRAM: Subroutine COMBLD computes the final member distortions, joint displacements, member-end forces and unbalanced joint loads for a dependent loading condition by combining the previously computed answers for previously processed loading conditions according to the specified linear combination.

CALLED BY: CLEAR, ALOCAT, NEWADR, RELEAS, UPADP

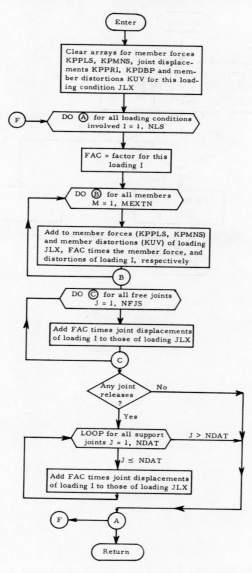

SUBROUTINE: NEWADR

TYPE: FORTRAN

DESCRIPTION OF PROGRAM: Subroutine NEWADR recomputes addresses of arrays used in BAKSUB and COMBLD after a memory reorganization or at the beginning of process. Arrays concerned are
KPPLS, KPMNS, KPPRI, KPDBP, KUV, KR.

CALLED BY: BAKSUB, COMBLD

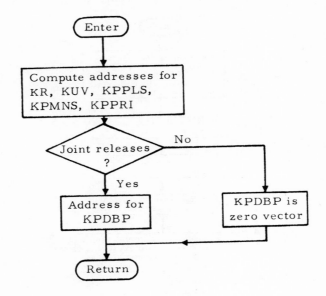

SUBROUTINE: STATCK

TYPE: FORTRAN

DESCRIPTION OF PROGRAM: STATCK sums all member-end forces at each joint (for each member incident to that joint) and stores the sum in array KR for the current loading condition. These sums represent the unbalanced joint loads as computed from the member-end forces. Member-end forces for each member are first rotated into joint coordinates.

CALLS: TRAMAT

CALLED BY: BAKSUB

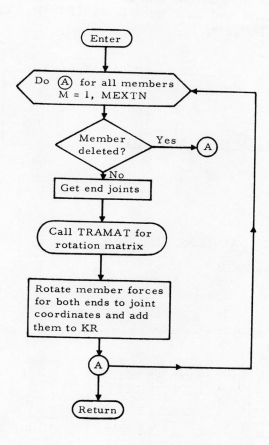

SUBROUTINE: DEFSUP

TYPE: FORTRAN

ARGUMENTS: None

DESCRIPTION OF PROGRAM: Subroutine DEFSUP computes the effect of the free joint displacements u' on the support joint displacements $\bar{\bar{u}}'$ at released support joints, if there are any free joints <u>and</u> released supports. The total displacements at support joints are

$$\bar{\bar{u}}' = \Omega_0 \{ \bar{\bar{\mathscr{P}}}' - \bar{\bar{A}}^t KAu' \}$$

where $\bar{\bar{\mathscr{P}}}'$ and u' reflect the effects of all types of loads. The part $\Omega_0 \bar{\bar{\mathscr{P}}}'$ of u' is computed by FOMOD and stored in KPDBP. Subroutine DEFSUP computes $\Omega_0 \bar{\bar{A}}^t KAu'$ and subtracts from KPDBP. The subscripts for the product are (L, I)*(I, J)*(J).

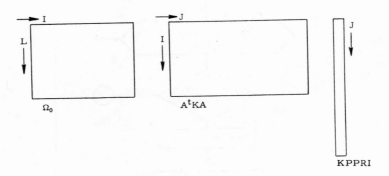

CALLS: ALOCAT, ADRESS, RELEAS.

CALLED BY: BAKSUB

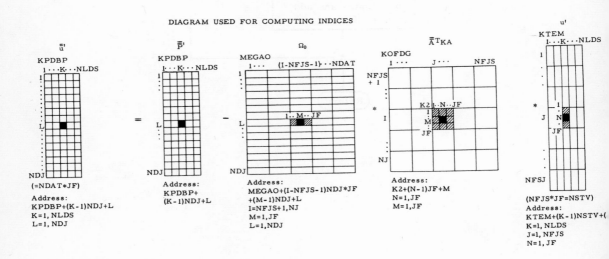

DIAGRAM USED FOR COMPUTING INDICES

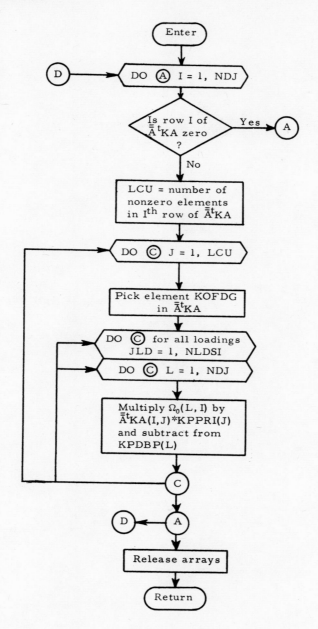

Appendix B

PROGRAM LISTING

The following program listing represents the status of the STRESS processor in November 1964. Various corrections, improvements, and extensions have since been made. Copies of the program containing these changes may be obtained from any of several commercial computer service bureaus which maintain STRESS.

```
        CLA     TOP                                                         ALOC0472
        SUB     NL                                                          ALOC0473
*       FAP                                                                 ALOC0000
*       DYNAMIC MEMORY ALLOCATOR, VERSION III, 9-12-63 LOGCHER             ALOC0001
*       CHAIN VERSION...13 SEPTEMBER 1963
*       REVISION OF DECEMBER 31,1963
        COUNT   800
        LBL     ALOCHN
        ENTRY   START                                                       0004
        ENTRY   DEFINE                                                      ALOC0004
        ENTRY   RELEAS                                                      ALOC0005
        ENTRY   ALOCAT                                                      ALOC0006
        ENTRY   DUMP                                                        ALOC0007
        ENTRY   REORG                                                       ALOC0008
START   SXA     IR1S,4                                                      ALOC0009
        SXA     IR1S+1,1                                                    ALOC0010
        CLA*    2,4                                                         ALOC0011
        ADD     ONE                                                         ALOC0012
        PDX     ,1
        STZ     U+1,1                                                       ALOC0013
        TIX     *-1,1,1                                                     ALOC0016
        CLA     *-2                                                         ALOC0017
        STA     TOP                                                         00018
        CLA*    1,4                                                         00019
        STO     TN                                                          ALOC0020
        CLS*    2,4                                                         ALOC0021
        ARS     18                                                          ALOC0022
        ADD     TOP                                                         ALOC0023
        STO     N1                                                          ALOC0024
        CLA     99                                                          ALOC0025
        ADD     ONE                                                         ALOC0026
        ARS     18                                                          ALOC0027
        STA     NT                                                          ALOC0028
        STA     99                                                          0C029
        SSM                                                                 ALOC0030
        ADD     N1                                                          ALOC0031
        ALS     18                                                          ALOC0032
        STO*    3,4                                                         ALOC0033
        CALL    FILES,ONE,TN,FN,ONE,TOP         INITIATE FILE TAPE          ALOC0034
        CLA     N1                                                          ALOC0035
        STO     NL                      INITIATE MEMORY ALLOCATION NO.      ALOC0036
IR1S    AXT     **,4                                                        ALOC0037
        AXT     **,1                                                        ALOC0038
        TRA     4,4                                                         ALOC0039
REORG   SXA     REORGF,1                                                    ALOC0040
        SXA     REORGF+1,4      MEMORY REORGANIZATION ROUTINE               ALOC0041
        STZ     NONEED                                                      ALOC0042
        STZ     NPERF                                                       ALOC0043
        CLA     N1                                                          ALOC0044
        STA     ADD1            DETERMINE AVAILABLE MEMORY                  ALOC0045
        STA     ADD2                                                        ALOC0046
ADD1    LXA     **,1                                                        ALOC0047
        TXL     NBR,1,0                                                     ALOC0048
ADD2    CAL*    **                                                          ALOC0049
        ADD     =1                                                          ALOC0050
        STA     N               SIZE OF ARRAY PLUS MEMORY WORD              ALOC0051
        CLA*    ADD2            MEMORY WORD (FIRST WORD IN ARRAY)           ALOC0052
        TPL     CONTU           NEEDED                                      ALOC0053
        ALS     1                                                           ALOC0054
        PBT                                                                 ALOC0056
```

```
        TRA     ORDN            ORDINARY ARRAY                          ALOC0057
        CLA     N               PERFERRED ARRAY                         ALOC0058
        ADD     NPERF                                                   ALOC0059
        STO     NPERF                                                   ALOC0060
        TRA     CONTU                                                   ALOC0061
ORDN    CLA     N                                                       ALOC0062
        ADD     NONEED                                                  ALOC0063
        STO     NONEED                                                  ALOC0064
CONTU   CLA     ADD1                                                    ALOC0065
        ANA     =0000000077777                                          ALOC0066
        SUB     N                                                       ALOC0067
        STA     ADD1                                                    ALOC0068
        STA     ADD2                                                    ALOC0069
        CAS     NL                                                      ALOC0070
        TRA     ADD1            TEST ANOTHER ARRAY                      ALOC0071
        CLA     NL                                                      ALOC0072
        ADD     NONEED                                                  ALOC0073
        SUB     NT                                                      ALOC0074
        SUB     NREQ                                                    ALOC0075
        SUB     =2                                                      ALOC0076
        TMI     PERFT                                                   ALOC0077
        CLA     SAV                                                     ALOC0078
        STA     SAV1                                                    ALOC0079
DUMPR   CLA     N1              DUMP AND COMPACT USING ORDINARY DALOC0079
        STO     NNL                                                     ALOC0080
        STO     K                                                       ALOC0081
        STA     ADD3                                                    ALOC0082
        STA     ADD4                                                    ALOC0083
        STA     ADD5                                                    ALOC0084
        STA     ADD6                                                    ALOC0085
        STA     ADD7                                                    ALOC0086
        STA     ERASE                                                   ALOC0087
        CLA*    ADD3            NEEDED, MOVE UP IN MEMORY               ALOC0088
SAV     TPL     MOVR                                                    ALOC0089
        STA     N                                                       ALOC0090
        NZT     N               NO BACK REFERENCE FOR DATA,PROCEDALOC0091
        TRA     NBRDP                                                   ALOC0092
        ALS     1                                                       ALOC0093
        PBT                                                             ALOC0094
        TRA     *+2             MOVE PERFERRED ARRAYS                   ALOC0095
SAV1    TRA     MOVR                                                    ALOC0096
        ALS     1                                                       ALOC0097
        PBT                                                             ALOC0098
        TRA     ERASE                                                   ALOC0100
ADD3    CLA*    **                                                      ALOC0101
        ANA     =0000000400000                                          ALOC0102
        TNZ     DPSLF                                                   ALOC0103
        XEC     ADD3                                                    ALOC0104
        STA     N                                                       ALOC0105
        ANA     =0000000777777                                          ALOC0106
        SSM
ADD4    STO*    **              WRITE FILE CONSISTING OF ARRAY          ALOC0107
        TSX     WFT,4           NEXT FILE NUMBER ASSIGNED               ALOC0108
        CLA     FN                                                      ALOC0110
ADD5    ORS*    **                                                      ALOC0111
FNNUM   CLA     K                                                       ALOC0112
        SUB     N                                                       ALOC0113
        SUB     =1                                                      ALOC0114
ENDTS   CAS     NL
        TRA     DUMPR+2
```

```
        NOP
        CALL    FILES,FIVE,ONE,ONE,ONE,ONE
        CLA     NNL
        STO     NL                                          ALOC0115
REORGF  AXT     **,1                                        ALOC0116
        AXT     **,4                                        ALOC0117
        TRA     1,4                                         ALOC0118
FIVE    OCT     5000000                                     ALOC0119
WFT     SXA     WTT,4
        LXA     N,4                                         ALOC0120
        SXD     NCOUNT,4                                    ALOC0121
        CLA     K                                           ALOC0122
        SUB     =1                                          ALOC0123
        STA     STARR                                       ALOC0124
        TSX     $FILES,4                                    ALOC0125
        TSX     TWO                                         ALOC0126
        TSX     TN                                          ALOC0127
        TSX     FN                                          ALOC0128
        TSX     NCOUNT                                      ALOC0129
STARR   TSX     **                                          ALOC0130
WTT     AXT     **,4                                        ALOC0131
        TRA     1,4                                         ALOC0132
NBR     CAL*    ADD2                                        ALOC0133
        ARS     18                                          ALOC0134
        ADD     =1                                          ALOC0135
        STA     N                                           ALOC0136
        TRA     ORDN                                        ALOC0137
PERFT   ADD     NPERF                                       ALOC0138
        TMI     *+4                                         ALOC0139
        CLA     SAV1-1                                      ALOC0140
        STA     SAV1                                        ALOC0141
        TRA     DUMPR                                       ALOC0142
        CLA     1,4                                         ALOC0143
        PAX     ,4                                          ALOC0144
        ZET     STORE                                       ALOC0145
        TSX     STER,1                                      ALOC0146
        CLA     ONE                                         ALOC0147
        STO     INORM                                           0149
        TSX     $CHAIN,4                                        0150
        TSX     ONE                                             0151
        TSX     FOUR                                            0152
NBRDP   CLA*    ADD3                                            0153
        ARS     18                                          ALOC0149
        STA     N                                           ALOC0150
        TRA     FNNUM                                       ALOC0151
ERASE   CAL*    **                                          ALOC0152
        ANA     =0000000777777                             ALOC0153
        SSM
        XEC     ADD4                                        ALOC0155
        STA     N                                           ALOC0156
        ANA     =0000000400000                             ALOC0157
        TNZ     ERSLBR                                      ALOC0158
        TRA     FNNUM                                       ALOC0159
ERSLBR  LXA     N,1                                         ALOC0160
ADD6    CLA     **,1                                        ALOC0161
        TMI     *+5                                         ALOC0162
        PDX     ,4                                          ALOC0163
        ALS     18                                          ALOC0164
        SSM                                                 ALOC0165
        STO     U+1,4                                       ALOC0166
                                                            ALOC0167
```

```
        TIX     ERSLBR+1,1,1                                    ALOC0168
        TRA     FNNUM                                           ALOC0169
DPSLF   XEC     ADD3                                            ALOC0170
        ALS     18
        STO     LP
        AXT     1,1
LP      TXH     LOOP+1,1,**
        CLA*    ADD6                                            ALOC0172
        TMI     LOOP                                            ALOC0173
        PDX     ,4                                              ALOC0174
        STA     N                                               ALOC0175
        SXA     K,4                                             ALOC0176
        ANA     =0C00000100000                                  ALOC0177
        TZE     ERSL                                            ALOC0178
        CLA     TOP                                             ALOC0179
        SUB     K                                               ALOC0180
        STO     K                                               ALOC0181
        CLA*    ADD6                                            ALOC0182
        ANA     =0000000777777
        SSM                                                     ALOC0183
        STO*    ADD6
        TSX     WFT,4                                           ALOC0185
        CLA     FN                                              ALOC0186
        ORS*    ADD6
        CLA     N                                               ALOC0188
        ALS     18                                              ALOC0189
        SSM                                                     ALOC0190
        STO*    K                                               ALOC0191
LOOP    TXI     LP,1,1                                          ALOC0192
        CLA     ADD3                                            ALOC0193
        STA     K                                               ALOC0194
        TRA     ADD3+3                                          ALOC0195
ERSL    CLA     N                                               ALOC0196
        ALS     18                                              ALOC0197
        SSM                                                     ALOC0198
        STO     U+1,4                                           ALOC0199
        CAL*    ADD6                                            ALOC0200
        ANA     =0000000777777
        SSM                                                     ALOC0202
        STO*    ADD6                                            ALOC0203
        TRA     LOOP                                            ALOC0204
MOVR    CLA     TOP                                             ALOC0205
        SUB     NNL                                             ALOC0206
        ALS     18                                              ALOC0207
ADD7    STO*    **
        XEC     ACD3                                            ALOC0209
        STA     N                                               ALOC0210
MOV     LXA     N,1                                             ALOC0211
        SXD     **+9,1                                            00212
        AXT     0,1                                             ALOC0213
        CLA     K                                               ALOC0214
        STA     **+7                                              00215
        CLA     NNL                                             ALOC0216
        STA     **+6                                              00217
        CAS     K
        TRA     **+2
        TRA     MOVCT
        TXH     **+4,1,**                                       ALOC0218
        CLA     **,1                                            ALOC0219
        STO     **,1                                            ALOC0220
```

```
         TXI     *-3,1,1                                                          ALOC0221
         CLA     NNL
         STA     UPDSL
         STA     *+1
         CLA*    **
         ANA     =0000000400000                                                   ALOC0223
         TNZ     UPDSL                                                            ALOC0224
 MOVCT   CLA     NNL                                                              ALOC0225
         SUB     N                                                                ALOC0226
         SUB     =1                                                               ALOC0226
         STO     NNL                                                              ALOC0227
         TRA     FNNUM                                                            ALOC0228
 UPDSL   CAL*    **                                                               ALOC0231
         PAX     ,1
         CLA     NNL                                                              ALOC0233
         STA     *+1
 LASCW   CLA     **,1                                                             ALOC0235
         TMI     *+6                                                              ALOC0236
         PDX     ,4                                                               ALOC0237
         PXA     ,1                                                               ALOC0238
         SSM                                                                      ALOC0239
         ADD     NNL                                                              ALOC0240
         STA     U+1,4                                                            ALOC0241
         TIX     LASCW,1,1                                                        ALOC0242
         TRA     MOVCT                                                            ALOC0243
 PTERR   CLA     1,4             FIRST ARGUMENT OF CALLING IN AC
         STA     *+8                       UNDEFINED CODEWORD ADDRESS
         SXA     *+5,4
         TSX     $PRER2,4        CALL ERROR PRINTING ROUTINE
         TSX     =017000000      ERROR MESSAGE NUMBER
         TSX     ONE             ARGUMENT NECESSARY FOR PRER2 BUT NOT USED
*                                HERE
         TSX     *+3
         AXT     **,4
         TRA     CHA2-3          TRANSFER TO SEQUENCE FOR RETURN FROM ALOCAT
         PZE                     SPACE FOR CODEWROD ADDRESS
 ALOCAT  SXA     CHA2-1,1        C(IR1) STORED IN RETURNING SEQUENCE              ALOC0244
         STZ     NREQ            CLEAR THIS LOCATION                              ALOC0245
         CLA     =2              PLACE A 2 IN THE RETURNING SEQUENCE TO           ALOC0246
         STA     CHA2                INDICATE RETURN FROM ALLOCATION TO
*                                LEVEL ARRAY
         CAL*    1,4             PLACE THE CODEWORD ITSELF IN THE AC              ALOC0248
         STA     NREQ            THE SIZE OF THE ARRAY IS STORED HERE             ALOC0249
         ANA     =0000000400000  TEST BIT 18 TO DETERMINE IF AN ARRAY OF          ALOC0250
*                                CODEWORDS OR OF DATA
         TNZ     SLCWAL          SINCE AC NOT ZERO, IT IS A SECOND LEVEL          ALOC0251
*                                ARRAY. GO TO STATEMTN SLCWAL
         CLA*    1,4             PLACE CODEWRD IN THE AC                          ALOC0252
         TZE     PTERR           IF CODEWRD IS ZERO, TRANSFER TO ERROR            ALOC0253
 CHA1    TPL     CLREL           IS THE CODEWORD POSITIVE OR NEGATIVE             ALOC0254
*                                POSITIVE MEANS THE ARRAY IS IN CORE
         CLA     NL              CODEWORD IS NEGATIVE AND THE ARRAY IS            ALOC0255
*                                ON TAPE OR BEING ALOCATED FOR THE FIRST TIM
*                                E.   PLACE THE LOCATION OF POOL IN THE AC
*                                START OF THE POOL IN THE AC
         STO     K               STORE THIS VALUE IN LOCATION CALLED K            ALOC0256
         SUB     NREQ            SUBTRACT THE AMOUNT PLUS ONE OF THE SPACE        ALOC0257
         SUB     =1              REQUIRED TO DETERMINE A NEW START OF POOL        ALOC0258
         CAS     NT              COMPARE WITH LOCATION OF BOTTOM OF POOL          ALOC0259
         TRA     OKFILE                    MEMORY AVAILABLE                       ALOC0260
```

222

```
          NOP                                                                  ALOC0261
          SXA       *+2,4         THERE IS NO SPACE AND A REORGANIZATION OF     ALOC0262
          TSX       REORG,4       MEMORY IS NECESSARY.                         ALOC0263
          AXT       **,4                                                       ALOC0264
          ZET       STORE         HAS THE CALLING SEQUENCE BEEN ALTERED FOR
*                                 SECOND LEVEL ARRAYS
*                                 REUQIRED IF A SECOND LEVEL ARRAY EXISTS
          TSX       STER,1        IT HAS THE TRANSFER RESTORES THE CALLING
*                                 SEQUENCE  AND PLACES ZEROS IN LOCATION STOR
          TRA       ALOCAT+4      ALLOCATE THE SPACE                           ALOC0265
OKFILE STO NL                     STORE THE NEW LOCATION OF THE START OF POOL  ALOC0266
          CLA*      1,4           PLACE CODEWORD IN THE AC                     ALOC0267
          ANA       =0177777000000 HAS THE CODEWORD BEEN ALOCATED BEFORE
          STO       FILE          YES, STORE THE DECREMENT IN LOCATION FILE    ALOC2
          TZE       OPEN                                                       ALOC0269
          CAL*      1,4           PLACE THE CODEWORD IN THE AC
          ANA       =0200000000000  HAS THE ARRAY BEEN REDEFINED WHILE ON TAPE
          TNZ       LCONT
          CLA       THREE         NO OPERATION NUMBER FOR NO SIZE CHANGE
          STO       MCNT
INALC CLA  TOP                    COMPUTE SUBSCRIPT OF U TO START OF ARRAY     ALOC0271
          SUB       K             AND PLACE IN CODEWORD DECREMENT              ALOC0272
          ALS       18                                                         ALOC0273
          STD*      1,4                                                        ALOC0274
          CLA       1,4           PLACE THE LOCATION OF THE CODEWORD IN THE AC ALOC0275
          ANA       =0000000077777  ERASE ALL BUT ADDRESS                     ALOC0276
          SLW*      K             STORE LOGICALLY AT THE HEAD OF THE ARRAY     ALOC0277
*                                 AS A BACK REFERENCE
*                                 E FIRST WORD OF THE ALLOCATED ARRAY
          CLA*      1,4           SET S AND 2 BITS OF CODEWORD TO ZERO         ALOC0278
          ANA       =0077777777777
          SLW*      1,4                                                        ALOC0280
          SXA       CHA2-4,4      STORE C(IR4)                                 ALOC0281
          LXA       NREQ,4        PLACE IN IR4 THE SIZE OF THE ARRAY           ALOC0282
          SXD       NCOUNT,4      C(IR4) PLACED IN DECREMENT OF NCOUNT         ALOC0283
          CLA       K             LOCATION OF FIRST WORD OF THE ARRAY PLACED   ALOC0284
*                                 IN THE AC
          SUB       =1            SUBTRACT ONE FROM THE ADRESS                 ALOC0285
          STA       *+6           STORE THIS SIX LOCATION FROM HERE            ALOC0286
          TSX       $FILES,4      GO TO SUBROUTINE WHICH HANDLES TRANSMISSION  ALOC0287
*                                 OF INFORMATION TO AND FROM TAPE
*                                 THE FOLLOWING ARE THE ARGUMENTS OF THE SUB.
          TSX       MCNT          INDICATES WHICH OPERATION TO BE DONE         ALOC0288
          TSX       TN            SCRATCH TAPE NUMBER                          ALOC0289
          TSX       FILE          TAPE FILE NUMBER                             ALOC0290
          TSX       NCOUNT        SIZE OF THE ARRAY OR EXPECTED SIZE           ALOC0291
          TSX       **            LOCATION OF FIRST WORD OF ARRAY              ALOC0292
          AXT       **,4          RESTORATION OF INDEX REGISTER 4              ALOC0293
          ZET       STORE         HAS THE CODEWORD BEEN RESTORED               ALOC0294
          TSX       STER,1        NO RESTORE IT                                ALOC0295
          AXT       **,1          YES RESTORED. NOW RESTORE IR1                ALOC0296
CHA2   TRA  **,4                  RESTORE IR4 AND RETURN TO CALLING SUBROUTIN  ALOC0297
LCONT CLA  TWO
          ADD       TWO           A FOUR  FOR OPERATION NUMBER FOR FILES
*                                 IMPLYING POSSIBLE SIZE CHANGE
          TRA       INALC-1       GO TO THAT WHICH PRECEDES CALLIN G OF FILES
OPEN  CLA  TOP                    PLACE 77462 IN THE AC  (ARRAY IN CORE)       ALOC0298
          SUB       K             SUBTRACT ENTERING START OF POOL              ALOC0299
          ALS       18            SHIFT ADDRESS INTO DECREMENT                 ALOC0300
          STD*      1,4           STORE IN DECREMENT OF CODEWORD               ALOC0301
```

```
        CAL*    1,4                     MAKE SIGN POSITIVE TO INDICATE NOT RELEASEDALOC0302
        STO*    1,4                                                             ALOC0303
        CLA     1,4                     PLACE ADDRESS OF CODEWORD IN AC          ALOC0304
        ANA     =0000000077777  ERASE ALL BUT ADDRESS                           ALOC0305
        SLW*    K                       STORE LOGICALLY IN THE FIRST WORD OF ALLOCAALOC0306
*                                       TED ARRAY
        NZT     NREQ                    IS THE ARRAY OF ZERO LENGTH              ALOC0307
        TRA     CHA2-3                  YES  GO TO RETURN SEQUENCE               ALOC0308
        LXA     NREQ,1                  NO. PLACE SIZE OF ARRAY IN IR1           ALOC0309
        CLA     K                       PLACE THE LOCATION OF THE FIRST WORD OF THEALOC0310
*                                       ARRAY IN THE AC
        STA     *+1                                                             ALOC0311
        STZ     **,1                    ZERO THE ARRAY OF LENGTH GIVEN BY C(NREQ) ALOC0312
        TIX     *-1,1,1                                                         ALOC0313
        TRA     CHA2-3                  GO TO RETURNING SEQUENCE                 ALOC0314
SLCWAL  CLA*    2,4                     THIS IS A SECOND LEVEL ARRAY9  PLACE IN THEALOC0315
*                                       AC THE SECOND WORD OF THE CALLING SEQUENCE
*                                       THIS NUMBER INDICATES WHICH ARRAY CONNECTED
*                                       WITH THE CODEWORD IS BEING ALLOCATED
        TZE     CHAR                    IF IT IS ZERO (THE CODEWORD ARAY),TRANSFER ALOC0316
        SXA     SVCNT-1,4
        CLA*    1,4                     PLACE CODEWORD IN THE AC                 ALOC0317
        TMI     *+4                     TRANSFER IF C(AC) ARE NEGATIVE
        PDX     ,1                      DECREMENT OF AC INTO IR1
        CLA     U+1,1                   PLACE FIRST WORD OF ARRAY IN AC
        TPL     SVCNT
        CLA     1,4                     PLACE ADRESS OF CODEWORD IN AC           ALOC0324
        STO     *+3                     STORE IN IN THE CALLING SEQUENCE FOLLOWING ALOC0325
        LXA     CHA2-1,1
        TSX     ALOCAT,4                CALL ALOCAT TO ALLOCATE THE ARRAY        ALOC0326
        TSX     **                      LOCATION OF CODEWORD                     ALOC0327
        TSX     ZERO                    PLACE A ZERO AS SECOND WORD OF CALLING SEQ ALOC0328
        AXT     **,4                    RESTORE IR4                              ALOC0329
SVCNT   TSX     SSLAD,1                 STORE CODEWORD AND PLACE A NEW VALUE IN ITSALOC0330
*                                       LOCATION
CHAR    CLA     =3                                                              ALOC0331
        STA     CHA2
        CLA*    1,4                     PLACE THE CODEWORD IN THE AC             ALOC0333
        STA     NREQ                    STORE THE ADDRESS PORTION WHICH SHOULD BE TALOC0334
*                                       HE SIZE OF THE ARRAY IN LOCATION NREQ
        TMI     CHA1+1                  A NEGATIVE SIGN MENAS FILE ON TAPE. TRNSFERALOC0335
        TZE     PTERR                   IF CODEWORD IS ZERO, TRANSFER TO ERROR   ALOC0336
CLREL   PDX     ,1                      C(AC) NON-ZERO AND ARRAY IS IN CORE,PLACE ALOC0337
*                                       DECREMENT OF AC IN IR1
        CLA     U+1,1                   CLEAR AC AND PLACE IN IT THE FIRST WORD OF ALOC0338
*                                       THE ARRAY
        SSP                             SET THE SIGN POSITIVE TO INDICATE IN CORE ALOC0339
        STO     U+1,1                   STORE IN THE FIRST WORD OF THE ARRAY     ALOC0340
        TRA     CHA2-3                  GO TO RETURNING SEQUENCE                 ALOC0341
DEFINE  CLA     6,4                                                             ALOC0342
        ARS     15                                                                00343
        LAS     =074000                                                           00344
        TRA     *+2                                                             ALOC0345
        TRA     SECCW                                                           ALOC0346
        CLA     =6                                                              ALOC0347
        STA     T
        CLA*    2,4                             N                               ALOC0349
        ARS     15                                                              ALOC0350
        ADD*    5,4                             U                               ALOC0351
        ARS     1                                                               ALOC0352
```

```
        ADD*    4,4                              ALOC0353
        ARS     1                                ALOC0354
        ADD*    3,4                    R         ALOC0355
CONTA   ARS     1
        SSM                                      ALOC0360
        STO     TEMP                             ALOC0361
        CLA*    1,4                              ALOC0362
        ANA     =0177777000000
        TZE     ENT
        CLA*    1,4
        TPL     TESTR
        STD     TEMP
        STP     TEMP
        CAL     =0600000000000
        ORS     TEMP
ENT     CAL     TEMP
        SLW*    1,4                              ALOC0367
ENDR    SXA     *+3,1                            ALOC0368
        ZET     STORE                            ALOC0369
        TSX     STER,1                           ALOC0370
        AXT     **,1                             ALOC0371
T       TRA     **,4
SECCW   CLA*    2,4                              ALOC0374
        TZE     FRSTCW                           ALOC0375
        SXA     *+2,1                            ALOC0376
        TSX     SSLAD,1                          ALOC0377
        AXT     **,1                             ALOC0378
FRSTCW  CLA     =7                               ALOC0379
        STA     T
        CLA*    3,4                              ALOC0381
        ARS     15                               ALOC0382
        ADD*    6,4                    U         ALOC0383
        ARS     1                                ALOC0384
        ADD*    5,4                              ALOC0385
        ARS     1                                ALOC0386
        ADD*    4,4                    R         ALOC0387
        TRA     CONTA                            ALOC0390
LGRR    PZE
LGR     PZE
SAM     CLA*    1,4
        SXA     SAMR,4                           ALOC0392
        SXA     SAMR+1,1                         ALOC0393
        ANA     =0000000400000                   ALOC0394
        TZE     SAMZ                             00395
        CLA*    1,4                              ALOC0396
        PDX     ,4                               ALOC0397
        PAX     ,1                               ALOC0398
        PXA     ,4                               ALOC0399
        SSM                                      ALOC0400
        ADD     TOP                              ALOC0401
        STA     *+1                              ALOC0402
        CLA     **,1                             ALOC0403
        TMI     *+5                              ALOC0404
        PDX     ,4                               ALOC0405
        ALS     18                               ALOC0406
        SSM                                      ALOC0407
        STO     U+1,4                            ALOC0408
        TIX     *-6,1,1                          ALOC0409
SAMZ    LXA     SAMR,4
        CLA     1,4
```

```
        STA     *+2
        TSX     $CLEAR,4
        TSX     **
SAMR    AXT     **,4                                            ALOC0410
        AXT     **,1                                            ALOC0411
        TRA     TESTR+4                                           00412
TESTR   ERA     TEMP                                            ALOC0413
        ANA     =0000000400000                                  ALOC0414
        STO     LGRR
        TNZ     SAM                                             ALOC0415
        CLA*    1,4                                             ALOC0416
        ANA     =0000000077777                                  ALOC0417
        SLW     NOLD                                            ALOC0419
        CAL     TEMP                                            ALOC0420
        ANA     =0000000077777                                  ALOC0421
        STO     NNEW                                            ALOC0422
        SUB     NOLD                                            ALOC0423
        STO     LGR
        TNZ     MOVER                                           ALOC0424
USEN    CAL*    1,4                                             ALOC0425
        ANA     =0077777000000                                  ALOC0426
        ACL     TEMP                                            ALOC0427
        STO*    1,4                                             ALOC0428
        TRA     ENDR                                            ALOC0429
MOVER   SXA     USNEW-2,1                                       ALOC0430
        CAL*    1,4                                             ALOC0431
        ARS     18                                              ALOC0432
        ANA     =0000000077777                                  ALOC0433
        SSM                                                     ALOC0434
        ADD     TOP                                             ALOC0435
        SUB     NOLD                                            ALOC0436
        SUB     =1                                              ALOC0437
        SUB     NL                                              ALOC0438
        TZE     NOMOV                                           ALOC0439
        CLA     NL                                              ALOC0440
        SUB     NT                                              ALOC0441
        SUB     NNEW                                            ALOC0442
        TMI     REORGI                                          ALOC0443
        CLA     LGR
        TMI     USNEW                                           ALOC0446
        PAX     ,1                                              ALOC0447
        CLA     NL                                              ALOC0448
        SUB     NOLD                                            ALOC0449
        STA     *+1                                             ALOC0450
        STZ     **,1                                            ALOC0451
        TIX     *-1,1,1                                         ALOC0452
        LXA     NOLD,1                                          ALOC0453
MVR     CAL*    1,4                                             ALOC0454
        ARS     18                                              ALOC0455
        ANA     =0000000077777                                  ALOC0456
        SSM                                                     ALOC0457
        ADD     TOP                                             ALOC0458
        STA     *+7                                             ALOC0459
        STA     *+3                                             ALOC0460
        CLA     NL                                              ALOC0461
        STA     *+2                                             ALOC0462
        CLA     **,1                                            ALOC0463
        STO     **,1                                            ALOC0464
        TIX     *-2,1,1                                         ALOC0465
        CLA     **                                              ALOC0466
```

```
        STO*    NL                                      ALOC0467
        CLA*    1,4                                     ALOC0468
        STZ*    *-3                                     ALOC0469
        ALS     18                                      ALOC0470
        SSM                                             ALOC0471
        STO*    *-6                                     ALOC0474
        ALS     18                                      ALOC0475
        STO*    1,4
        CLA     LGRR
        TNZ     CHNL
        CLA*    1,4
        ANA     =0400000
        TZE     CHNL
        NZT     NOLD
        TRA     CHNL
        SXA     CHNL-1,4
        LXA     NOLD,1
        CLA     NL
        STA     *+1
        CLA     **,1
        TMI     *+6
        PDX     ,4
        PXA     ,1
        SSM
        ADD     NL
        STA     U+1,4
        TIX     *-7,1,1
        AXT     **,4
CHNL    CLA     NL                                      ALOC0476
        SUB     NNEW                                    ALOC0477
        SUB     =1                                      ALOC0478
        STO     NL                                      ALOC0479
        AXT     **,1                                    ALOC0480
        TRA     USEN                                    ALOC0481
USNEW   CAL*    1,4
        ANA     =0400000
        TZE     *+2
        TSX     DTRYBR,1
        LXA     NNEW,1                                  ALOC0482
        TRA     MVR                                     ALOC0483
REORGI  SXA     REIR,4                                  ALOC0484
        CLA     NNEW                                    ALOC0485
        STO     NREG                                    ALOC0486
        TSX     REORG,4                                 ALOC0487
REIR    AXT     **,4                                    ALOC0488
        CLA     STORE
        TZE     *+3
        TSX     STER,1
        TSX     SSLAD,1
        CLA*    1,4                                     ALOC0489
        TPL     MVR-10                                  ALOC0490
        TRA     ENDR-6
NOMOV   CAL*    1,4
        ANA     =0400000
        TZE     *+4
        CLA     LGR
        TPL     *+2
        TSX     DTRYBR,1
        CLA     NL                                      ALOC0492
        ADD     =1
```

```
          STA      RELEAS-3
          SUB      =1
          SUB      NNEW                                                    ALOC0493
          ADD      NOLD                                                    ALOC0494
          CAS      NT                                                      ALOC0495
          NOP                                                              ALOC0496
          TRA      ZENM                                                    ALOC0497
          CLA      NNEW                                                    ALOC0498
          SUB      NOLD                                                    ALOC0499
          STO      NREQ                                                    ALOC0500
          SXA      *+2,4                                                   ALOC0501
          TSX      REORG,4                                                 ALOC0502
          AXT      **,4                                                    ALOC0503
          CLA      STORE
          TZE      *+3
          TSX      STER,1
          TSX      SSLAD,1
          CLA*     1,4
          TMI      ENT-4
          TRA      NOMOV+6
DTRYBR    SXA      DTLOP+1,1                                               ALOC0504
          SXA      DTLOP+2,4
          CLA*     1,4
          PAX      ,1
          SXD      DTLP,1
          PDX      ,4
          PXA      ,4
          SSM
          ADD      TOP
          STA      DTLP+1
          CLA      NNEW
          ADD      =1
          PAX      ,1
DTLP      TXH      DTLOP+1,1,**
          CLA      **,1
          TMI      DTLOP
          PDX      ,4
          STZ      U+1,4
          ALS      18
          SSM
          STO      U+1,4
DTLOP     TXI      DTLP,1,1
          AXT      **,1
          AXT      **,4
          TRA      1,1
ZENM      STA      NL                                                     ALOC0505
          CLA      NNEW
          SUB      NOLD
          TMI      USEN
          PAX      ,1
          STZ      **,1
          TIX      *-1,1,1
          TRA      USEN
RELEAS    SXA      IR11,1                 CALL RELEAS(NAME)    OR          ALOC0506
          CAL*     1,4                    CALL RELEAS(NAME,U,V)    OR      ALOC0507
          ANA      =0000000400000         CALL RELEAS(SLNAME,J)    OR      ALOC0508
          TNZ      SECL                   CALL RELEAS(SLNAME,J,U,V)        ALOC0509
          CLA      =3                                                     ALOC0510
          STA      A34                                                    ALOC0511
          CLA      =4                                                     ALOC0512
                                                                         ALOC0513
```

```
        STA      SXRT-1                                          ALOC0514
        CLA      =2                                              ALOC0515
        STA      A23                                             ALOC0516
        STA      A24                                             ALOC0517
        STA      ONEAR                                           ALOC0518
A23     CLA      **,4                                            ALOC0519
        ARS      15                                                00520
        LAS      =074000                                           00521
        TRA      ONEAR                                           ALOC0522
        TRA      *+2                                             ALOC0523
        TRA      ONEAR                                           ALOC0524
        CLA*     1,4                                             ALOC0525
        TPL      A24                                             ALOC0526
        XEC      SXRT-1                                          ALOC0527
        XEC      SXRT                                            ALOC0528
        TRA      IR11-2                                          ALOC0529
A24     CLA*     **,4                                            ALOC0530
        ALS      1                                               ALOC0531
A34     ADD*     **,4                                            ALOC0532
        ARS      3                                               ALOC0533
        STO      NNEW                                            ALOC0534
        CAL*     1,4                                             ALOC0535
        ANA      =0777777477777                                  ALOC0536
        ACL      NNEW                                            ALOC0537
        SLW*     1,4                                             ALOC0538
        ALS      18                                              ALOC0539
        AXT      **,1                                            ALOC0540
SXRT    SXA      RETN,1                                          ALOC0541
        SSM                                                      ALOC0542
        STO      NNEW                                            ALOC0543
        CAL*     1,4                                             ALOC0544
        PDX      ,1                                              ALOC0545
        CAL      NNEW                                            ALOC0546
        STP      U+1,1                                           ALOC0547
        ZET      STORE                                           ALOC0548
        TSX      STER,1                                          ALOC0549
IR11    AXT      **,1                                            ALOC0550
RETN    TRA      **,4                                            ALOC0551
ONEAR   AXT      **,1                                            ALOC0552
        CLA*     1,4                                             ALOC0553
        TMI      A24-2                                           ALOC0554
        ALS      18                                              ALOC0555
        TRA      SXRT                                            ALOC0556
SECL    CLS*     2,4                                             ALOC0557
        TZE      FIRSTL                                          ALOC0558
        TSX      SSLAD,1                                         ALOC0559
FIRSTL  CLA      =4                                              ALOC0560
        STA      A34                                             ALOC0561
        CLA      =5                                              ALOC0562
        STA      SXRT-1                                          ALOC0563
        CLA      =3                                              ALOC0564
        TRA      A23-3                                           ALOC0565
DUMP    CLA      =2                                              ALOC0566
        STA      RETDU                                           ALOC0567
        CLA      1,4                                             ALOC0568
        STA      CWLOC                                           ALOC0569
        CLA*     1,4                                             ALOC0570
        TPL      CONTB                                           ALOC0571
        CLA      1,4                                             ALOC0572
        STA      *+9                                             ALOC0573
```

```
        CLA     2,4                                      ALOC0574
        STA     *+5                                      ALOC0575
        STA     *+7                                      ALOC0576
        SXA     *+7,4                                    ALOC0577
        TSX     ALOCAT,4                                 ALOC0578
CWLOC   TSX     **                                       ALOC0579
        NOP     **                                       ALOC0580
        TSX     RELEAS,4                                 ALOC0581
        TSX     **                                       ALOC0582
        NOP     **                                       ALOC0583
        AXT     **,4                                     ALOC0584
        CLA*    CWLOC                                    ALOC0585
CONTB   ANA     =0000000400000                           ALOC0586
        TNZ     SECLD                                    ALOC0587
        STZ     TEMP                                     ALOC0588
        CLA*    CWLOC                                    ALOC0589
        STD     TEMP                                     ALOC0590
        ALS     18                                       ALOC0591
        STD     NCOUNT                                   ALOC0592
        SXA     RETDU-1,4                                ALOC0593
        CALL    DUMPER,TEMP,NCOUNT,CWLOC                 ALOC0594
        AXT     **,4                                     ALOC0595
RETDU   TRA     **,4                                     ALOC0596
SECLD   CLA     =3                                       ALOC0597
        STA     RETDU                                    ALOC0598
        CLA*    2,4                                      ALOC0599
        TZE     CONTB+2                                  ALOC0600
        CAL*    1,4                                      ALOC0601
        ANA     =0077777000000                           ALOC0602
        ADD*    2,4                                      ALOC0603
        ARS     18                                       ALOC0604
        SSM                                              ALOC0605
        ADD     TOP                                      ALOC0606
        STA     CWLOC                                    ALOC0607
        CLA*    CWLOC                                    ALOC0608
        TPL     CONTB+2                                  ALOC0609
        CLA     CWLOC                                    ALOC0610
        TRA     DUMP+7                                   ALOC0611
SSLAD   CLA     1,4                                      ALOC0612
        STO     STORE                                    ALOC0613
        CAL*    1,4                                      ALOC0614
        ANA     =0077777000000                           ALOC0615
        ADD*    2,4                                      ALOC0616
        ANA     =0077777000000                           ALOC0617
        ARS     18                                       ALOC0617
        SSM                                              ALOC0618
        ADD     TOP                                      ALOC0619
        STA     1,4                                      ALOC0620
        TRA     1,1                                      ALOC0621
STER    CLA     STORE                                    ALOC0622
        STO     1,4                                      ALOC0623
        STZ     STORE                                    ALOC0624
        TRA     1,1                                      ALOC0625
STORE   PZE                                              ALOC0626
FN      PZE                                              ALOC0627
K       PZE                                              ALOC0628
N       PZE                                              ALOC0629
NONEED  PZE                                              ALOC0630
NPERF   PZE                                              ALOC0631
NNL     PZE                                              ALOC0632
```

```
NOLD     PZE                                                                      ALOC0633
NNEW     PZE                                                                      ALOC0634
NCOUNT   PZE                                                                      ALOC0635
 TEMP    PZE                                                                      ALOC0636
 FILE    PZE                                                                      ALOC0637
 MCNT    PZE
 ZERO    PZE                                                                      ALOC0638
 ONE     OCT     000001000000                                                     ALOC0639
 TWO     OCT     000002000000            TWO                                      ALOC0640
 THREE   OCT     000003000000                                                     ALOC0641
 FOUR    OCT     000004000000                                                        0647
 U       COMMON  110                                                                 0648
 INORM   COMMON  9                                                                   0649
 TOP     COMMON  1               TOP OF COMMON PLUS ONE                           ALOC0643
 N1      COMMON  1               TOP OF POOL                                      ALOC0644
 NL      COMMON  1               NEXT MEMORY LOCATION TO BE ASSIGNALOC0645
 NT      COMMON  1               BOTTOM OF POOL                                   ALOC0646
 NREQ    COMMON  1                                                                ALOC0647
 TN      COMMON  1                 SCRATCH TAPE FOR FILING                        ALOC0648
         END                                                                      ALOC0649
*        LIST8
*        LABEL
CFILEA5     ROUTINE FOR READING AND WRITING FILE TAPES 1-17-64 LOGCHER -A5
C        REVISED VERSION FOR PROCESSOR - CHAIN TAPE USED AS SYSTEM TAPE A5
         SUBROUTINE FILES(NOP,TN,NFILE,NCOUNT,ARRAY)
         COMMON U,ISCAN,ISUCC,FIL,LFILE,FIL2,NTAPE,NXREC
         DIMENSION ARRAY(100),U(112),FIL(7),FIL2(5),ISCAN(5)
         DIMENSION NTAPE(5),NXREC(5),BUF(254),LBUR(254)
         EQUIVALENCE (BUF,LBUR)
         XNTF(L)=((L-1)/20+1)-(4*((L-1)/ 80))
         XNRNF(L)=(L-1)/80 +1
         XLFNF(L)=((L-1)/20)*20+21
         XFIBF(L)=L-20*((L-1)/20)
  5      GO TO (10,20,30,30,19),NOP
C        SCRATCH TAPES USED NOW ARE B1, B2, B3, A4
  10     NTAPE(1)=5
         NTAPE(2)=6
         NTAPE(3)=7
         NTAPE(4)=8
         LFILE=1
         NBUF=0
         DO 11 I=1,4
         NT=NTAPE(I)
         NXREC(I)=1
  11     REWIND NT
  12     RETURN
C        DUMP BUFFER ON TAPE AND RETURN
  19     NS1=3
         IF(NBUF)12,12,234
  20     IF(NCOUNT-252)231,231,21
C        IS BUFFER EMPTY
  21     IF(NBUF)25,25,22
  22     NS1=1
C        EMPTY BUFFER
 234     NIT=XNTF(LFILE-1)
         NT=NTAPE(NIT)
         NIR=XNRNF(LFILE-1)
 230     IF(NIR-NXREC(NIT))23,24,26
  23     PRINT 1,LFILE
         GO TO 12
```

```
  1      FORMAT(39H1ERROR IN TAPE WINDING FOR WRITING FILE,I6)
 24      WRITE TAPE NT,(BUF(I),I=1,254)
         NBUF=0
         LFILE=XLFNF(LFILE-1)
         NXREC(NIT)=NXREC(NIT)+1
         GO TO (25,235,13),NS1
 25      NIT=XNTF(LFILE)
         NT=NTAPE(NIT)
         NIR=XNRNF(LFILE)
         NS2=2
 29      IF(NIR-NXREC(NIT))23,27,240
 26      NS2=1
C        FORWARD SPACING ROUTINE
240      READ TAPE NT,NOLD
         NXREC(NIT)=NXREC(NIT)+1
         GO TO (230,29,37),NS2
C        WRITE ARRAY ON TAPE - IT EXCEEDS BUFFER SIZE
 27      WRITE TAPE NT,LFILE,NCOUNT,(ARRAY(I),I=1,NCOUNT)
         NFILE=LFILE
         LFILE=XLFNF(LFILE)
         NXREC(NIT)=NXREC(NIT)+1
 13      IF(65400-LFILE)14,12,12
 14      PRINT 15
         INORM=1
         CALL CHAIN(1,4)
 15      FORMAT(50H1TAPE CAPACITY EXCEEDED.   PROBLEM CANNOT CONTINUE.)
C        WILL ARRAY FIT IN BUFFER
231      IF(NBUF+NCOUNT-252)232,232,233
C        NO - EMPTY BUFFER
233      NS1=2
         GO TO 234
C        IS THIS THE FIRST FILE
232      IF(XFIBF(LFILE)-1)235,237,235
C        YES  -  DUMP BUFFER IF NON EMPTY AND -235- PUT ARRAY IN BUFFER
237      IF(NBUF)235,235,233
C        NO - 235 - PUT ARRAY IN BUFFER
235      IBUF=NBUF+2
         LBUR(IBUF-1)=LFILE
         LBUR(IBUF)=NCOUNT
         DO 236 I=1,NCOUNT
         I1=IBUF+I
236      BUF(I1)=ARRAY(I)
         NBUF=I1
         NFILE=LFILE
         LFILE=LFILE+1
         GO TO 12
 30      LB=XFIBF(NFILE)-1
         IF(LBUR(1)-NFILE+LB)33,74,33
 33      NIR=XNRNF(NFILE)
         NIT=XNTF(NFILE)
         NT=NTAPE(NIT)
         NS2=3
 37      IF(NIR-NXREC(NIT))31,35,240
 31      N=NXREC(NIT)-NIR
         IF(NXREC(NIT)/4-N)45,45,60
 60      DO 32 I=1,N
 32      BACKSPACE NT
 35      NXREC(NIT)=NIR+1
         IF(NOP-3)22,36,40
 45      REWIND NT
```

```
        NXREC(NIT)=1
        GO TO 37
36      IF(NCOUNT-252)70,70,38
38      READ TAPE NT,NOLD,N,(ARRAY(I),I=1,NCOUNT)
        IF(NFILE-NOLD)44,12,44
44      PRINT 2,NFILE
        ISUCC=2
        ISCAN=2
        CALL CHAIN(1,4)
70      READ TAPE NT,(BUF(I),I=1,254)
74      NBF=2
        IF(LB)44,77,78
78      DO 71 I=1,LB
71      NBF=NBF+2+LBUR(NBF)
77      IF(LBUR(NBF)-NCOUNT)75,72,72
72      DO 73 I=1,NCOUNT
        J=NBF+I
73      ARRAY(I)=BUF(J)
        GO TO 12
75      N=N+1
        DO 76 I=N,NCOUNT
76      ARRAY(I)=0.0
        NCOUNT=N-1
        GO TO 72
40      IF(XFIBF(NFILE)-1)44,41,70
41      READ TAPE NT,NOLD,N
        BACKSPACE NT
        IF(NCOUNT-N)38,38,42
42      N=N+1
        DO 43 I=N,NCOUNT
43      ARRAY(I)=0.0
        NCOUNT=N-1
        GO TO 38
2       FORMAT(25H1TAPE READING ERROR, FILE,I6)
        END
*       LIST8
*       LABEL
CDUMPER    ALLOCATOR VERSION I DUMPER
        SUBROUTINE DUMPER(K,N,LOCW)
        COMMON U,IU
        DIMENSION U(100),IU(100)
        EQUIVALENCE (U,IU)
        PRINT 1,LOCW
        PRINT 2
        DO 10 I=1,N
        J=K+I
10      PRINT 3,U(J),IU(J),U(J),U(J)
        RETURN
1       FORMAT (52H1DUMP OF ARRAY ASSOCIATED WITH CODEWORD AT LOCATION ,O
        15)
2       FORMAT (12H-   FLOATING,6X,5HFIXED,6X,3HBCD,6X,5HOCTAL)
3       FORMAT (E14.4,I9,A10,O14)
        END
*       FAP
        COUNT    80
        LBL      PACKW
* PACKING AND UNPACKING ROUTINES FOR STRESS VERSION 3   JULY 24,1963
        ENTRY    PACKW
        ENTRY    UPACW
        ENTRY    PADP
```

```
            ENTRY   UPADP
*SUBROUTINE PACKW(A,I,J,K,L,M)
*STRESS PROGRAMMING SYSTEM
*   S.J.FENVES, 3/22/63
*PACKS I,J,K,L,M INTO A
*S,1,2 I
*3-17  J
*18-23 K
*24-29 L
*30-35 M
 PACKW STZ*    1,4                    CLEAR A
       CLA*    2,4                    I
       ALS     15
       STP*    1,4                    STORE I
       CLA*    3,4
       STD*    1,4                    STORE J
       CLA*    4,4
       ANA     M2
       ARS     6
       ORS*    1,4                    STORE K
       CLA*    5,4
       ANA     M2
       ARS     12
       ORS*    1,4                    STORE L
       CLA*    6,4
       ANA     M2
       ARS     18
       ORS*    1,4                    STORE M
       TRA     7,4
*SUBROUTINE UPACW(A,I,J,K,L,M)
*S.J.FENVES 3/22/63
*UNPACKS A ORIGINALLY PACKED BY PACKW
 UPACW CAL*    1,4                    A TO ACCUMULATOR
       STZ*    2,4
       STZ*    3,4
       STZ*    4,4
       STZ*    5,4
       STZ*    6,4
       STD*    3,4                    STORE J
       ARS     15                     I TO DECREMENT
       STD*    2,4                    STORE I
       CAL*    1,4                    A TO AC
       ANA     MASK                   CLEAR LEFT HALF
       ALS     6                      K
       STD*    4,4                    STORE K
       ANA     MASK
       ALS     6
       STD*    5,4                    STORE L
       ANA     MASK
       ALS     6
       STD*    6,4                    STORE M
       TRA     7,4
 MASK  OCT     000000777777
 M2    OCT     000077000000
* SUBROUTINE PADP(A,I,J,K)
*   R.D. LOGCHER  7/24/63
*   PACKS I INTO PREFIX OF A
*   PACKS J INTO THE DECREMENT OF A
*   PACKS K INTO THE ADDRESS OF A
 PADP  CLA*    2,4
```

```
        STZ*     1,4
        ALS      15
        STP*     1,4
        CLA*     3,4
        STD*     1,4
        CLA*     4,4
        ARS      18
        STA*     1,4
        TRA      5,4
*  SUBROUTINE UPADP(A,I,J,K)
*  R.D.LOGCHER  24/7/63
*  UNPACKS WORD GENERATED BY SUBROUTINE PADP
*  I FROM PREFIX OF A
*  J FROM DECREMENT OF A
*  K FROM ADDRESS OF A
 UPADP  CLA*     1,4
        STZ*     2,4
        STZ*     3,4
        STZ*     4,4
        STD*     3,4
        ALS      18
        STD*     4,4
        CAL*     1,4
        ARS      15
        STD*     2,4
        TRA      5,4
        END
*       FAP
        COUNT    200                                                CHAIC010
        SST                                                         CHAIC020
*   32K 709/7090 FORTRAN LIBRARY        9CHN                        CHAIC030
*   32K 709/7090 FORTRAN LIBRARY    MIT VERSION.  MARCH 28,1962     MICHAIC040
        TTL     MCNITOR CHAIN ROUTINE / 9CHN FOR STRESS III PROCESSOR   CHAIC050
*   ALL CARDS SEQUENCE NUMBERED WITH MI ARE ONLY FOR MIT MONITOR
*   ALL CARDS SEQUENCE NUMBERED WITH STRES ARE FOR STRESS PROCESSOR
        LBL     STCHA5                                             STRESCHAIC060
*   MAY 16 1963 REVISED FOR TSS BACKGROUND                          MICHAIGC70
*   AUG 22 1963 * REVISED FOR 7094                                  MICHAIC071
*   NOV 20 1963 * REVISED FOR STRESS III                           STRESCHAIC072
*   JAN 15 1964 * REVISED FOR STRESS III PROCESSOR TAPE ON A5      STRESCHAIC076
        ENTRY   CHAIN                                               CHAIC080
 CHAIN  LTM                                                         CHAIC090
        EMTM                                   7094                 MICHAI0092
        ENB     =0400000        DISABLE EVERY TRAP BUT CLOCK 7094   MICHAI0094
        SXA     *+2,4                                               CHAIC120
        XEC*    $(TES)                                              CHAIC130
        AXT     **,4                                                CHAI0140
        CLA*    1,4                                                 CHAI0150
        STD     CHWRD                                               CHAIC180
 CHA    AXT     1093,4          SET FOR                             CHAIC490
        SXA     XA3+1,4         A5                                  CHAIC500
        TXI     *+1,4,-448      IRC = 1205 OCTAL                    CHAIC510
        AXT     4,2                                                 CHAIC520
        SXA     CHWRD,2                                             CHAIC530
        AXT     2048,2                                              CHAIC540
        PXD     0,2                                                 CHAIC550
        ORS     XCI+8           SET 5L                              MICHAIC560
        ALS     1                                                   CHAIC570
        ORS     XCI+16          SET A                               MICHAIC580
        CLM                                                         CHAIC590
```

```
        LDQ     XXA             TCOA                                              CHAI0600
CHAB    SXA     CHD,4           AC=PREFIX +OR-,MQ=TCO  AC=2202,2203,1204  BSCHAI0610
        SXA     XLA,4           BSR                                               CHAI0620
        SXA     CHE+3,4         REW                                               CHAI0630
        TXI     *+1,4,16        SET IRC FOR BIN                                   CHAI0640
        SLQ     CHC+2                                                            MICHAI0645
        SXA     CHC+3,4         RTB                                              MICHAI0650
        STP     CHC+4           RCH                                              MICHAI0660
        SLQ     CHC+5           TCO                                              MICHAI0670
        STP     CHC+6           TRC                                              MICHAI0680
        STP     CHC+7           TEF                                              MICHAI0690
        SXA     XLC,4           RTB                                               CHAI0700
        STP     XLC+1           RCH                                               CHAI0710
        SLQ     XLD             TCO                                               CHAI0720
        STP     XLD+1           TRC                                               CHAI0730
        STP     XLD+2           SCH                                               CHAI0740
        STP     CHC-1           TEF                                               CHAI0750
        AXT     XL(0)-XLA+1,1 MOVE                                                CHAI0760
        AXT     0,2             LOADER TO UPPER                                   CHAI0770
        CLA     XLA,2                                                             CHAI0780
        STO     .XLA,2                                                            CHAI0790
        TXI     *+1,2,-1                                                          CHAI0800
        TIX     *-3,1,1                                                           CHAI0810
        TEFA    *+1                                                               CHAI0820
CHC     AXT     5,1                                                               CHAI0830
        CLA     CHWRD                                                             CHAI0840
        TCOB    *                                                                MICHAI0845
        RDS     **              2                                                 CHAI0850
        RCHB    CHSEL1          3                                                 CHAI0860
        TCOB    *               4                                                 CHAI0870
        TRCB    CHD             5                                                 CHAI0880
        TEFB    CHE             6                                                 CHAI0890
        LXA     LBL,2                                                             CHAI0900
        TXH     *+2,2,4                                                           CHAI0910
        TXH     *+3,2,1                                                           CHAI0920
        AXT     4,2             ANY TAPE EXCEPT 2 OR 3 MAKE 4                     CHAI0930
        SXA     LBL,2                                                             CHAI0940
        SUB     LBL             IS THIS THE CORRECT LINK                          CHAI0950
        TNZ     CHC             NOT THIS ONE                                      CHAI0960
        CLA     LKRCW                                                             CHAI0970
        STO     .XLZ            CONTROL WORD FOR READING LINK                     CHAI0980
        STA     .XL(0)                                                           MICHAI0985
        CAL     PROG                                                              CHAI0990
        ERA     XGO                                                               CHAI1000
        PDX     ,1                                                                CHAI1010
        TXH     .XXA,1,0        MAKE SURE 3RD WORD IS TRA TO SOMEWHERE            CHAI1020
        TZE     .XXA            IF NOT, COMMENT BAD TAPE                          CHAI1030
        CLA     PROG                                                              CHAI1040
        STO     .XGO            TRANSFER TO LINK                                  CHAI1050
        LXD     .XLZ,1                                                       STRESCHAI1051
        PXA     ,1                                                           STRESCHAI1052
        ADD     =100                                                         STRESCHAI1053
        STO     TEMP
        ZET     INORM                                                        STRESCHAI1055
        TRA     *+5                                                          STRESCHAI1056
        SUB     NL                                                           STRESCHAI1057
        TMI     *+3                                                          STRESCHAI1058
        STO     NREQ                                                         STRESCHAI1059
        TSX     $REORG,4                                                     STRESCHAI1060
        CLA     TEMP
```

```
        STO     NT
        AXT     5,1                                         CHAI1061
        TRA     .XLC                                        CHAI1070
TEMP    PZE
CHD     BSRB    **                                          CHAI1080
        TIX     CHC+1,1,1                                   CHAI1090
        TRA     .XXA                                        CHAI1100
CHE     NZT     CHAIN                                       CHAI1110
        TRA     NGD         SECOND EOF NO GOOD              CHAI1120
        STZ     CHAIN                                       CHAI1130
        REWA    **                                          CHAI1140
        TRA     CHC                                         CHAI1150
NGD     LXA     XA3+1,1                                     CHAI1160
        SXD     COM1+3,1                                    CHAI1170
        TCOA    *                                           MICHAI1175
        WTDA    PRTAPE                                      CHAI1180
        RCHA    CHSEL2                                      CHAI1190
        TRA*    $EXIT                                       CHAI1200
CHWRD   PZE                                                 CHAI1210
TCOB    TCOB    *                                           CHAI1220
CHSEL1  IORT    LBL,,3                                      CHAI1230
LBL                                                         CHAI1240
LKRCW                                                       CHAI1250
PROG                                                        CHAI1260
CHSEL2  IORT    COM1,,7                                     CHAI1270
COM1    BCI     7,1 LINK NOT ON TAPE    .  JOB TERMINATED.  CHAI1280
        REM                                                 CHAI1290
XLA     BSRA    **              REDUNDANCY, TRY AGAIN       CHAI1300
        IOT                                                 CHAI1310
        NOP                                                 CHAI1320
        TIX     .XLC,1,1                                    CHAI1330
XXA     TCOA    .XXA                                        CHAI1340
        WTDA    PRTAPE          COMMENT BAD TAPE            CHAI1350
        RCHA    .XLX                                        CHAI1360
XXA1    TCOA    .XXA1                                       MICHAI1365
        WPDA                                                CHAI1370
        RCHA    .XCIL                                       CHAI1390
XXA2    TCOA    .XXA2                                       MICHAI1400
        REWA    1                                           CHAI1470
        REWA    4                                           CHAI1480
        REWB    1                                           CHAI1490
        REWB    2                                           CHAI1500
        REWB    3                                           CHAI1510
Z1      HTR     .Z1+1                                       MICHAI1520
        ENK                                                 CHAI1530
        TQP     .XLB                                        CHAI1540
        BSFA    2               REPEAT JOB                  CHAI1550
XLB     RTBA    1               READ IN 1 TO CS             CHAI1560
        RCHA    .XLY                                        CHAI1570
Z2      TCOA    .Z2                                         MICHAI1575
        RTBA    1                                           CHAI1580
Z3      TCOA    .Z3                                         MICHAI1585
        RTBA    1                                           CHAI1590
        TRA     1               GO TO SIGN ON               CHAI1600
        REM                                                 CHAI1610
XLC     RDS     **              READ IN LINK                CHAI1620
        RCHB    .XLZ                                        CHAI1630
XLD     TCOB    .XLD                                        CHAI1640
        TRCB    .XLA                                        CHAI1650
        SCHB    .XSCH                                       CHAI1660
```

```
        LAC     .XSCH,2                                                  CHAI1670
        CLA*    .XL(0)                                                 MICHAI1680
        STD     .XLD1                                                    CHAI1690
XLD1    TXI     .XLD1+1,2,**                                             CHAI1700
        TXH     .XXA,2,0        WAS RECORD READ IN UP TO PROGRAM BREAK   CHAI1710
XGO     TRA     **              OK, GO TO THE LINK                       CHAI1720
        REM                                                              CHAI1730
XCI     OCT     000100001200,0  9L                                    MICHAI1740
        OCT     000000400000,0  8L COL 6=1,2                           MICHAI1750
        OCT     000200000000,0  7L CHANNEL A,B                         MICHAI1760
        OCT     000000004004,0  6L COL 7=2,3,4                         MICHAI1770
        OCT     000441000440,0  5L TAPE NO.                            MICHAI1780
        OCT     041000000100,0  4L BAD XX DEPRESS                      MICHAI1790
        OCT     000000010000,0  3L KEY S TO RERUN                      MICHAI1800
        OCT     200032100002,0  2L                                     MICHAI1810
        OCT     100000000010,0  1L                                     MICHAI1820
        OCT     000030510100,0  0L                                     MICHAI1830
        OCT     000302005254,0  11L                                    MICHAI1840
        OCT     351441000402,0  12L                                    MICHAI1850
XA3     BCI     1,1 BAD                                                   CHAI1860
        BCI     1,                                                       CHAI1870
XLX     IORT    .XA3,,2                                                  CHAI1880
XCIL    IOCT    .XCI,,24                                               MICHAI1890
XLY     IOCP    0,,3                                                     CHAI1910
        TCH     0                                                        CHAI1920
XLZ     IORT    **,,**                                                   CHAI1930
XSCH    PZE                                                              CHAI1940
XL(0)   PZE                                                              CHAI1950
        REM                                                              CHAI1960
PRTAPE  EQU     3                                                        CHAI1970
.X      COMMON  0                                                        CHAI1980
.XLA    EQU     .X+2                                                     CHAI1990
.XXA    EQU     .XLA+4                                                   CHAI2000
.XXA1   EQU     .XXA+3                                                 MICHAI2010
.XXA2   EQU     .XXA1+3                                                  CHAI2020
.Z1     EQU     .XXA2+6                                                MICHAI2025
.XLB    EQU     .Z1+4                                                  MICHAI2030
.Z2     EQU     .XLB+2                                                 MICHAI2033
.Z3     EQU     .Z2+2                                                  MICHAI2036
.XLC    EQU     .Z3+3                                                  MICHAI2040
.XLD    EQU     .XLC+2                                                   CHAI2050
.XLD1   EQU     .XLD+6                                                   CHAI2060
.XGO    EQU     .XLD1+2                                                  CHAI2070
.XCI    EQU     .XGO+1                                                   CHAI2080
.XA3    EQU     .XCI+24                                                MICHAI2090
.XLX    EQU     .XA3+2                                                   CHAI2100
.XCIL   EQU     .XLX+1                                                   CHAI2110
.XLY    EQU     .XCIL+1                                                MICHAI2120
.XLZ    EQU     .XLY+2                                                   CHAI2130
.XSCH   EQU     .XLZ+1                                                   CHAI2140
.XL(0)  EQU     .XSCH+1                                                  CHAI2150
U       COMMON  110                                                 STRESCHAI2161
INORM   COMMON  11                                                  STRESCHAI2162
NL      COMMON  1                                                   STRESCHAI2163
NT      COMMON  1                                                   STRESCHAI2164
NREQ    COMMON  1                                                   STRESCHAI2165
        END                                                              CHAI2170
*       LIST8
*       LABEL
        SUBROUTINE PRER2(J,N,A)
```

```
      DIMENSION IT(4),BETA(7),BUFA(85),B(8),SYSFIL(32),PROFIL(27),
     1 ARRFIL(14),U(5),IU(5)
      COMMON U,IU,INDEX,IT,BETA,BUFA,B,JJC,JMC,JLD,
     1 CHECK,NMAX,INCRM,ISOLV,ISCAN,IIII,IMOD,JJJJ,ICONT,ISUCC,IMERG,
     2 TOP,N1,NL,NT,NREQ,TN,LFILE,SYSFIL,NJ,NB,NDAT,ID,JF,NSQ,NCORD,
     3 IMETH, NLDS,NFJS,NSTV,NMEMV,IPSI,NMR,NJR,PROFIL,NAME,KXYZ,KJREL,
     4 JPLS,JMIN,MTYP,KPSI,MEMB,LOADS,MODN,KS,KMKST,KSTDB,KATKA,KPPLS,
     5 KPMNS,KUV,KPPRI,KR,KSAVE,KSRTCH,ARRFIL,JEXT
      EQUIVALENCE(U,IU,INDEX),(IT,IU(2),U(2))
C     KENNETH F. REINSCHMIDT, ROOM 1-255, EXT. 2117
C     STRESS...STRUCTURAL ENGINEERING SYSTEM SOLVER
C     VERSION III...13 AUGUST 1963
      GO TO (1,2,3,4,5,6,7,8,9,10,11,12,13,14,15,16,17,18,19),J
    1 PRINT 801,N
      GO TO 500
  801 FORMAT (7H MEMBER,I5,29H SINGULAR FLEXIBILITY MATRIX. )
    2 PRINT 802,N
      GO TO 500
  802 FORMAT (46H OVERFLOW INVERTING FLEXIBILITY MATRIX, MEMBER,I5,1H.)
    3 PRINT 803,N,A
      GO TO 500
  803 FORMAT (26H COMPUTED LENGTH OF MEMBER,I5,38H NOT EQUAL TO THE SUM
     1OF THE SEGMENTS.  ,G5,28H SEGMENTS HAVE BEEN DELETED. )
    4 PRINT 804,N
  500 RETURN
  804 FORMAT (10H TIME USED ,I6,9H SECONDS. )
    5 PRINT 805,N
      GO TO 500
  805 FORMAT (7H MEMBER,I5,29H UNSTABLE, TOO MANY RELEASES. )
    6 PRINT 806,N
      GO TO 500
  806 FORMAT (5H PART,I5,22H OF PROBLEM COMPLETED. )
    7 PRINT 807
      GO TO 500
  807 FORMAT(48H STRUCTURE FOUND UNSTABLE DUE TO JOINT RELEASES. )
    8 CALL ALOCAT(JEXT)
      NN=JEXT+N
      NN=IU(NN)
      PRINT 808,NN,A
      CALL RELEAS(JEXT)
      GO TO 500
  808 FORMAT(6H JOINT,I5,13H IS UNSTABLE.,G12.4)
    9 PRINT 809,A
      GO TO 500
  809 FORMAT (11H THE WORD *,A6,23H* CANNOT BE TRANSLATED. )
   10 PRINT 810,N,A
      GO TO 500
  810 FORMAT(5H THE ,I3,22HTH LOAD DATA FOR JOINT,I4,15H ARE INCORRECT.)
   11 PRINT 811,N,A
      GO TO 500
  811 FORMAT (4H THE ,I5,31H TH SEGMENT HAS BEEN ALTERED TO ,G15.5 )
   12 PRINT 812,N,A
      GO TO 500
  812 FORMAT (51H MEMBER TYPE AND LOAD TYPE ARE INCOMPATIBLE, MEMBER ,
     1 I5,10H LOAD TYPE ,G5)
   13 PRINT 813,N,A
      GO TO 500
  813 FORMAT (15H LOAD ON MEMBER ,I5,18H LOADING CONDITION ,G5,
     1 34H INCOMPATIBLE WITH STRUCTURE TYPE. )
   14 PRINT 814,N,A
```

```
        GO TO 500
814     FORMAT(62H MODULUS OF ELASTICITY (OR SHEAR MODULUS) NOT GIVEN FOR
       1MEMBER,I5,14H VALUE USED IS,G15.5)
   15 PRINT 815,A
        GO TO 500
  815 FORMAT (47H ALLOCATION OF UNDEFINED CODEWORD IN LOCATION   ,O5)
  16    PRINT 816,N,A
        GO TO 500
  816   FORMAT(13H JOINT NUMBER,I5,25H FOR INCIDENCES OF MEMBER,I5,16H DOE
       1S NOT EXIST.)
  17    PRINT 817,N
        GO TO 500
  817   FORMAT(14H MEMBER NUMBER,I5,61H IS DELETED AS A RESULT OF DELETION
       1 OF ONE OF ITS END JOINTS.)
  18    PRINT 818,A,N,N,A
        GO TO 500
  818   FORMAT(21H LOADING COMBINATION I4,25H INCLUDES LOAD CONDITION I4,3
       18H WHICH IS NOT SPECIFIED YET.EFFECT OF I4,4H ON I4,16H WILL BE DE
       2LETED)
  19    PRINT 819,N
        GO TO 500
  819   FORMAT (6H JOINT,I5,69H IS A FREE JOINT.  RELEASES SPECIFIED FOR I
       1T ARE THEREFORE INCORRECT.)
        END
*       FAP                                                              FRM70000
*   MODIFIES 9/24/63 FOR CLOCK ENABLED AFTER TRAPPING ON REG RUNS       FRM70001
*(TME) MERWIN C043-25                                                   FRM70010
*       CLOCK MODIFIED FEB. 3, 1964 FOR STRESS FOR USE WITH CHAIN
*       LOCATION CNTRL MOVED TO LOWER COMMON  -  75
        LBL     STIME
        COUNT   220                                                     FRM70030
*       AUGUST 27,1962                                                  FRM70040
        ENTRY   (TIME)                                                  FRM70050
        ENTRY   TIMER                                                   FRM70060
        ENTRY   JOBTM                                                   FRM70070
        ENTRY   RSCLCK                                                  FRM70080
        ENTRY   STOPCL                                                  FRM70090
        ENTRY   KILLTR                                                  FRM70100
        ENTRY   TIMLFT                                                  FRM70110
        ENTRY   RSTRTN                                                  FRM70120
        ENTRY   (FRM7)                                                  FRM70130
*                                                                       FRM70140
*                RSCLCK...RESET CLOCK TO TIME,,ZERO,,                   FRM70150
*                STOPCL        GIVES BACK TIME SINCE LAST RSCLCK.       FRM70160
*                JOBTM         GIVES BACK TIME SINCE START OF JOB.      FRM70170
*                TIMER         SETS CLOCK TO ACT LIKE ALARM CLOCK.      FRM70180
*                KILLTR        RESETS ALARM CLOCK BEFORE IT GOES OFF.   FRM70190
*                TIMLFT        GIVES BACK TIME LEFT FOR JOB TO RUN.     FRM70200
*                RTNRST        RETURN TO INTERRUPTED PROGRAM            FRM70210
*                              WITH MACHINE CONDITIONS RESTORED AS OFFRM70220
*                              LAST INTERRUPT                           FRM70230
*                CALLING SEQUENCE...CALL RSCLCK                         FRM70240
*                CALLING SEQUENCE...CALL STOPCL(I)                      FRM70250
*                 CALLING SEQUENCE... CALL TIMLFT(I)                    FRM70260
*                CALLING SEQUENCE...CALL TIMER(I,STMT)                  FRM70270
*                CALLING SEQUENCE...CALL KILLTR                         FRM70280
*                CALLING SEQUENCE...CALL JOBTM(I)                       FRM70290
*                CALLING SEQUENCE...CALLRSTRTN                          FRM70300
*                                                                       FRM70310
*                                                                       FRM70320
```

```
*              REGISTERS IN LOWER CORE                            FRM70330
*     7  TTR   CLKL                                              FRM70340
* CLKL STL    CLKIND                                            FRM70350
*         TTR*  6                                               FRM70360
*CLKIND                                                         FRM70370
*JOBTIM                                                         FRM70380
*                                                               FRM70390
 CLKRD SYN    5                                                 FRM70400
CLKLOC SYN    6                                                 FRM70410
CLKTRA SYN    7                                                 FRM70420
 CNTRL SYN    75              OLD TIME LEFT. BASE POINT
LSTSET SYN    76
TMLEFT SYN    77
(JBTM) SYN    91                                                FRM70430
JOBTIM SYN    (JBTM)                                            FRM70440
(CKIN) SYN    92                                                FRM70450
CLKIND SYN    (CKIN)                                            FRM70460
*                                                               FRM70470
*             (TIME)                                            FRM70480
(TIME) SYN    *                                                 FRM70490
       SXA    TM2,1                                             FRM70500
 TM1   CAL    CLKRD                                             FRM70510
       STZ    CLKRD                                             FRM70520
       COM                                                      FRM70530
       STO    TMLEFT                                            FRM70540
       STO    LSTSET                                            FRM70550
       ANA    =0777777                                          FRM70560
       ORA    STRT                                              FRM70570
       LXA    NTBL,1                                            FRM70580
       SLW    TMTBL,1                                           FRM70590
       TXI    *+1,1,-1     OPCODE VAR.FIELD                     FRM70600
       SXA    NTBL,1                                            FRM70610
       CAL    TRA7                                              FRM70620
       SLW    CLKTRA                                            FRM70630
       CAL    TMLEFT                                            FRM70640
       COM                                                      FRM70650
       STO    CLKRD                                             FRM70660
       NZT    CLKIND                                           1FRM70670
       TTR    TM2                                              2FRM70680
       CALL   EXIT                                             3FRM70690
 TM2   AXT    **,1                                              FRM70700
       TTR    1,4                                               FRM70710
 TRA7  TTR    FROM7                                             FRM70720
 STRT  PON    0,0,GOMON     VECTOR ENTRY FOR JOB TIME UP.       FRM70730
*                                                               FRM70740
*             GOMON SETS CLOCK INDICATOR NON-ZERO AND THEN TO EXIT.  FRM70750
*                                                               FRM70760
 GOMON STL    CLKIND                                            FRM70770
       TTR    $EXIT                                             FRM70780
*                                                               FRM70790
********************************************************************FRM70800
*             JOBTM                                             FRM70810
*             CALLING SEQUENCE... TSX JOBTM,4                   FRM70820
*                         PZE TIMEX        RESULT IN DECR.      FRM70830
*                         (TXH)            RESULT IN ADDR.      FRM70840
*                                                               FRM70850
 JOBTM SYN    *                                                 FRM70860
       CAL    CLKRD                                             FRM70870
       COM                                                      FRM70880
       STO    TMET                                              FRM70890
```

```
         CAL     1,4                                              FRM70900
         STP     PRFX                                             FRM70910
         CLA     JOBTIM                                           FRM70920
         SUB     TMLEFT                                           FRM70930
         ADD     LSTSET                                           FRM70940
         SUB     TMET                                             FRM70950
         NZT     PRFX                                             FRM70960
         ALS     18                                               FRM70970
         STO*    1,4                                              FRM70980
         TTR     2,4                                              FRM70990
  TMET                           TEMPORARY                        FRM71000
*******************************************************************FRM71010
 *               TIMLFT                                           FRM71020
 *               CALLING SEQUENCE...TSX TIMLFT,4                  FRM71030
 *                         PZE LEFT       RESULT IN DECR.         FRM71040
 *                         (TXH) LEFT     RESULT IN ADDR.         FRM71050
 *                                                                FRM71060
 TIMLFT  SYN     *                                                FRM71070
         SXA     TML1,4                                           FRM71080
         TSX     JOBTM,4                                          FRM71090
         TXH     RUNR                                             FRM71100
  TML1   AXT     **,4                                             FRM71110
         CAL     1,4                                              FRM71120
         STP     PRFX                                             FRM71130
         CLA     JOBTIM                                           FRM71140
         SUB     RUNR                                             FRM71150
         NZT     PRFX                                             FRM71160
         ALS     18                                               FRM71170
         STO*    1,4                                              FRM71180
         TTR     2,4                                              FRM71190
  RUNR   PZE                                                      FRM71200
*******************************************************************FRM71210
 *                                                                FRM71220
*******************************************************************FRM71230
 *               RSCLCK                                           FRM71240
 *               CALLING SEQUENCE...TSX RSCLCK,4                  FRM71250
 RSCLCK  SYN     *                                                FRM71260
         SXA     RSC1,4                                           FRM71270
         TSX     JOBTM,4                                          FRM71280
         TXH     CNTRL                                            FRM71290
  RSC1   AXT     **,4                                             FRM71300
         TTR     1,4                                              FRM71310
 *               STOPCL                                           FRM71320
 *               CALLING SEQUENCE...CALL STOPCL(J)                FRM71330
 *                                                                FRM71340
 STOPCL  SYN     *                                                FRM71350
         SXA     STP1,4                                           FRM71360
         TSX     JOBTM,4                                          FRM71370
         TXH     NEWT                                             FRM71380
  STP1   AXT     **,4                                             FRM71390
         CAL     1,4                                              FRM71400
         STP     PRFX                                             FRM71410
         CLA     NEWT                                             FRM71420
         SUB     CNTRL                                            FRM71430
         NZT     PRFX                                             FRM71440
         ALS     18              0=FORTRAN TYPE                   FRM71450
         STO*    1,4             NON-ZERO=MAD TYPE                FRM71460
         TTR     2,4                                              FRM71470
  NEWT                                                            FRM71490
*******************************************************************FRM71500
```

```
*                   TIMER                                                FRM71510
*                   CALLING SEQUENCE...CALL TIMER(I,N)                   FRM71520
*                                 ...EXECUTE TIMER.(A,B)                 FRM71530
*                   WHERE I GIVES TIME IN 60THS OF SECOND FOR CLOCK TO RUN FRM71540
*                   BEFORE ALARM. N CONTAINS LOCATION WHERE CONTROL IS SENT FRM71550
*                   AFTER ALARM CLOCK GOES OFF.                          FRM71560
*                                                                        FRM71570
  TIMER SYN         *                                                    FRM71580
        SXA         CLK10,1                                              FRM71580
        CAL         1,4                                                  FRM71590
        STP         PRFX                                                 FRM71600
        CAL*        1,4               TIME INTERVAL IN SECONDS           FRM71610
        NZT         PRFX                                                 FRM71620
        ARS         18                0=FORTRAN. NON-ZERO=MAD            FRM71630
        SLW         NEWINT            TIME INTERVAL IN 60THS OF A SECOND FRM71640
*                   COMPUTE ELAPSED TIME FROM LAST READING               FRM71650
        LXA         NTBL,1            NO. OF OPENINGS IN TABLE           FRM71660
        TXL         CLK10,1,0         TABLE FULL                         FRM71670
        SXD         CLK2,1                                               FRM71680
        CAL         CLKRD                                                FRM71690
        STZ         CLKRD                                                FRM71700
        COM                                                              FRM71710
        XCL                                                              FRM71720
        XCA                           -TIME LEFT IN AC                   FRM71730
        ADD         LSTSET            LAST SETTING-TIME LEFT             FRM71740
        STO         TMUSED            =TIME USED IN 60THS OF SECOND.     FRM71750
        AXT         N,1                                                  FRM71760
  CLK1  CLA         TMTBL,1           DECREASE TIME LIMITS               FRM71770
        SUB         TMUSED            ON ALL ENTRIES                     FRM71780
        STO         TMTBL,1           IN THE TABLE                       FRM71790
        TXI         *+1,1,-1                                             FRM71800
  CLK2  TXH         CLK1,1,**                                            FRM71810
        CAL         2,4               PICK UP LOCATION                   FRM71820
        ANA         =077777                                              AFRM71830
        ALS         18                TO TRANSFER TO WHEN CLOCK GOES OFF FRM71840
        ORA         NEWINT                                               FRM71850
        SLW         TEMP                                                 FRM71860
*                   SEARCH FOR PROPER POSITION IN TABLE.                 FRM71870
        LXA         NTBL,1                                               FRM71880
        SXD         CLK5,1                                               FRM71890
        SXD         CLK8,1                                               FRM71900
        AXT         N,1                                                  FRM71910
  CLK3  CAL         TMTBL,1                                              FRM71920
        ANA         =0777777                                             FRM71930
        LAS         NEWINT                                               FRM71940
        TRA         CLK4                                                 FRM71950
        NOP                                                              FRM71960
        TRA         SMLR                                                 FRM71970
  CLK4  TXI         *+1,1,-1                                             FRM71980
  CLK5  TXH         CLK3,1,**                                            FRM71990
        LDQ         TEMP              IF NEW INTERVAL IS LESS            FRM72000
        STQ         TMTBL,1           THAN REST THAT ARE                 FRM72010
        TRA         CLK9              IN THE TABLE                       FRM72020
*                                                                        FRM72030
*                   FOUND A LARGER VALUE. INSERT NEW VALUE               FRM72040
*                   IN TABLE AND PUSH DOWN THE REST.                     FRM72050
  SMLR  LDQ         TMTBL,1                                              FRM72060
        CAL         TEMP                                                 FRM72070
        SLW         TMTBL,1                                              FRM72080
        TRA         CLK7                                                 FRM72090
  CLK6  CAL         TMTBL,1                                              FRM72100
```

```
        STQ     TMTBL,1                                          FRM72110
        XCL                                                      FRM72120
CLK7    TXI     *+1,1,-1                                         FRM72130
CLK8    TXH     CLK6,1,**                                        FRM72140
        STQ     TMTBL,1                                          FRM72150
CLK9    TXI     *+1,1,-1                                         FRM72160
        SXA     NTBL,1                                           FRM72170
*               SET CLOCK FOR NEW VALUE                          FRM72180
        CLA     TMLEFT                                           FRM72190
        SUB     TMUSED                                           FRM72200
        SUB     CLKRD                                            FRM72210
        STO     TMLEFT          ADJUST TOTAL TIME COUNT          FRM72220
        XCL                     LOWEST INTERVAL IN MQ            FRM72230
        ANA     =0777777                                         FRM72240
        SLW     LSTSET                                           FRM72250
        COM                                                      FRM72260
        STO     CLKRD           RESET C(5) TO NEW VALUE          FRM72270
CLK10   AXT     **,1                                             FRM72280
        TRA     3,4             EXIT                             FRM72290
TMUSED                                                           FRM72300
NEWINT                                                           FRM72310
TEMP                                                             FRM72320
PRFX                                                             FRM72330
N       EQU     10                                               FRM72350
TMTBL   BES     N                                                FRM72360
NTBL    PZE     N                                                FRM72370
*****************************************************************FRM72380
*               KILLTR                                           FRM72390
*               CALLING SEQUENCE...CALL KILLTR                   FRM72400
*               KILLS LAST TIMER ALARM CLOCK SETTING             FRM72410
*               EXCEPT IF LAST SET FOR JOB TIME SETTING.         FRM72420
*                                                                FRM72430
KILLTR  SYN     *                                                FRM72440
        SXA     CL1,1                                            FRM72450
        LXA     NTBL,1                                           FRM72460
        TXI     *+1,1,1                                          FRM72470
        CAL     TMTBL,1                                          FRM72480
        STP     PRFX                                             FRM72490
        ZET     PRFX                                             FRM72500
TRARTN  TTR     RTN                                              FRM72510
        STZ     KILLOV                                           FRM72520
        TNO     *+2                                              FRM72530
        STL     KILLOV                                           FRM72540
        CAL     CLKRD                                            FRM72550
        COM                                                      FRM72560
        XCL                                                      FRM72570
        XCA                                                      FRM72580
        ADD     LSTSET                                           FRM72590
        STO     LSTSET                                           FRM72600
        CAL     TRARTN                                           FRM72610
        ALS     18                                               FRM72620
        STD     TMTBL,1                                          FRM72630
        NZT     KILLOV                                           FRM72640
        TOV     *+1                                              FRM72650
        LXA     CL1,1                                            FRM72660
        TTR     FROM7           TREAT AS IF TRAP FROM 7.         FRM72670
KILLOV                                                           FRM72680
*                                                                FRM72690
*               JOB TIME SETTING CANNOT BE KILLED.               FRM72700
RTN     LXA     CL1,1                                            FRM72710
```

```
        TTR     1,4                                                 FRM72720
*                                                                   FRM72730
*****************************************************************FRM72740
*               RSTRTN                                               FRM72750
*               CALLING SEQUENCE...CALL RSTRTN                       FRM72760
RSTRTN  SYN     *                                                   FRM72770
        CAL     *                                                   FRM72780
        ALS     2                                                   FRM72790
        CLA     SPQ                                                 FRM72800
        ALS     2                                                   FRM72810
        ORA     LOGAC                                               FRM72820
        NZT     OVFL                                                FRM72830
        TOV     *+1                                                 FRM72840
        LDQ     MQ                                                  FRM72850
        LXA     CL1,1                                               FRM72860
        LXA     CL2,2                                               FRM72870
        LXA     CL4,4                                               FRM72880
        TTR*    6                                                   FRM72890
*****************************************************************FRM72900
*               EXTERNAL REGISTERS REFERENCED                        FRM72910
*               CLKRD           5                                   FRM72920
*               ENTRY FROM 7 AFTER CLOCK HAS GONE OFF.              FRM72930
*                                                                   FRM72940
(FRM7)  SYN     *                                                   FRM72950
 FROM7  SYN     *                                                   FRM72960
        SXA     CL1,1                                               FRM72970
        SXA     CL2,2                                               FRM72980
        SXA     CL4,4                                               FRM72990
        STZ     OVFL                                                FRM73000
        STQ     MQ                                                  FRM73010
        SLW     LOGAC           SAVE LOGICAL ACCUMULATOR            FRM73020
        ARS     2                                                   FRM73030
        STO     SPQ                                                 FRM73040
        TNO     *+2                                                 FRM73050
        STL     OVFL                                                FRM73060
        LXA     NTBL,1                                              FRM73070
        TXI     *+1,1,1                                             FRM73080
        CAL     TMTBL,1                                             FRM73090
        PDX     0,2             SET TRANSFER                        FRM73100
        SXA     TRAD,2          LOCATION.                           FRM73110
        ALS     2                                              AFRM73120
        PBT                                                    BFRM73130
        TTR     *+2                                            CFRM73140
        TTR     CLKN            GO TO MONITOR SETTING. DONT RESET CLOCK DFRM73150
        SXA     NTBL,1                                              FRM73160
        SXD     CLKM,1                                              FRM73170
        AXT     N,1                                                 FRM73180
CLKL    CLA     TMTBL,1         REVISE PREVIOUS                     FRM73190
        SUB     LSTSET          ENTRIES IN THE TABLE.               FRM73200
        STO     TMTBL,1                                             FRM73210
        TXI     *+1,1,-1                                            FRM73220
CLKM    TXH     CLKL,1,**                                           FRM73230
        ANA     =0777777        USE LAST ENTRY IN TABLE             FRM73240
        SLW     NEWINT          FOR NEW SETTING                     FRM73250
        CLA     TMLEFT          REVISE TOTAL TIME                   FRM73260
        SUB     LSTSET          LEFT FOR JOB.                       FRM73270
        STO     TMLEFT                                              FRM73280
        CAL     NEWINT          UPDATE LSTSET AND                   FRM73290
        SLW     LSTSET          CHANGE SETTING OF                   FRM73300
        COM                     LOCATION 5 TO NEXT                  FRM73310
```

```
        STO     CLKRD           ENTRY IN THE TABLE.              FRM73320
*                       RESTORE MACHINE CONDITIONS              FRM73330
  CLKN  SYN     *                                              AFRM73340
        ALS     2       TURN ON OVERFLOW LIGHT                  FRM73350
        CLA     SPQ                                             FRM73360
        ALS     2       RESTORE +, Q, AND S                     FRM73370
        ORA     LOGAC                                           FRM73380
        NZT     OVFL                                            FRM73390
        TOV     *+1                                             FRM73400
        LDQ     MQ                                              FRM73410
        RCT                                                     FRM73415
  CL1   AXT     **,1            RESTORE INDEX REGISTERS         FRM73420
  CL2   AXT     **,2                                            FRM73430
  CL4   AXT     **,4                                            FRM73440
  TRAD  TTR     **                                              FRM73450
  OVFL                                                          FRM73460
*                                                               FRM73470
  SPQ                                                           FRM73480
  LOGAC                                                         FRM73490
  MQ                                                            FRM73500
        END                                                     FRM73520
*       FAP
*       DUMMY INPUT-OUTPUT ROUTINE FOR IBM MONITOR
*       CHANGES PRINT STATEMENT TO WRITE OUTPUT TAPE 2
*       CHANGES READ STATEMENT TO READ INPUT TAPE 4
        LBL     DUMIO
        ENTRY   (CSH)
        ENTRY   (SPH)
(SPH)   CLA     TWO
        TRA*    $(STHM)
(CSH)   CLA     FOUR
        TRA*    $(TSHM)
TWO     OCT     2000000
FOUR    OCT     4000000
        END
*       LIST8
*       SYMBOL TABLE
*       LABEL
CMAIN1/6
C    STRESS III  6 LINK CHAIN VERSION
     DIMENSION  Y(6,6),T(6,6),Q(6,6),U(36),IU(36),SYPA(40),FILL(5)
     COMMON U,T,Q,CHECK,NMAX,INORM,ISOLV,ISCAN,IIII,IMOD,ILINK,ICONT,
    1ISUCC,SYPA,NJ,NB,NDAT,ID,JF,NSQ,NCORD,IMETH,NLDS,NFJS,NSTV
    2,NMEMV,IPSI,NMR,NJR,ISODG,NDSQ,NDJ,SDJ,NPR,NBB,NFJS1,JJC,JDC,
    3JMIC,JMPC,JLD,JEXTN,MEXTN,LEXTN,JLC,NLDSI,IYOUNG,ISHER,IEXPAN,
    4IDENS,NBNEW,FILL,
    5NAME,KXYZ,KJREL,JPLS,JMIN,MTYP,KPSI,MEMB,
    6LOADS, INPUT,KS,KMKST,KSTDB,KATKA,KPPLS,KPMIN,KUV,KPPRI,KR,KMK,
    7KV,K33,LA2R ,LA2RT,NV    ,NTP   ,NSCR7,NSCR8,NSCR9,NSCR10,KDIAG,KOFD
    8G,IOFDG,LDNM,MEGAO,JEXT,JINT,KUDBP,KMEGA,KPDBP,JTYP,MTYP1,KB,MLOAD
    9,JLOAD,KATR,LEXT,KYOUNG,KSHER,KEXPAN,KDENS
     EQUIVALENCE(U(1),IU(1),Y(1))
     IF(INORM)33,15,33
15   IF(ISOLV-1)40,36,75
33   CALL PRERR(2)
34   IF(ISOLV-1)36,36,35
35   IF(ICONT)36,10,36
36   CALL PRERR(3)
     ISCAN=2
40   CALL PHAS1A
```

```
         ISCAN=ISCAN
         GO TO (50,10),ISCAN
 50      ISUCC=ISUCC
         GO TO (70,60),ISUCC
 60      IF(ICONT)36,10,36
 70      CALL CHAIN(2,A4)
 75      ISUCC=ISUCC
         GO TO (80,78),ISUCC
 78      CALL PRERR(5)
         GO TO 60
 80      IF(ICONT)85,82,85
 85      CALL PRER2(6,ICONT,0)
         GO TO 40
 82      CALL PRERR(6)
 10      CALL STOPCL(ITIME)
         ITIME=ITIME/60
         CALL PRER2(4,ITIME,0.)
         GO TO 40
         END
*        LIST8
*        LABEL
CPHAS1A    VERSION 3   PHAS1A  FENVES - LOGCHER - MAZZOTTA JAN 17, 1964
         SUBROUTINE PHAS1A
         DIMENSION LABL(12),BETA(85),SYSFIL(27),PRBFIL(3),CWFIL(49),U(99),
        1IU(2)
         DIMENSION B(6)
         COMMON U,IU,A,IA,LABL,BETA,K,ITABLE,J,NE,ITS,IB,IS,IL,INDEX,IN,
        1CHECK,NMAX,INORM,ISOLV,ISCAN,III,IMOD,JJJJ,ICONT,ISUCC,IMERG,TOP,N
        21,NL,NT,NREQ,TN,LFILE,TOLER,IPRG,IRST,IRLD,IRPR,SYSFIL,
        3NJ,NB,NDAT,ID,JF,NSQ,NCORD,IMETH,NLDS,NFJS,NSTV,NMEMV,IPSI,NMR,NJR
        4,ISODG,NDSC,NDJ,IPDBP,IUDBP,NBB,NFJS1,JJC,JDC,JMIC,JMPC,JLD,JEXTN,
        5MEXTN,LEXTN,JLC,NLDSI,IYOUNG,ISHEAR,IEXPAN,IDENS,INNN,NLDG
         COMMON JTSTAB,PRBFIL,
        6NAME,KXYZ,KJREL,JPLS,JMIN,MTYP,KPSI,MEMB,LOADS,MODN,KS,KMKST,KSTDB
        7,KATKA,KPPLS,KPMNS,KUV,KPPRI,KR,KSAVE,KS1,KS2,KS3,KS4,KS5,KS6,KS7,
        8KS8,KS9,KS10,KDIAG,KOFDG,KAD,LOADN,MEGAO,JEXT,JINT,KUDBP,KMEGA,
        9KPDBP,JTYP,MTYP1,KB,MLOAD,JLOAD,KATR,LINT,KYOUNG,KSHEAR,KEXPAN
         COMMON KDENS,CWFIL
         EQUIVALENCE(U,IU,A,IA),(U(2),IU(2),LABL),(U(14),BETA),(U(99),K,BK)
F        LIST1,LIST2,LIST4,LIST7,LIST8,LIST10,LIST11
B        B(1)=000000000001
B        B(2)=CC0000000002
B        B(3)=C00000000004
B        B(4)=000000000010
B        B(6)=000000000017
         FFLAG=0.
         CALL PRERR(1)
C        FIRST WORD
C
 10      IX=0
         J=0
         M=1
 11      I1=MATCH(LIST1,K,M)
         GO TO (12,14,97,96,15,16),I1
 12      IX=IX+1
         IF(IX-12) 13,10,10
 13      M=0
         GO TO 11
 14      J=K
         IF(ITABLE-1) 98,100,200
```

```
   15    K1=K
         GO TO 17
   16    IF(K1-4) 17,17,95
   17    ITABLE=0
         GO TO (100,200,300,400,500,600,700,800,900,1000,1100,1200,1300,140
        10,1500,1600,1700,1800,1900,2000),K1
C
C        MEMBER
C        SECOND WORD FOR MEMBER
  100    CALL MEMDAT
         NE=NE
         GO TO(10,48,87,91,92,96,97,98,99,85,95,94,90),NE
C
C        SECOND WORD FOR JOINT
  200    CALL JTDAT
         NE=NE
         GO TO(10,48,90,91,92,93,98,99,96,97),NE
C
C        FIRST WORD NUMBER
  300    I2=MATCH(LIST4,K ,4)
         GO TO (99,98,97,96,301,95),I2
  301    K2=K
         I3=MATCH(0,K ,2)
         GO TO (99,302,97,96,94,95),I3
  302    K3=K
  304    GO TO (305,10),ISCAN
  305    K4=1
         CALL SIZED(K2,K3,K4)
         IF(K4-2) 10,90,90
C
C        FIRST WORD TABULATE
C
C        READ WORDS
  400    FLAG=0.0
  401    I2=MATCH(LIST2,K,0)
         GO TO (403,98,97,96 ,402,95),I2
  402    IF(K-5)420,1720,430
C        RESULTS REQUESTED
  430 IF(K-6)401,420,401
B420     FLAG=FLAG+B(K)
         GO TO 401
C        PROCESS REQUESTS
  403    GO TO (404,10),ISCAN
  404    IF(ICONT) 4041,405,4041
 4041    IRPR=1
         CALL ALOCAT(LOADS,JLC)
         GO TO (91 ,411,410,415),IMOD
C        NORMAL
  405    IF(JLC)410,407,409
B407     FFLAG=FFLAG+FLAG
         GO TO 10
  409    CALL ALOCAT(LOADS,JLC)
  410    JS=LOADS+JLC
         JT=IU(JS)+1
B        U(JT)=U(JT)+FLAG
         GO TO 421
C        CHANGE
  411    JS=LOADS+JLC
         JT=IU(JS)+1
B        U(JT)=(U(JT)*777777700000)+FLAG
```

```
            GO TO 421
C           DELETE
  415    JS=LOADS+JLC
         JT=IU(JS)+1
B        U(JT)=-(-(U(JT))+FLAG)
  421    CALL RELEAS(LOADS,JLC)
         GO TO 10
C
C           LOADING
  500    GO TO (524,10),ISCAN
  524    IF(ICONT) 501,502,501
  501    GO TO (91,506,511,506),IMOD
  502    IF(NLDS) 503,86,503
  503    JLD=JLD+1
         IF(JLD-NLDS) 504,504,85
  504    CALL DEFINE(LOADS,JLD,24,0,0,1)
         CALL ALOCAT(LOADS,JLD)
         NLDG=NLDG+1
         L=LOADS+JLD
         IK=IU(L)+1
B        U(IK)=U(IK)+FFLAG+100000000000
         IU(IK+1)=10
         I2=MATCH(LIST1,K,0)
         IF(I2-5)513,520,513
  520    IF(K-19)513,521,513
  521    IU(IK)=IU(IK)+32768
  513    IK=IK+1
         DO 505 IX=1,12
         M=IK+IX
         NM=KBUF+IX
  505    IU(M)=IU(NM)
         JLC=JLD
  530    CALL RELEAS(LOADS,JLC)
         GO TO 10
  506    I2=MATCH(0,K,2)
         IF(I2-2) 85,507,85
  507    JLC=K
         IF(IMOD-2)508, 10,508
  508    CALL ALOCAT(LOADS,JLC)
         L=LOADS+JLC
         IK=IU(L)+1
         CALL UPADP(U(IK),LTYP,NLD,A)
         IU(IK)=0
         IF(LTYP-2)522,523,522
  522    DO 510 IX=1,NLD
         M=LOADS+JLC
         M=IU(M)+14+IX
         IF(IU(M)) 509,510,509
  509    CALL UPADP(IU(M),INDEX,IPL,JA)
         GO TO (5091,5092,5092,5092,5091),INDEX
 5091    CALL ALOCAT(JLOAD,JA)
         CALL RELEAS(JLOAD,JA)
         LL=JLOAD+JA
         GO TO 5093
 5092    CALL ALOCAT(MLOAD,JA)
         CALL RELEAS(MLOAD,JA)
         LL=MLOAD+JA
 5093    LM=IU(LL)+2
         IU(LM)=IU(LM)-1
         LM=IU(LL)+IPL
```

```
          IU(LM)=0
   510    CONTINUE
   523    CALL DEFINE(LOADS,JLC,1,0,0,1)
          NLDG=NLDG-1
          GO TO 530
   511    JLD=JLD+1
          IF(JLD-LEXTN) 504,504,512
   512    CALL SIZED(4,JLD,2)
          GO TO 504
C
C         STRUCTURE
C
C         STORE PROBLEM TITLE
   600    CALL RSCLCK
          CALL START(5,301,NMAX)
          IMOD=1
          ISCAN=1
          ISOLV=1
          I=MATCH(KBUF,0,5)
          CALL DEFINE(NAME,12,0,0,1)
          CALL ALOCAT(NAME)
          DO 601 I=1,12
          JS=NAME+I
          JT=KBUF+I
   601    U(JS)=U(JT)
          CALL RELEAS(NAME)
C         INITIAL CHECK
B         CHECK=000006774100
          GO TO 10
C
C         METHOD
C
C         READ SECOND WORD AND SET IMETH
   700    I2=MATCH(LIST7,K,0)
          GO TO (99,98,97,96,701,95),I2
   701    GO TO (96,702,703,96,96),K
   702    IMETH=1
          GO TO 704
   703    IMETH=2
C         TESTS
   704    GO TO (706,90,90,90),IMOD
B706      CHECK=CHECK+000000000040
          GO TO 10
C
C         TYPE
C
C         READ TWO WORDS AND FIND TYPE, SET NCORD,JF
   800    I2=MATCH(LIST11,K,0)
          GO TO (99,98,97,96,801,95),I2
   801    ISODG = 0
          GO TO (802,803,96,96,96),K
   802    JT=0
          NCORD=2
          GO TO 804
   803    JT=3
          NCORD=3
   804    I3=MATCH(LIST11,K,0)
          GO TO (99,98,97,96,805,95),I3
   805    GO TO (96,96,806,807,808),K
   806    JT=JT+1
```

```
            ISODG = 1
            GO TO 809
   807      JT=JT+2
            GO TO 809
   808      JT=JT+3
   809      GO TO (810,811,811,811,812,89),JT
   810      JF=2
            GO TO 813
   811      JF=3
            GO TO 813
   812      JF=6
C          TESTS
   813      GO TO (820,814,90,90),IMOD
C          CHANGE TEST TO BE INSERTED HERE
   814      GO TO 820
   820      GO TO (821,10),ISCAN
   821      ID=JT
            IF (ISODG) 823,822,823
   822 ISODG = JF
   823 NSQ=JF*JF
B          CHECK=CHECK+000000000020
            GO TO 10
C
C          MODIFICATION
C
C          READ THIRD WORD, TEST, CALL RESTORING SUBROUTINE
   900      I2=MATCH(LIST10,K,4)
            GO TO (99,98,97,96,901,95),I2
   901      GO TO (903,10),ISCAN
   903      GO TO (904,905,96),K
C          FIRST-RESTORE
   904      CALL RESTOR
            IRST=1
            GO TO 906
C          LAST-REALLOCAT
   905      CALL ALOCAT(KXYZ)
            CALL ALOCAT(JPLS)
            CALL ALOCAT(JMIN)
            CALL ALOCAT(KPSI)
            CALL ALOCAT(JTYP)
            CALL ALOCAT(MLOAD,0)
            CALL ALOCAT(JLOAD,0)
            CALL ALOCAT(LOADS,0)
            IRPR=0
            IRST=0
            IRLD=0
C          STORE NAME
   906      I=MATCH(KBUF,0,5)
            CALL DEFINE(MODN,12,0,0,1)
            CALL ALOCAT(MODN)
            DO 907 I=1,12
            JS=MODN+I
            JT=KBUF+I
   907      U(JS)=U(JT)
            CALL RELEAS(MODN)
            ISOLV=1
            GO TO 10
C
C          MODIFIERS
C
```

```
C      CHANGE
 1000 IMOD=2
      GO TO 10
C      ADDITION
 1100 IMOD=3
      GO TO 10
C      DELETION
 1200 IMOD=4
      GO TO 10
C
C      SOLVE
C
C      READ THIRD WORD
 1300 IF(ISOLV-1) 1310,1311,1310
 1310 ISUCC=2
      GO TO 1308
 1311 I2=MATCH(LIST10,K,4)
      GO TO (1302,1301,1301,1301,1305,1301),I2
C      ERROR INTERPRETED AS BLANK
 1301 CALL PRERR(7)
 1302 GO TO (1303,1308),ISCAN
 1303 ICONT=0
 1304 CALL PHAS1B
 1308 RETURN
C      SOLVE THIS PART
 1305 GO TO (1301,1301,1306),K
 1306 ICONT=ICONT+1
      GO TO (1304,10),ISCAN
C
C      STOP
C
 1400 CALL EXIT
C      PROBLEM FINISHED, RETURN AND END
 1500 ISCAN=2
      GO TO 1308
C      OUTPUT MODE
 1600 IMOD = 2
      GO TO 10
C      SELECTIVE PRINTING
 1700 I2 = MATCH(LIST2,K,0)
      GO TO (10,1704,97,96,1701,1700),I2
 1701 IF(K-5)1702,1720,1700
 1702 K2=5-K
      IF(KPPRI)1700,88,1700
C      PRINT DATA
 1720 GO TO (1721,10),ISCAN
 1721 CALL DPRINT
      CALL PRERR(1)
      GO TO 10
 1704 GO TO (1703,1700),ISCAN
 1703 CALL SELOUT(JLC,K2,K)
      GO TO 1700
C      CONSTANTS
 1800 IF(IMOD-4)1807,90,90
 1807 IM=0
 1808 I2 = MATCH(LIST8,K,0)
      GO TO (10,1830,1801,96,1802,1808),I2
 1801 VALUE = BK
      GO TO 1808
 1802 GO TO (1803,1804,1806,1806,1806,1806),K
```

```
1803 U(IJ) = VALUE
     GO TO 1808
1804 GO TO (1841,1808),ISCAN
1841 CALL ALOCAT(U(IM))
     CALL RELEAS(U(IM))
     DO 1805 I = 1,MEXTN
     IL = IU(IM)+I
1805 U(IL) = U(IJ)
     IU(IJ) = 1
     GO TO 1808
1806 IJ = K+188
     IM = IJ+57
     CALL DEFINE(U(IM),MEXTN,0,0,1)
     GO TO 1808
1830 IF(IM) 98,98,1831
1831 GO TO (1832,1808),ISCAN
1832 IF(K-MEXTN) 1835,1835,1833
1833 IF(ICONT) 1834,1836,1834
1834 CALL SIZED(2,K,2)
     CALL DEFINE(U(IM),MEXTN,0,0,1)
1835 CALL ALOCAT(U(IM))
     CALL RELEAS(U(IM))
     IU(IJ) = 1
     IL = IU(IM)+K
     U(IL) = VALUE
     GO TO 1808
1836 CALL PRERR(31)
     GO TO 1808
C    COMBINATION LOADING
1900 IF(JLC) 1912,86,1912
1912 GO TO (1901,10),ISCAN
1901 CALL ALOCAT(LOADS,JLC)
     L=LOADS+JLC
     IK=IU(L)+1
     CALL UPADP(U(IK),LTYP,NLD,A)
       IF ( LTYP-2)84,1902,84
1902 IF(ICONT)1925,1903,1925
1903 IT=0
     IB=0
1907 I2=MATCH(LIST11,K,0)
     GO TO (1920,1904,1908,1907,1907,1907),I2
1904 IF(JLC-K) 1950,1950,1953
1950 CALL PRER2(18,JLC,K)
     IF(IB-IT) 1951,1951,1952
1951 I2=MATCH(LIST11,K,0)
     GO TO (1920,1904,1907,1951,1951,1951),I2
1952 IB=IB-1
     GO TO 1907
1953 IT=IT+1
     IL=IT
1911 NLS=IU(IK+1)
     IF(2*(NLD+IL)-NLS)1906,1906,1905
1905 NLS=NLS+10
     IU(IK+1)=NLS
     CALL DEFINE(LOADS,JLC,NLS+14,0,0,1)
     L=LOADS+JLC
     IK=IU(L)+1
     IF(IB-IT)1906,1906,1909
1906 IS=IK+12+2*(NLD+IT)
     IU(IS)=K
```

```
            GO TO 1907
      1908  IB=IB+1
            IF(IB-IT)1909,1909,1910
      1909  IS=IK+13+2*(NLD+IB)
            U(IS)=BK
            GO TO 1907
      1920  IU(IK)=IU(IK)+IT
      1921  CALL RELEAS(LOADS,JLC)
            GO TO 10
      1910  IL=IB
            GO TO 1911
      1925  GO TO (91,1937,1903,1937),IMOD
      1937  HOLD=0
      1926  I2=MATCH(LIST11,K,0)
            GO TO (1921,1927,1933,1926,1926,1926),I2
      1927  DO 1928 I=1,NLD
            IT=IK+12+2*I
            IF(K-IU(IT))1928,1939,1928
      1928  CONTINUE
      1939  IF(IMOD-2)91,1929,1932
      1929  IF(HOLD)1931,1930,1931
      1930  IHOLD=2
            GO TO 1926
      1931  IHOLD=1
            U(IT+1)=HOLD
            GO TO 1926
      1932  IU(IT)=0
            GO TO 1926
      1933  IF(IHOLD-2)1935,1934,1935
      1934  U(IT+1)=BK
            GO TO 1926
      1935  HOLD=BK
            GO TO 1926
C     CHECK JOINT STABILITY
C     SET PARAMETER
      2000  JTSTAB=1
            IF(ICONT) 2001,10,2001
      2001  GO TO (91,10,10,2002),IMOD
      2002  JTSTAB=0
            GO TO 10
C
C     ERROR MESSAGES PRINTED WITH SUBROUTINE PRERR
        99  I=14
            GO TO(50,50,50,10),IMOD
        98  I=9
            GO TO 50
        97  I=12
            GO TO 50
        96  CALL PRER2(9,0,BK)
        48  I=7
            GO TO 50
        95  I=8
            GO TO 50
        94  I=10
      B50   CHECK=CHECK*777777773777
        51  CALL PRERR(I)
            GO TO 10
        93  I=30
      B     CHECK=CHECK*777777677777
            GO TO 51
```

```
 92    I=31
B      CHECK=CHECK*777777577777
       GO TO 51
 91    I=11
       GO TO 52
 90    I=29
B52    CHECK=CHECK*777777737777
       GO TO 51
 89    I=13
B      CHECK=CHECK*777777767777
       GO TO 51
 86    I=18
B      CHECK=CHECK*777777377777
       GO TO 51
 85    I=32
B      CHECK=CHECK*777777757777
       GO TO 51
   87  I=33
B      CHECK=CHECK*777777777577
       GO TO 51
   88  I=21
       GO TO 51
 84    I=28
B      CHECK=CHECK*777777757777
       GO TO 50
       END
*      FAP
       LBL      LISTS
       COUNT    20
       ENTRY    LIST1
       ENTRY    LIST2
       ENTRY    LIST3
       ENTRY    LIST4
       ENTRY    LIST5
       ENTRY    LIST6
       ENTRY    LIST7
       ENTRY    LIST8
       ENTRY    LIST10
       ENTRY    LIST11
LIST1  DEC      20
       BCI      9,MEMBEROJOINTNUMBERTABULALOADINSTRUCTMETHODOOTYPEMODIFI
       BCI      9,CHANGEADDITIDELETIOSOLVEOOSTOPFINISHSELECTOPRINTCONSTA
       BCI      2,COMBINOCHECK
LIST2  DEC      8
       BCI      8,DISPLADISTORREACTIFORCESOODATAOOOALLMEMBEROJOINT
LIST3  DEC      6
       BCI      6,COORDIRELEASOLOADSOOLOADDISPLANUMBER
LIST4  DEC      4
       BCI      4,JOINTSMEMBERSUPPORLOADIN
LIST5  DEC      37
       BCI      9,00000X00000Y00000Z0FORCEMOMENTDISPLAROTATIO0000A00000I
       BCI      9,000ENDOSTARTCONCENUNIFORLINEAROOOCOPOCOOOWOOOOWAOOOOWB
       BCI      8,COOOLAOOOOLBOOOOOSOOOCOFSUPPOROOFREEOOAREAINERTI
       BCI      8,0COOOOLLENGTHOSTEELOOBETAOOOOAXOOOOAYOOOOAZOOOOIX
       BCI      3,COOOIYOOOOIZDISTOR
LIST6  DEC      16
       BCI      9,00FROMOOGOESNUMBERINCIDEPROPERRELEASCONSTRDISTORCOLOAD
       BCI      7,OLOADSPRISMASTIFFNFLEXIBVARIABOSTEELOOOEND
LIST7  DEC      5
       BCI      5,PRISMASTIFFNFLEXIBOSTEELVARIAB
```

```
      LIST8 DEC       6
            BCI       6,000ALLO00BUT00000E00000G000CTEDENSIT
      LIST10 DEC      3
            BCI       3,0FIRST00LAST00PART
      LIST11 DEC      5
            BCI       5,0PLANE0SPACE0TRUSS0FRAME00GRID
            END
*         LIST8
*         LABEL
CMEMDAT   OF PHASIA  STRESS III  FENVES-LOGCHER-MAZZOTA  NOV. 19,1963
          SUBROUTINE MEMDAT
          DIMENSION LABL(12),BETA(85),SYSFIL(27),PRBFIL(6),CWFIL(49),U(99),
         1IU(2)
          COMMON U,IU,A,IA,LABL,BETA,K,ITABLE,J,NE,ITS,IB,IS,IL,INDEX,IN,
         1CHECK,NMAX,INORM,ISOLV,ISCAN,III,IMOD,JJJJ,ICONT,ISUCC,IMERG,TOP,N
         21,NL,NT,NREQ,TN,LFILE,TOLER,IPRG,IRST,IRLD,IRPR,SYSFIL,
         3NJ,NB,NDAT,ID,JF,NSQ,NCORD,IMETH,NLDS,NFJS,NSTV,NMEMV,IPSI,NMR,NJR
         4,ISODG,NDSQ,NDJ,IPDBP,IUDBP,NBB,NFJS1,JJC,JDC,JMIC,JMPC,JLD,JEXIN,
         5MEXTN,LEXTN,JLC,NLDSI,IYOUNG,ISHEAR,IEXPAN,IDENS,PRBFIL,
         6NAME,KXYZ,KJREL,JPLS,JMIN,MTYP,KPSI,MEMB,LOADS,MODN,KS,KMKST,KSIDB
         7,KATKA,KPPLS,KPMNS,KUV,KPPRI,KR,KSAVE,KS1,KS2,KS3,KS4,KS5,KS6,KS7,
         8KS8,KS9,KS10,KDIAG,KOFDG,KAD,LOADN,MEGA0,JEXT,JINT,KUDBP,KMEGA,
         9KPDBP,JTYP,MTYP1,KB,MLOAD,JLOAD,KATR,LINT,KYOUNG,KSHEAR,KEXPAN
          COMMON KDENS,CWFIL
          EQUIVALENCE(U,IU,A,IA),(U(2),IU(2),LABL),(U(14),BETA),(U(99),K,BK)
          DIMENSION IHOLD(12),HOLD(64)
F         LIST5,LIST6,LIST11
          NE=1
          IF(ITABLE-1) 100,115,96
  100     I2=MATCH(LIST6,K,0)
          GO TO (1281,101,97,96,102,104),I2
  101     J=K
          GO TO 100
  102     IF(K-3) 103,100,103
  103     K2=K
  104     IF(J)114,105,114
  105     ITABLE=1
          JIB=0
          NSS=0
          KN=K2
          GO TO (99,99,99,10,106,10,10,10,10,10,158,158,158,158,158,10),K2
  106     DO 107 I=1,12
  107     IHOLD(I)=0
          DO 108 I=1,64
  108     HOLD(I)=0.
          JHOLD=0
          GO TO 100
  114     JIB=0
          NSS=0
          KN=K2
  115     IF(J-MEXTN) 118,118,116
  116     IF(ICONT) 92,92,117
  117     CALL SIZED(2,J,2)
  118     GO TO(120,120,96,120,109,130,131,150,151,151,158,158,158,158,158,
         1 154),KN
  109     I2=MATCH(LIST6,K,0)
          GO TO (99,98,97,96,110,158),I2
  110     K2=K
          IF(K-10) 96,96,158
C         MEMBER INCIDENCES
```

```
120    I3=MATCH(0,K3,2)
       IF(I3-2) 128,121,96
121    I3=MATCH(0,K4,2)
       IF(I3-2) 99,122,96
122    GO TO (123,10),ISCAN
123    IF(ICONT) 126,124,126
124    JMIC=JMIC+1
125    IJ=JPLS+J
       IU(IJ)=K3
       IJ=JMIN+J
       IU(IJ)=K4
       GO TO 10
126    IRST=1
       GO TO (91,125,124,127),IMOD
127    JMIC=JMIC-1
       JMPC=JMPC-1
       CALL ALOCAT(MTYP)
       IJ=MTYP+J
       IU(IJ)=0
       CALL RELEAS(MTYP)
       CALL DEFINE(MEMB,J,0,0,0,0)
       GO TO 10
1281   IF(J) 128,10,128
128    GO TO (99,99,99,127),IMOD
C      MEMBER RELEASES AND CONSTRAINTS
130    IX=1
       GO TO 132
131    IX=2
132    IG=2
       CALL READ(IG,J)
       IF(ITS) 98,1320,98
1320   GO TO(133,10),ISCAN
133    IF(IG) 134,48,134
134    IF(K2-6) 97,136,135
135    IA=IA*4096
136    CALL ALOCAT(MTYP)
       CALL RELEAS(MTYP)
       IF(ICONT)138,137,138
137    NMR=1
       IJ=MTYP+J
B      U(IJ)=U(IJ)+A
       GO TO 10
138    IRST=1
       GO TO (91,145,137,146),IMOD
145    IF(IX-1) 99,143,144
B143   U(IJ)=U(IJ)*770000777777
       GO TO 137
B144   U(IJ)=U(IJ)*007777777777
       GO TO 137
146    IJ=MTYP+J
B      U(IJ)=-(-(U(IJ))+A)
       GO TO 10
C      MEMBER DISTORTIONS AND LOADS
150    INDEX=4
       GO TO 152
151    INDEX=2
152    IG=3
       CALL READ(IG,J)
       GO TO (153,10),ISCAN
153    IF(IG) 10,48,10
```

```
  154 INDEX = 3
      IF (ITABLE-1) 155,152,155
  155 I3 = MATCH(LIST6,K,0)
      GO TO 152
C     MEMBER PROPERTIES
  158 IMLOT=K2-10
      IG=1
      GO TO (160,170,170,180,190),IMLOT
  160 CALL READ(IG,J)
      IF(ITS) 98,1600,98
 1600 IF(IG-4) 162,161,162
  161 K2=15
      GO TO 158
  162 N=6
      GO TO (163,10),ISCAN
  163 IF(IG) 164,48,164
  164 IF(ITABLE-1) 1665,165,85
  165 IF(J) 167,166,167
  166 DO 1651 I=1,IB
 1651 HOLD(I)=BETA(I)
      IF(IL) 1654,1653,1654
 1653 IL=12
 1654 DO 1652 I=1,IL
 1652 IHOLD(I)=LABL(I)
      JIB=IB
      GO TO 10
 1665 JIB=0
  167 CALL DEFINE(MEMB,J,N,0,0,1)
      CALL ALOCAT(MEMB,J)
      L=MEMB+J
      IF(JIB) 1671,1675,1671
 1671 DO 1674 I=1,JIB
      IF(IHOLD(I)-7) 1672,1673,1673
 1672 IJ=IHOLD(I)+IU(L)
      U(IJ)=HOLD(I)
      GO TO 1674
 1673 IJ=KPSI+J
      U(IJ)=HOLD(I)
 1674 CONTINUE
 1675 IF(IB) 1676,199,1676
 1676 DO 1679 I=1,IB
      IF(LABL(I)-7) 1677,1678,1678
 1677 IJ=LABL(I)+IU(L)
      U(IJ)=BETA(I)
      GO TO 1679
 1678 IJ=KPSI+J
      U(IJ)=BETA(I)
 1679 CONTINUE
      GO TO 199
  170 N=NSQ
      DO 171 KKI=1,NSQ
      I3=MATCH(LIST5,K,0)
      GO TO (179,98,171,170,172,172),I3
  171 BETA(KKI)=BK
  172 I4=MATCH(0,K,0)
      GO TO (179,98,173,96,94,95),I4
  173 IF(J) 174,175,174
  174 L=KPSI+J
      U(L)=BK
      GO TO 170
```

```
        175   HOLD(1)=BK
              JHOLD=2
              GO TO 170
        179   IF(J) 1793,10,1793
        1793  CALL DEFINE(MEMB,J,NSQ,0,0,1)
              CALL ALOCAT(MEMB,J)
              L=MEMB+J
              DO 1791 I=1,NSC
              IJ=IU(L)+I
        1791  U(IJ)=BETA(I)
              IF(JHOLD-2) 199,1792,199
        1792  L=KPSI+J
              U(L)=HOLD(1)
              GO TO 199
C             STEEL SECTION
C             VARIABLE SECTICN
        180   IF(NSS)185,184,185
        184   I3=MATCH(LIST11,K,2)
              IF(I3-2)187,181,94
        187   IF(ITABLE)99,99,10
        181   IF(J) 1861,186,1861
        186   NSS=1
        1861  NS=K
              I3=MATCH(LIST11,K,0)
        185   N=7*NS
              IF(J) 182,160,182
        182   CALL CEFINE(MEMB,J,N,0,0,1)
              CALL ALOCAT(MEMB,J)
              L=MEMB+J
              IF(ITABLE-1) 183,1821,183
        1821  IF(JIB) 1822,1831,1822
        1822  NM=JIB/NS
              IF(NM-7)1828,1828,87
        1828  I=0
              IF(IHOLD(1)-8) 1824,1823,1824
        1823  IJ=KPSI+J
              U(IJ)=HOLD(1)
              I=1
        1824  IF(NM) 1841,1831,1841
        1841  DO 1825 IJK=1,NS
              DO 1825 KK=1,NM
              IIT=KK+I
              IJ=IHOLD(IIT)+IU(L)+7*(IJK-1)
              IK=KK+NM*(IJK-1) +I
        1825  U(IJ)=HOLD(IK)
              IF(I) 1831,1826,1831
        1826  IF(JIB-NM*NS) 1831,1831,1827
        1827  IJ=KPSI+J
              U(IJ)=HOLD(JIB)
              GO TO 1831
        183   JIB=0
        1831  CALL READ(IG,J)
              IF(ITS) 98,1830,98
        1830  GO TO(1832,10),ISCAN
        1832  IF(IG) 1833,48,1833
        1833  IF(IB) 1834,199,1834
        1834  NM=IB/NS
              IF(NM-7)1829,1829,87
        1829  I=0
              IF(LABL(1)-8) 1836,1835,1836
```

```
     1835 IJ=KPSI+J
          U(IJ)=BETA(1)
          I=1
     1836 IF(NM) 1842,199,1842
     1842 DO 1837 IJK=1,NS
          DO 1837 KK=1,NM
          IIT=KK+I
          IJ=LABL(IIT)+IU(L)+7*(IJK-1)
          IK=KK+NM*(IJK-1)+I
     1837 U(IJ)=BETA(IK)
          IF(I) 199,1838,199
     1838 IF(IB-NM*NS) 199,199,1839
     1839 IJ=KPSI+J
          IB=IB
          U(IJ)=BETA(IB)
          GO TO 199
     190  IG=1
          CALL READ(IG,J)
          PRINT 191
          GO TO 199
     191  FORMAT(35H STEEL SECTIONS CANNOT YET BE READ.)
     199  IRST=1
          GO TO(1991,1992,1991,90),IMOD
     1991 JMPC=JMPC+1
     1992 IF(J) 1993,10,1993
     1993 CALL RELEAS(MEMB,J)
          CALL ALOCAT(MTYP)
          L=MTYP+J
          CALL PACKW(A,0,U(L),IMLOT,N,0)
          U(L)=A
          CALL RELEAS(MTYP)
          L=KPSI+J
          IF(U(L))1994,10,1994
     1994 IPSI=1
     10   RETURN
     48   NE=2
          GO TO 10
     87   NE=3
          GO TO 10
     91   NE=4
          GO TO 10
     92   NE=5
          GO TO 10
     96   NE=6
          GO TO 10
     97   NE=7
          GO TO 10
     98   NE=8
          GO TO 10
     99   NE=9
          GO TO 10
     85   NE=10
          GO TO 10
     95   NE=11
          GO TO 10
     94   NE=12
          GO TO 10
     90   NE=13
          GO TO 10
          END
```

```
*      LIST8
*      LABEL
CJTDAT    OF PHAS1A STRESS III FENVES-LOGCHER-MAZZOTA  NOV. 19,1963
       SUBROUTINE JTDAT
       DIMENSION LABL(12),BETA(85),SYSFIL(27),PRBFIL(6),CWFIL(49),U(99),
      1IU(2)
       COMMON U,IU,A,IA,LABL,BETA,K,ITABLE,J,NE,ITS,IB,IS,IL,INDEX,IN,
      1CHECK,NMAX,INORM,ISOLV,ISCAN,III,IMOD,JJJJ,ICONT,ISUCC,IMERG,TOP,N
      21,NL,NT,NREQ,TN,LFILE,TOLER,IPRG,IRST,IRLD,IRPR,SYSFIL,
      3NJ,NB,NDAT,ID,JF,NSQ,NCORD,IMETH,NLDS,NFJS,NSTV,NMEMV,IPSI,NMR,NJR
      4,ISODG,NDSQ,NDJ,IPDBP,IUDBP,NBB,NFJS1,JJC,JDC,JMIC,JMPC,JLD,JEXTN,
      5MEXTN,LEXTN,JLC,NLDSI,IYOUNG,ISHEAR,IEXPAN,IDENS,PRBFIL,
      6NAME,KXYZ,KJREL,JPLS,JMIN,MTYP,KPSI,MEMB,LOADS,MODN,KS,KMKST,KSTDB
      7,KATKA,KPPLS,KPMNS,KUV,KPPRI,KR,KSAVE,KS1,KS2,KS3,KS4,KS5,KS6,KS7,
      8KS8,KS9,KS10,KDIAG,KOFDG,KAD,LOADN,MEGAO,JEXT,JINT,KUDBP,KMEGA,
      9KPDBP,JTYP,MTYP1,KB,MLOAD,JLOAD,KATR,LINT,KYOUNG,KSHEAR,KEXPAN
       COMMON KDENS,CWFIL
       EQUIVALENCE(U,IU,A,IA),(U(2),IU(2),LABL),(U(14),BETA),(U(99),K,BK)
F      LIST3
       NE=1
       IF(ITABLE-1) 200,96,206
  200  I2=MATCH(LIST3,K,0)
       GO TO (101,201,97,96,202,204),I2
  101  GO TO (99,99,99,2170),IMOD
  201  J=K
       GO TO 200
  202  GO TO (203,203,203,203,203,200),K
  203  K2=K
  204  IF(J) 93,205,206
  205  ITABLE=2
       GO TO 10
  206  IF(J-JEXTN) 209,209,207
  207  IF(ICONT) 93,93,208
  208  CALL SIZED(1,J,2)
  209  GO TO (210,220,230,230,250,96),K2
C
C      JOINT COORDINATES
  210  IG=1
       CALL READ(IG,J)
       IF(ITS) 98,2110,98
 2110  L=JTYP+J
       IF(ICONT) 216,211,216
  211  IF(IS-1) 212,212,214
  212  JJC=JJC+1
  213  IU(L)=1
       GO TO 2151
  214  JDC=JDC+1
  215  IU(L)=2
C      STORE COORDINATES
 2151  IF(3     -IB) 2152,2152,2153
 2152  N=3
       GO TO 2154
 2153  N=IB
 2154  IJ=KXYZ+3*(J-1)
       IF(IB)10,10,2156
 2156  DO 2155 I=1,N
       JT=IJ+LABL(I)
 2155  U(JT)=BETA(I)
       GO TO 10
  216  IRST=1
```

```
      GO TO (91,2161,211,2171),IMOD
2161  IF(IS) 2162,2151,2162
2162  IF(IU(L)-IS) 2163,2151,2163
2163  IF(IS-1) 2173,2173,2165
2165  JJC=JJC-1
      GO TO 214
2170  L=J+JTYP
2171  IF(IU(L)-1) 2173,2172,2173
2172  JJC=JJC-1
      GO TO 2174
2173  JDC=JDC-1
      CALL ALOCAT(KJREL)
      IJ=KJREL+1
      NR=5*IU(IJ)
      DO 2177 IK=1,NR,5
      M=IK+IJ
      IF(IU(M)-J) 2177,2178,2177
2177  CONTINUE
      GO TO 2174
2178  IU(M+1)=0
2174  CALL RELEAS(KJREL)
      GO TO (91,212,91,2179),IMOD
2179  L=JTYP+J
      IU(L)=3
      DO 2175  N=1,3
      IJ=KXYZ+3*(J-1)+N
2175  U(IJ)=0.
      GO TO 10
C
C     JOINT RELEASES
 220  IG=2
      CALL READ(IG,J)
      IF(ITS) 98,2200,98
2200  GO TO(221,10),ISCAN
221   IF(IG) 222,48,222
222   IF(ICONT)227,223,227
 223  NJR=1
      CALL ALOCAT(KJREL)
      IJ=KJREL+1
      NR=IU(IJ)
      IMR=5*NR
      IF(NR) 2232,2232,2231
2231  DO 2291 IK=1,IMR,5
      M=IK+IJ
      IF(IU(M)-J) 2291,224,2291
2291  CONTINUE
2232  IU(IJ)=NR+1
      M=5*NR+6
      CALL DEFINE(KJREL,M,0,0,1)
      M=KJREL+5*NR+2
      IU(M)=J
B224  U(M+1)=U(M+1)+A
      CALL RELEAS(KJREL)
      IF (IB) 10,10,226
226   DO 225 IX=1,IB
      NM=M+IX+1
225   U(NM)=BETA(IX)
      GO TO 10
227   GO TO (91,228,223,228),IMOD
228    CALL ALOCAT(KJREL)
```

```
       CALL RELEAS(KJREL)
       IJ=KJREL+1
       NR=IU(IJ)
       DO 2285 IX=1,NR
       M=KJREL+5*IX-3
       IF(IU(M)-J) 2285,2286,2285
  2285 CONTINUE
       GO TO 90
  2286 IF(IMOD-2) 90,2281,229
  2281 IU(M+1)=0
       NJR=1
       GO TO 224
B229   U(M+1)=-(-(U(M+1))+A)
       GO TO 10
C      JOINT LOADS
  230  INDEX=1
  231  IG=3
       CALL READ(IG,J)
       GO TO (232,10),ISCAN
  232  IF(IG) 10,48,10
C      JOINT DISPLACEMENTS
  250  INDEX=5
       GO TO 231
  10   RETURN
  48   NE=2
       GO TO 10
  90   NE=3
       GO TO 10
  91   NE=4
       GO TO 10
  93   NE=6
       GO TO 10
  96   NE=9
       GO TO 10
  97   NE=10
       GO TO 10
  98   NE=7
       GO TO 10
  99   NE=8
       GO TO 10
       END
*      LIST8
*      LABEL
CDPRINT  SUBROUTINE FOR PRINTING DATA LOGCHER MAZZOTTA NOV)1963
       SUBROUTINE DPRINT
       DIMENSION LABL(12),BETA(86),SYSFIL(27),PRBFIL(6),CWFIL(49),U(2),
      1IU(2)
       COMMON U,IU,A,IA,LABL,BETA,ITABLE,J,NE,ITS,IB,IS,IL,INDEX,IN,
      1CHECK,NMAX,INORM,ISOLV,ISCAN,III,IMOD,JJJJ,ICONT,ISUCC,IMERG,TOP,N
      21,NL,NT,NREQ,TN,LFILE,TOLER,IPRG,IRST,IRLD,IRPR,SYSFIL,
      3NJ,NB,NDAT,ID,JF,NSQ,NCORD,IMETH,NLDS,NFJS,NSTV,NMEMV,IPSI,NMR,NJR
      4,ISODG,NDSQ,NDJ,IPDBP,IUDBP,NBB,NFJS1,JJC,JDC,JMIC,JMPC,JLD,JEXTN,
      5MEXTN,LEXTN,JLC,NLDSI,IYOUNG,ISHEAR,IEXPAN,IDENS,PRBFIL,
      6NAME,KXYZ,KJREL,JPLS,JMIN,MTYP,KPSI,MEMB,LOADS,MODN,KS,KMKST,KSTDB
      7,KATKA,KPPLS,KPMNS,KUV,KPPRI,KR,KSAVE,KS1,KS2,KS3,KS4,KS5,KS6,KS7,
      8KS8,KS9,KS10,KDIAG,KOFDG,KAD,LOADN,MEGAO,JEXT,JINT,KUDBP,KMEGA,
      9KPDBP,JTYP,MTYP1,KB,MLOAD,JLOAD,KATR,LINT,KYOUNG,KSHEAR,KEXPAN
       COMMON KDENS,CWFIL
       EQUIVALENCE(U,IU,A,IA),(U(2),IU(2),LABL)
       EQUIVALENCE (B,IBB),(C,IC)
```

```
          DIMENSION FMT(3)
   8      FMT(1)=606074013020
   8      FMT(3)=305434606060
   2      CALL ALOCAT(NAME)
   3      K1=NAME+1
          K2=NAME+12
          PRINT 99,(U(I),I=K1,K2),ICONT
          CALL RELEAS(NAME)
  99      FORMAT(35H1PROBLEM DATA FROM INTERNAL STORAGE//1X,12A6//
         12H *,I3,36H*TH MODIFICATION OF INITIAL PROBLEM.//)
  98      FORMAT(16H STRUCTURAL DATA//5H TYPE)
          PRINT 98
          IF(ID) 9,15,9
   9      GO TO (10,11,12,13,14),ID
  10      PRINT 96
          GO TO 15
  11      PRINT 95
          GO TO 15
  12      PRINT 94
          GO TO 15
  13      PRINT 93
          GO TO 15
  14      PRINT 92
  96      FORMAT(1H+,6X,11HPLANE TRUSS)
  95      FORMAT(1H+,6X,11HPLANE FRAME)
  94      FORMAT(1H+,6X,10HPLANE GRID)
  93      FORMAT(1H+,6X,11HSPACE TRUSS)
  92      FORMAT(1H+,6X,11HSPACE FRAME)
  15      PRINT 91
  91      FORMAT(8HOMETHOD )
          IF(IMETH) 8,18,8
   8      GO TO (16,17),IMETH
  16      PRINT 90
          GO TO 18
  17      PRINT 89
  90      FORMAT(1H+,7X,9HSTIFFNESS)
     89 FORMAT(1H+,7X,11HFLEXIBILITY)
  18      PRINT 88,NJ,NB,NDAT,NLDS
  88      FORMAT(17H NUMBER OF JOINTS,I6/11X,7HMEMBERS,I5/11X,8HSUPPORTS,I4
         1/11X,8HLOADINGS,I4,/)
C         JOINT COORDINATES
          IF(NJ) 7,39,7
   7      PRINT 87
  87      FORMAT(18HOJOINT COORDINATES//6H JOINT,7X,1HX,10X,1HY,10X,1HZ,10X,
         16HSTATUS)
          CALL ALOCAT(JTYP)
          CALL ALOCAT(KXYZ)
          DO 23  I=1,JEXTN
          IJ=JTYP+I
          IF(IU(IJ))19,23,19
  19      IF(IU(IJ)-3)20,23,20
  20      I1=KXYZ+1+3*(I-1)
          I2=I1+NCORD-1
          PRINT 86,I,(U(L),L=I1,I2)
  86      FORMAT(I5,3F11.3)
          IF(IU(IJ)-2) 23,21,23
  21      PRINT 85
  85      FORMAT(1H+,45X,7HSUPPORT)
  23      CONTINUE
C         JOINT RELEASES
```

```
        IF(NJR)27,39,27
27      PRINT 84
        CALL ALOCAT(KJREL)
84      FORMAT(15HOJOINT RELEASES//7H  JOINT,4X,5HFORCE,4X,6HMOMENT,
       17X,5HTHETA/7H NUMBER,4X,7HX  Y  Z,2X,7HX  Y  Z,7X,1H1,7X
       21H2,7X,1H3)
        I=KJREL+1
        NR=IU(I)
        DO 32  K=1,NR
        LRC=KJREL+3+5*(K-1)
        IA=IU(LRC)
B       A=A*000077000000
        IF(IA)28,32,28
28      LJ=LRC-1
        PRINT 83,IU(LJ)
83      FORMAT(I5)
        IS=64
        DO 31  IK=1,6
        IS=IS/2
        INO=IA/IS
        IF(INO-1) 31,29,29
29      N=28-3*IK
        IBB=N/10
        IC=N-IBB*10
        IBB=IBB*64
B       FMT(2)=730000677301+C+B
        PRINT FMT
        IA=IA-IS
31      CONTINUE
        II=LRC+1
        IE=LRC+3
        PRINT 82,(U(I),I=II,IE)
82      FORMAT(1H+,29X,3F8.2)
32      CONTINUE
        CALL RELEAS(KJREL)
C       MEMBER INCIDENCES
C       MEMBER PROPERTIES
39      IF(NB) 40,38,40
40      PRINT 820
        CALL ALOCAT(JPLS)
        CALL ALOCAT(JMIN)
        CALL ALOCAT(MTYP)
        CALL ALOCAT(KPSI)
820     FORMAT(25HOMEMBER  START  END  TYPE,15X,7HSEGMENT,5X,2HAX,6X,
       112HAY,6X,2HAZ,7X,2HIX,8X,2HIY,8X,2HIZ,8X,1HL,6X,4HBETA)
        DO 43  I=1,MEXTN
        IK=MTYP+I
        IF(IU(IK)) 41,43,41
41      CALL UPACW(U(IK),K1,K2,K3,K4,K5)
C       K3=IMLOT
C       K4=N
        IP=JPLS+I
        IM=JMIN+I
        PRINT 81,I,IU(IP),IU(IM)
81      FORMAT(2X,I3,4X,I3,3X,I3)
        GO TO(42,430,44,45,46),K3
42      PRINT 80
80      FORMAT(1H+,20X,9HPRISMATIC)
        K4=1
        LO=6
```

```
         GO TO 47
430      PRINT 79
79       FORMAT(1H+,20X,15HSTIFFNESS GIVEN)
         GO TO 47
44       PRINT 78
78       FORMAT(1H+,20X,17HFLEXIBILITY GIVEN)
         GO TO 47
45       PRINT 77
77       FORMAT(1H+,20X,16HVARIABLE SECTION)
         LO=7
         K4=K4/7
         GO TO 47
C        STEEL SECTIONS NOT ACCOUNTED FOR HERE
46       GO TO 47
47       CALL ALOCAT(MEMB,I)
         IT=MEMB+I
         ITT=IU(IT)
         IF(K3-1)48,49,48
48       IF(K3-4)52,49,52
49       DO 50  J=1,K4
         JFK=ITT+LO*(J-1)+1
         LBJ=JFK+LO-1
         PRINT 76, J,(U(II),II=JFK,LBJ)
76       FORMAT(40X,I3,5X,3F8.3,3F10.2,F7.1)
50       CONTINUE
54       CALL RELEAS(MEMB,I)
         IF(IPSI)51,43,51
51       IJ=KPSI+I
         PRINT 475,U(IJ)
475      FORMAT(1H+,110X,F12.6)
         GO TO 43
52       DO 53  IQ=1,JF
         JFK=ITT+1+(IQ-1)*JF
         LBJ=JFK+JF-1
   53    PRINT 474,(U(II),II=JFK,LBJ)
  474    FORMAT(40X,6F12.4)
         GO TO 54
43       CONTINUE
C        MEMBER RELEASES
         IF(NMR) 55,61,55
   55    PRINT 473
  473    FORMAT(16HOMEMBER RELEASES//7H MEMBER,7X,5HSTART,14X,3HEND/
        110X,5HFORCE,3X,6HMOMENT,4X,5HFORCE,3X,6HMOMENT/
        29X,7HX    Y    Z,9H  X    Y    Z,9H   X    Y    Z,9H   X    Y    Z)
         DO 60  I=1,MEXTN
         IW=MTYP+I
B        IF(U(IW)*0077777000000) 56,60,56
56       CALL UPACW(IU(IW),K1,IRC,K3,K4,K5)
57       IS=4096
         PRINT 72,I
72       FORMAT(I5)
         DO 59  J=1,12
         IS=IS/2
         INO=IRC/IS
         IF(INO-1) 59,58,58
58       IF(J-6) 73,73,74
73       N=23
         JC=J
         GO TO 75
74       N=41
```

```
         JC=J-6
  75     N=N-3*(JC-1)
         IBB=N/10
         IC=N-IBB*10
         IBB=IBB*64
  B      FMT(2)=730000677301+B+C
         PRINT FMT
         IRC=IRC-IS*INO
  59     CONTINUE
  60     CONTINUE
  61     CONTINUE
         DO 620  I=1,4
         IAD=190+I
         IF(U(IAD))601,620,601
 601     GO TO(605,604,603,602),I
 602     PRINT 699
 699     FORMAT(17HOMEMBER DENSITIES)
         GO TO 606
 603     PRINT 698
 698     FORMAT(34HOCOEFFICIENTS OF THERMAL EXPANSION)
         GO TO 606
 604     PRINT 697
 697     FORMAT(13HOSHEAR MODULI)
         GO TO 606
 605     PRINT 696
 696     FORMAT(15HOYOUNG'S MODULI)
 606     IF(IU(IAD)-1) 610,607,610
 607     PRINT 695
 695     FORMAT(14H MEMBER  VALUE)
         CALL ALOCAT(IU(IAD+57))
         K=IU(IAD+57)
         DO 609 J=1,MEXTN
         IP=MTYP+J
         M=K+J
         IF(IU(IP)) 608,609,608
 608     PRINT 694,J,U(M)
 694     FORMAT(2X,I4,3X,F14.2)
 609     CONTINUE
         CALL RELEAS(IU(IAD+57))
         GO TO 620
 610     PRINT 693,U(IAD)
 693     FORMAT(1X,F14.2,22H VALUE FOR ALL MEMBERS)
 620     CONTINUE
         CALL RELEAS(MTYP)
  38     IF(NLDS) 171,170,171
 170     PRINT 300
 300     FORMAT(37HOALL LOADINGS DELETED IN MODIFICATION)
         GO TO 399
 171     PRINT 299
         DO 400 IN=1,LEXTN
 299 FORMAT (13H-LOADING DATA// 37H GIVEN IN TABULAR FORM WITHOUT LABEL
     1S)
         CALL ALOCAT(LOADS,IN)
         IL=LOADS+IN
         IL1=IU(IL)+1
         CALL UPACW(IU(IL1),J1,J2,J,J,J5)
         IF(J1) 172,400,172
 172     II=IU(IL)+3
         IE=II+11
         PRINT 301,(IU(IP),IP=II,IE)
```

```
  301   FORMAT(1H0,12A6)
        IF(J5) 173,182,173
C       WHAT INFORMATION WANTED
  173   PRINT 302
  302   FORMAT(9HOTABULATE)
        IF(J5-15) 175,174,175
  174   PRINT 303
  303   FORMAT(1H+,9X,3HALL)
        GO TO 182
  175   N=16
        DO 181   IC=1,4
        N=N/2
        INF=J5/N
        IF(INF) 176,181,176
  176   J5=J5-N
        GO TO(177,178,179,180),IC
  177   PRINT 304
  304   FORMAT(9X,6HFORCES)
        GO TO 181
  178   PRINT 305
  305   FORMAT(9X,9HREACTIONS)
        GO TO 181
  179   PRINT 306
  306   FORMAT(9X,11HDISTORTIONS)
        GO TO 181
  180   PRINT 307
  307   FORMAT(9X,13HDISPLACEMENTS)
  181   CONTINUE
  182   IF(J1-1) 185,185,183
C       COMBINATION LOADING
  183   PRINT 308
  308   FORMAT(32H LOADING CONDITION     PERCENTAGE)
        J22=2*J2
        DO 184   KK=1,J22,2
        KKK=IE+KK
        KKKK=KKK+1
        PRINT 309,IU(KKK), U(KKKK)
  309   FORMAT(10X,I5,11X,F6.2)
  184   CONTINUE
        GO TO 400
C       REGULAR LOADS
  185   DO 398   LL=1,J2
        IW=IE+LL
        IF(IU(IW)) 1851,398,1851
 1851   CALL UPADP(IU(IW),K1,K2,K5)
        GO TO (186,191,189,190,187),K1
C       JOINT LOADS
  186   PRINT 310
  310   FORMAT(12X,5HLOADS)
        GO TO 188
  187   PRINT 3101
 3101   FORMAT(12X,12HDISPLACEMENT)
  188   PRINT 311,K5
  311   FORMAT(1H+,5HJOINT,I5)
        CALL ALOCAT(JLOAD,K5)
        CALL RELEAS(JLOAD,K5)
        IWH=JLOAD
        GO TO 220
C       MEMBER LOADS
  189   PRINT 312
```

```
       312  FORMAT(12X,8HEND LOAD)
            GO TO 202
       190  PRINT 313
       313  FORMAT(12X,11HDISTORTIONS)
            GO TO 202
       191  PRINT 314
       314  FORMAT(12X,4HLOAD)
            CALL ALOCAT(MLOAD,K5)
            CALL RELEAS(MLOAD,K5)
            IM=MLOAD+K5
            IMM=IU(IM)+K2+1
            CALL UPACW(IU(IMM),L,L,L3,L4,L)
            IF(L4-2)192,193,194
       192  PRINT 3141
      3141  FORMAT(1H+,25X,12HCONCENTRATED)
            GO TO 195
       193  PRINT 315
       315  FORMAT(1H+,25X,7HUNIFORM)
            GO TO 195
       194  PRINT 316
       316  FORMAT(1H+,25X,6HLINEAR)
       195  IF(L3/4) 197,197,196
       196  PRINT 317
       317  FORMAT(1H+,16X,6HMOMENT)
            L3=L3-3
            GO TO 198
       197  PRINT 318
       318  FORMAT(1H+,16X,5HFORCE)
       198  IF(L3-2) 199,200,201
       199  PRINT 319
       319  FORMAT(1H+,23X,1HX)
            GO TO 202
       200  PRINT 320
       320  FORMAT(1H+,23X,1HY)
            GO TO 202
       201  PRINT 321
       321  FORMAT(1H+,23X,1HZ)
       202  PRINT 322,K5
            CALL ALOCAT(MLOAD,K5)
            CALL RELEAS(MLOAD,K5)
       322  FORMAT(1H+,6HMEMBER,I4)
            IWH=MLOAD
       220  LP=IWH+K5
            LPL=IU(LP)+K2+1
            CALL UPACW(IU(LPL),L,L2,L,L,L)
             ISTR=LPL+1
            IEND=LPL+L2-2
            PRINT 323,(U(IK),IK=ISTR,IEND)
       323  FORMAT(1H+,38X,8F11.4/(39X,8F11.4))
       398  CONTINUE
       400  CALL RELEAS(LOADS,IN)
       399  RETURN
            END
     *      LIST8
     *      LABEL
     CSIZED    VERSION 3   LOGCHER-FENVES  JULY 23,1963
     C        PROCESSES SIZE DESCRIPTORS
     C        STRESS
     C        STRUCTURAL ENGINEERING SYSTEMS SOLVER
            SUBROUTINE SIZED(J,K,L)
```

```
      DIMENSION LABL(12),BETA(90),SYSFIL(27),PRBFIL(11),CWFIL(55),U(2),
     1IU(2)
      COMMON U,IU,A,IA,LABL,BETA,IB,IS,IL,INDEX,IN,
     1CHECK,NMAX,INORM,ISOLV,ISCAN,III,IMOD,JJJJ,ICONT,ISUCC,IMERG,TOP,N
     21,NL,NT,NREQ,TN,LFILE,TOLER,IPRG,IRST,IRLD,IRPR,SYSFIL,
     3NJ,NB,NDAT,ID,JF,NSQ,NCORD,IMETH,NLDS,NFJS,NSTV,NMEMV,IPSI,NMR,NJR
     4,ISODG,NDSQ,NDJ,IPDBP,IUDBP,NBB,NFJS1,JJC,JDC,JMIC,JMPC,JLD,JEXTN,
     5MEXTN,LEXTN,JLC,PRBFIL,
     6NAME,KXYZ,KJREL,JPLS,JMIN,MTYP,KPSI,MEMB,LOADS,MODN,KS,KMKST,KSTDB
     7,KATKA,KPPLS,KPMNS,KUV,KPPRI,KR,KSAVE,KS1,KS2,KS3,KS4,KS5,KS6,KS7,
     8KS8,KS9,KS10,KDIAG,KOFDG,KAD,LOADN,MEGAO,JEXT,JINT,KUDBP,KMEGA,
     9KPDBP,JTYP,MTYP1,KB,MLOAD,JLOAD,CWFIL
      EQUIVALENCE(U,IU,A,IA),(U(2),IU(2),LABL)
      GO TO (100,110),L
  100 IF(ICONT) 190,101,105
  101 GO TO (200,300,400,500),J
  105 IF(IMOD-2) 190,106,190
  106 GO TO (250,350,401,550),J
  110 GO TO (201,301,191,501),J
  190 L=2
  191 RETURN
C
C     NO. OF JOINTS
  200 NJ=K
B     CHECK=CHECK+000000000001
      JEXTN=0
      CALL DEFINE(KJREL,1,0,0,1)
  201 N=K*3
      CALL DEFINE(KXYZ,N,0,0,1)
      CALL DEFINE(JTYP,K,0,1,1)
      CALL DEFINE(JLOAD,0,K,1,0,1)
      CALL ALOCAT(KXYZ)
      CALL ALOCAT(JTYP)
      CALL ALOCAT(JLOAD,0)
      JX=JEXTN+1
      DO 202 I=JX,K
  202 CALL DEFINE(JLOAD,I,2,0,0,1)
      JEXTN=K
      GO TO 191
  250 NJ=K
      IF(NJ-JEXTN)600,600,201
C     NO. OF MEMBERS
  300 NB=K
      MEXTN=0
B     CHECK=CHECK+000000000004
  301 CALL DEFINE(JPLS,K,0,1,1)
      CALL DEFINE(JMIN,K,0,1,1)
      CALL DEFINE(MTYP,K,0,0,1)
      CALL DEFINE(MTYP1,K,0,0,1)
      CALL DEFINE(KPSI,K,0,0,1)
      CALL DEFINE(MEMB,0,K,1,0,1)
      CALL DEFINE(MLOAD,0,K,1,0,1)
      CALL ALOCAT(JPLS)
      CALL ALOCAT(JMIN)
      CALL ALOCAT(KPSI)
      CALL ALOCAT(MEMB,0)
      CALL ALOCAT(MLOAD,0)
      JX=MEXTN+1
      DO 302 I=JX,K
  302 CALL DEFINE(MLOAD,I,2,0,0,1)
```

```
      MEXTN=K
      GO TO 191
  350 NB=K
      IF(NB-MEXTN)600,600,301
C     NO. OF SUPPORTS
B400  CHECK=CHECK+000000000002
  401 NDAT=K
      GO TO 191
C
C     NO. OF LOADINGS
  500 NLDS=K
B     CHECK=CHECK+000000000010
  501 CALL DEFINE(LOADS,0,K,1,0,1)
  510 CALL ALOCAT(LOADS,0)
      LEXTN=K
      GO TO 191
  550 NLDS=K
      IF(NLDS-LEXTN)600,600,501
  600 IRST=1
      GO TO 191
      END
*     LIST8
*     LABEL
CREAD     SUBROUTINE FOR READING JOINT AND MEMBER LABELS FENVES-LOGCHER
      SUBROUTINE READ(IG,J)
      DIMENSION KL(36)
      DIMENSION LABL(12),BETA(86),SYSFIL(27),PRBFIL(6),CWFIL(49),U(2),
     1IU(2)
      COMMON U,IU,A,IA,LABL,BETA,ITABLE,J,NE,ITS,IB,IS,IL,INDEX,IN,
     1CHECK,NMAX,INORM,ISOLV,ISCAN,III,IMOD,JJJJ,ICONT,ISUCC,IMERG,TOP,N
     21,NL,NT,NREQ,TN,LFILE,TOLER,IPRG,IRST,IRLD,IRPR,SYSFIL,
     3NJ,NB,NDAT,ID,JF,NSQ,NCORD,IMETH,NLDS,NFJS,NSTV,NMEMV,IPSI,NMR,NJR
     4,ISODG,NDSQ,NDJ,IPDBP,IUDBP,NBB,NFJS1,JJC,JDC,JMIC,JMPC,JLD,JEXTN,
     5MEXTN,LEXTN,JLC,NLDSI,IYOUNG,ISHEAR,IEXPAN,IDENS,PRBFIL,
     6NAME,KXYZ,KJREL,JPLS,JMIN,MTYP,KPSI,MEMB,LOADS,MODN,KS,KMKST,KSTDB
     7,KATKA,KPPLS,KPMNS,KUV,KPPRI,KR,KSAVE,KS1,KS2,KS3,KS4,KS5,KS6,KS7,
     8KS8,KS9,KS10,KDIAG,KOFDG,KAD,LOADN,MEGAO,JEXT,JINT,KUDBP,KMEGA,
     9KPDBP,JTYP,MTYP1,KB,MLOAD,JLOAD,KATR,LINT,KYOUNG,KSHEAR,KEXPAN
      COMMON KDENS,CWFIL
      EQUIVALENCE(U,IU,A,IA),(U(2),IU(2),LABL)
      EQUIVALENCE (K,BK)
F     LIST5
      IB=0
      IL=0
      I1=0
      I2=0
      IN=1
      IS=0
      ITS=0
  100 I3=MATCH(LIST5,K,0)
      GO TO (130,101,102,96,103,104),I3
  101 ITS=K
      GO TO 100
  102 IB=IB+1
      BETA(IB)=BK
      GO TO 100
  103 KL(IN)=K
      GO TO 105
  104 K=KL(IN)
  105 GO TO (106,106,106,107,108,107,108,107,108,109,110,111,111,111,112
```

```
        1,112,112,113,114,115,116,117,116,117,107,108,118,118,121,122,1055,
        21055,1055,1056,1056,1056,107),K
1055    I2=0
        I1=-3C
        GO TO 106
1056    I2=3
        I1=-33
106     IA=K+I1+I2
        GO TO 119
107     I2=0
        GO TO 120
108     I2=3
        GO TO 120
109     I1=0
        GO TO 1101
110     I1=6
1101    INDEX=3
        GO TO 120
111     IA=K-11
        GO TO 119
112     IA=1
        GO TO 119
113     IA=2
        GO TO 119
114     IA=3
        GO TO 119
118     GO TO(1181,1181,114),IG
1181    IA=7
        GO TO 119
122     IA=8
        GO TO 119
121     IG=4
        GO TO 500
115     IA=4
119     IL=IL+1
        LABL(IL)=IA
120     IN=IN+1
        GO TO 100
116     IS=2
        GO TO 100
117     IS=1
        GO TO 100
130     GO TO (131,97),ISCAN
131     IF(IL)134,132,134
132     IF(IB-12)1322,1322,1321
1321    N=12
        GO TO 1323
1322    N=IB
        IL=N
1323    IF(IB)134,134,1324
1324    DO 133 I=1,N
133     LABL(I)=I+I1
134     GO TO (500,135,137),IG
135     IA=0
        DO 136 I=1,IL
136     IA=IA+2**(LABL(I)-1)
        GO TO 500
137     IF(JLD) 1371,95,1371
B1371   CHECK=CHECK+000000000200
        CALL ALOCAT(LOADS,JLC)
```

```
            IF(ICONT)200,138,200
     138    L=LOADS+JLC
            IK=IU(L)+1
            CALL UPADP(IU(IK),X,NLD,Y)
            NLS=IU(IK+1)
            IU(IK)=IU(IK)+1
            IF(NLD-NLS) 140,139,139
     139    NLS=NLS+10
            IU(IK+1)=NLS
            CALL DEFINE(LOADS,JLC,NLS+14,0,0,1)
     140    GO TO(141,170,142,171,141),INDEX
     141    N=8
            CALL ALOCAT(JLOAD,J)
            L=JLOAD+J
            CALL UPADP(IU(L),X,Y,NSZ)
            IK=IU(L)+1
            IF(NSZ-2)148,148,149
     148    IU(IK)=3
            IU(IK+1)=0
     149    NKPL=IU(IK)
            NBL=IU(IK+1)
            IU(IK)=NKPL+8
            IU(IK+1)=NBL+1
            CALL DEFINE(JLOAD,J,NKPL+7,0,0,1)
            CALL RELEAS(JLOAD,J)
            L=JLOAD+J
            GO TO 144
     142    N=14
     143    CALL ALOCAT(MLOAD,J)
            L=MLOAD+J
            CALL UPADP(IU(L),X,Y,NSZ)
            IK=IU(L)+1
            IF(NSZ-2)146,146,147
     146    IU(IK)=3
            IU(IK+1)=0
     147    NKPL=IU(IK)
            NBL=IU(IK+1)
            NM=NKPL+N-1
            IU(IK)=NKPL+N
            IU(IK+1)=NBL+1
            CALL DEFINE(MLOAD,J,NM,0,0,1)
            CALL RELEAS(MLOAD,J)
            L=MLOAD+J
     144    M=LOADS+JLC
            IJ=IU(M)+NLD+15
            CALL PADP(IU(IJ),INDEX,NKPL,J)
     145    IJ=IU(L)+NKPL
            CALL PADP(IU(IJ),INDEX,JLC,NLD+1)
            IJ=IJ+1
            CALL PACKW(IU(IJ),0,N,LABL(1),LABL(2),0)
     150    GO TO (151,153,151,151,151),INDEX
     151    DO 152 M=1,IB
            NM=LABL(M)
            KKL=IJ+NM
     152    U(KKL)=BETA(M)
            GO TO 400
     153    IF(IL-2)156,156,154
     154    DO 155 M=1,IB
            NM=LABL(M+2)
            KKL=IJ+NM
```

```
155      U(KKL)=BETA(M)
         GO TO 400
156      IF(LABL(2)-2)158,157,160
157      LABL(5)=4
158      LABL(4)=3
159      LABL(3)=1
         GO TO 154
160      DO 161 M=1,4
161      LABL(M+2)=M
         GO TO 154
170      N=6
         GO TO 143
171      N=8
         GO TO 143
200      IRLC=1
         GO TO (97,201,138,201),IMOD
201      CALL PADP(KKL,INDEX,0,J)
         IX=1
         L=LOADS+JLC
         IK=IU(L)+1
         NLD=IU(IK)
         IK=IK+13
         IF(ITS) 2022,2021,2022
2021 ITS=1
2022 DO 204 I=1,NLD
         M=IK+I
B        A=U(M)*700000777777
         IF(IA-KKL)204,202,204
202      IF(IX-ITS)203,205,203
203      IX=IX+1
204      CONTINUE
         PRINT 98
         GO TO 97
98       FORMAT(32H LOAD SPECIFIED CANNOT BE FOUND.)
205      GO TO (97,206,97,209),IMOD
206      NKPL=IU(M)
         NLD=I
         GO TO (207,2081,2082,2083,207),INDEX
207      CALL ALOCAT(JLOAD,J)
         CALL RELEAS(JLOAD,J)
         N=8
         L=JLOAD+J
         GO TO 145
2081     N=6
         GO TO 208
2082     N=14
         GO TO 208
2083     N=8
208      CALL ALOCAT(MLOAD,J)
         CALL RELEAS(MLOAD,J)
         L=MLOAD+J
         GO TO 145
209      I=IU(M)
         IU(M)=0
         GO TO (210,211,211,211,210),INDEX
210      CALL ALOCAT(JLOAD,J)
         CALL RELEAS(JLOAD,J)
         L=JLOAD+J
         GO TO 212
211      CALL ALOCAT(MLOAD,J)
```

```
        CALL RELEAS(MLOAD,J)
        L=MLOAD+J
  212   NM=IU(L)+XABSF(I)
        N=IU(L)+2
        IU(N)=IU(N)-1
        IU(NM)=0
        GO TO 400
  95    CALL PRERR(36)
  B     CHECK=CHECK*777777777377
        GO TO 97
  96    CALL PRER2(9,0,BK)
  97    IG=0
  400   CALL RELEAS(LOADS,JLC)
  500   RETURN
        END
*       FAP                                                                           01
*       SUBROUTINE MATCH  R.D.LOGCHER  VERSION CORRECTED 1-15-64                      005
        COUNT   385                                                                  0010
        LBL     MATCH                                                                015
        ENTRY   MATCH                                                                0020
MATCH   SXA     BACK,4                                                               0030
        SXA     BACK+2,2                                                             C040
        SXA     BACK+3,1                                                             0050
        LDQ     SAVE                                                                 060
WORD    AXT     **,1                                                                 0070
LETTER  AXT     **,2                                                                 C080
  .     PXA                                                                          0081
        LGL     36                                                                   083
        LAS     =0607777777777                                                       086
        TRA     *+2                                                                  089
        TRA     SAL                                                                  092
        LGR     36                                                                   095
CHKT    CLA*    3,4                                                                  098
        SUB     ONE                                                                  0100
        TZE     NCARD                       NEW CARD                                 0110
        STZ     DICK                                                                 115
        TMI     NF                          NEW FIELD                                0120
        SUB     THREE                                                                0130
        TMI     NUMB+1                      NUMBER ONLY EXPECTED                     0140
        TNZ     BUFAD                       BUFFER ADDRESS FOR CARD TRANSMISSION     0150
        TSX     STB,4                                                                0160
STF     PXA                                 STRIP ONE LOGICAL FIELD                  C170
        LGL     6                                                                    0180
        LAS     BLANK                                                                0190
        TRA     *+2                                                                  0200
        TRA     BLK1                                                                 C210
        LAS     COMMA                                                                0220
        TRA     *+2                                                                  0230
        TRA     BLK1                                                                 0240
        TIX     STF,2,1                                                              0250
        TSX     NWD,4                                                                0260
        TRA     STF                                                                  0270
BLK1    LGR     6                                                                    0280
NF      TSX     STB,4                       EXAMINE ONE FIELD                        C290
        STZ     TEMP                                                                 0300
        PXA                                                                          0310
        LGL     6                                                                    0320
        LAS     =00C0000000040              MINUS                                    C330
        TRA     *+2                                                                  0340
        TRA     NUMB                        NUMBER NEXT                              0350
```

```
        LAS     =000000000033          DECIMAL                           0360
        TRA     *+2                                                      0370
        TRA     NUMB                                                     0380
        LAS     =000000000020                                            0390
        TRA     *+3                                                      0400
        NOP                                                              0410
        TRA     NUMB                                                     0420
        AXT     6,4                                                      0430
LPNF    LGR     6                                                        0440
        CAL     TEMP                                                     0450
        LGL     6                                                        0460
        SLW     TEMP                                                     0470
INLP    TIX     *+4,2,1                                                  0480
        SXA     *+2,4                                                    0490
        TSX     NWD,4                                                    0500
        AXT     **,4                                                     0510
        PXA                                                              0520
        LGL     6                                                        0530
        LAS     BLANK                                                    0540
        TRA     *+2                                                      0550
        TRA     BLK2                                                     0560
        LAS     COMMA                                                    0570
        TRA     *+2                                                      0580
        TRA     BLK2                                                     0590
        TIX     LPNF,4,1                                                 0600
        TRA     INLP                                                     0610
BLK2    LGR     6                                                        0620
        SXA     WORD,1                                                   0630
        SXA     LETTER,2                                                 0640
        LXA     BACK,4                                                   0650
        CLA*    1,4                                                      0660
        STA     *+1                                                      0670
ADLIST  CLA     **                                                      0680
        PAX     ,1                                                       0690
        ADD     *-2                                                      0700
        ADD     =1                                                       0710
        STA     *+8                                                      0720
        CAL     TEMP                                                     0730
        LAS     =000000002446                                            0740
        TRA     *+2                                                      0750
        TRA     DITTO                  FIELD CONTAINS DITTO              0760
        LAS     =0002431636346                                           0770
        TRA     *+2                                                      0780
        TRA     DITTO                  FIELD CONTAINS DITTO              0790
        LAS     **,1                                                     0800
        TRA     *+2                                                      0810
        TRA     FOUND                                                    0820
        TIX     *-3,1,1                                                  0830
        SLW*    2,4                                                      0840
        CLA     FOUR                                                     0850
        TRA     BACK+1                                                   0860
FOUND   PXA     ,1                                                       0870
        SSM                                                              0880
        ADD     =1                                                       0890
        ADD*    ADLIST                                                   0900
        ALS     18                                                      0910
        STO*    2,4                                                      0920
        CLA     FIVE                                                     0930
        TRA     BACK+1                                                   0940
DITTO   CLA     SIX                                                      0950
```

```
        TRA     BACK+1                                                    0960
NCARD   ZET     DICK                                                       965
        TRA     CONTER-1                                                   970
        TXH     READC,1,1                                                  975
        TXL     READC,2,5                                                 0980
        PXA                          NEW CARD ALREADY READ                0990
        LGL     6                                                         1CC0
        LAS     =054                 * FOR COMMENT CARD                   1010
        TRA     *+2                                                       1020
        TRA     READC                                                     1030
        LAS     =053                 $ FOR CONTINUATION CARD              1C40
        TRA     *+2                                                       1050
        TRA     CONTER               CONTINUATION CARD ERROR-NOT RIGHT    1060
        LGR     6                                                         1070
        TRA     NF                                                        1080
        STZ     DICK
CONTER  CAL     =0532551514651                                           1090
        LXA     BACK,4                                                   11C0
        SLW*    2,4                                                      1110
        TNX     *+1,2,1                                                  1120
        SXA     WORD,1                                                   1130
        SXA     LETTER,2                                                 1140
        CLA     FOUR                                                     1150
BACK    AXT     **,4                                                     1160
        STQ     SAVE                                                     1170
        AXT     **,2                                                     1180
        AXT     **,1                                                     1190
        TRA     4,4                                                      1200
NUMB    LGR     6                                                        1210
RFLN    CLA     =1                                                       1220
        STO     FLCO                                                     1230
        STZ     PTPLC                                                    1240
        STZ     BCON                                                     1250
        STZ     BCON+1                                                   1260
        STZ     SIGN                                                     1270
        AXT     0,4                                                      1280
        TRA     INTO                                                     1290
ONCON   TIX     INTO,2,1                                                 13C0
        SXA     *+2,4                                                    1310
        TSX     NWD,4                                                    1320
        AXT     **,4                                                     1330
        PXA                                                              1335
        LGL     36                                                       1340
        LAS     =0607777777777                                           1350
        TRA     *+2                                                      1360
        TRA     STRD-1                                                   1370
        LGR     36                                                       1380
INTO    PXA                                                              1390
        LGL     6                                                        1400
        LAS     =0000000000040                                           1410
        TRA     *+2                                                      1420
        TRA     ADSN                                                     1430
        LAS     =0000000000014                                           1440
        TRA     TSTPT                                                    1450
        TRA     ADSN                                                     1460
        TRA     NBD                                                      1470
ADSN    CLA     =1                                                       1480
        STO     SIGN                                                     1490
        TRA     ONCON                                                    15C0
NBD     LAS     =0000000000012                                           1510
```

```
        TRA     ONCON                                              1520
        TRA     ONCON                                              1530
        TXI     *+1,4,1                                            1540
        LGR     6                                                  1550
        PXA     ,4                                                 1560
        SUB     =7                                                 1570
        TPL     SEWD                                               1580
        CAL     BCDN                                               1590
        LGL     6                                                  1600
        SLW     BCDN                                               1610
        TRA     ONCON                                              1620
SEWD    SUB     =2                                                 1630
        TPL     PTOUT                                              1640
        CAL     BCDN+1                                             1650
        LGL     6                                                  1660
        SLW     BCDN+1                                             1670
        TRA     ONCON                                              1680
PTOUT   AXT     8,4                                                1690
        TRA     TESNU                                              1700
TSTPT   LAS     =0000000000033                                     1710
        TRA     BLK7                                               1720
        TRA     SPT                                                1730
        TRA     BLK7                                               1740
SPT     SXA     PTPLC,4                                            1750
        STZ     FLOO                                               1760
        TRA     ONCON                                              1770
BLK7    PXA     ,4                                                 1780
        TZE     ONCON                                              1790
        TRA     STRD+1                                             1800
TESNU   PXA                                                        1810
        LGL     6                                                  1820
        LAS     =0000000000011                                     1830
        TRA     STRD                                               1840
        NOP                                                        1850
INTB    TIX     TESNU,2,1                                          1860
        SXA     *+2,4                                              1870
        TSX     NWD,4                                              1880
        AXT     **,4                                               1890
        TRA     TESNU                                              1900
        LGR     30                                                 1910
STRD    LGR     6                                                  1920
        TXL     NFND,4,0                                           1930
        PXA     ,4                                                 1940
        SUB     PTPLC                                              1950
        STO     PTPLC                                              1960
        STQ     SAVE                                               1970
        SXA     LETTER,2                                           1980
        STZ     FXANS                                              1990
        AXT     0,2                                                2000
NUMD    PXA     ,4                                                 2010
        SUB     =7                                                 2020
        TMI     INLOP                                              2030
        ADD     =1                                                 2040
        PAX     ,4                                                 2050
        CLA     =1                                                 2060
        STO     DIV                                                2070
SECD    ADD     ABCDN                                              2080
        STA     PUT                                                2090
PUT     CAL     **                                                 2100
        LDQ     =0                                                 2110
```

```
            LRS     6                              2120
            SLW*    PUT                            2130
            PXA                                    2140
            LRS     29                             2150
            MPY     NUM,2                          2160
            XCA                                    2170
            ADD     FXANS                          2180
            STO     FXANS                          2190
            TXI     TST3,2,1                       2200
TST3        TIX     PUT,4,1                        2210
            CLA     DIV                            2220
            TZE     FINISH                         2230
            AXT     6,4                            2240
INLOP       STZ     DIV                            2250
            PXA                                    2260
            TRA     SECD                           2270
GTFX        CLA     FXANS                          2280
            ALS     18                             2290
            TRA     RESULT+1                       2300
FINISH      CLA     FLCO                           2310
            TNZ     GTFX                           2320
            CLA     FXANS                          2330
            ADD     FLO                            2340
            STO     TEMP                           2350
            CLM                                    2360
            FAD     TEMP                           2370
            STO     TEMP                           2380
            CLA     PTPLC                          2390
            TZE     TEND                           2400
            PAX     ,4                             2410
            CLA     TEMP                           2420
            FDP     DIV,4                          2430
RESULT      XCA                                    2440
            ZET     SIGN                           2450
            SSM                                    2460
IR2         LXA     LETTER,2                       2470
            LDQ     SAVE                           2480
            TIX     *+2,2,1                        2490
            TSX     NWD,4                          2500
            SXA     WORD,1                         2510
            SXA     LETTER,2                       2520
            LXA     BACK,4                         2530
            STO*    2,4                            2540
            PXA                                    2550
            NZT     FLCO                           2560
            CLA     ONE                            2570
            ADD     TWO                            2580
            TRA     BACK+1                         2590
TEND        LDQ     TEMP                           2600
            TRA     RESULT                         2610
NFND        TSX     NXTCD,4                        2620
            TRA     NUMB+1                         2630
            LXA     BACK,4                         2640
            STQ*    2,4                            2650
            TRA     BRET                           2660
STB         SXA     STBF,4                         2670
            AXT     6,4                            2680
            PXA                                    2690
            LGL     6                              2700
            LAS     COMMA                          2710
```

```
          TRA      *+2                                                    2720
          TRA      CMB                                                    2730
          LAS      BLANK                                                  2740
          TRA      *+2                                                    2750
          TRA      BLK3                                                   2760
          LGR      6                                                      2770
STBF      AXT      **,4                                                   2780
          TRA      1,4                                                    2790
BLK3      TIX      *+6,2,1                                                2800
          SXA      *+4,4                                                  2810
          TSX      NXTCD,4                                                2820
          TRA      *+2                                                    2830
          TRA      BRET                                                   2840
          AXT      **,4                                                   2850
          TIX      STB+2,4,1                                              2860
BRET      SXA      LETTER,2                                               2870
          SXA      WORD,1                                                 2880
          CLA      ONE                                                    2890
          TRA      BACK                                                   29C0
CMB       AXT      7,4                                                    2900
          TRA      BLK3                                                   2910
READC     TSX      READ,4                                                 2920
          TRA      NCARD+2                                                2930
SAL       TSX      NXTCD,4                                                2940
          TRA      *+2                                                    2943
          TRA      NFND+2                                                 2944
          CLA      =1                                                     2945
          STO      DICK                                                   2946
          TRA      CHKT                                                   2947
NXTCD     TXL      NWD,1,11                                               2948
          CLA      =1                                                     2950
          STO      DICK                                                   2960
READ      SXA      READR,4                                                2970
          TSX      $(CSH),4                                               2980
          PZE      FMT1                                                   2990
          AXT      1,1                                                    3C00
          STR                                                             3010
          STQ      BUF+1,1                                                3020
          TXI      *+1,1,1                                                3030
          TXL      *-3,1,12                                               3040
          TSX      $(RIN),4                                               3050
          TSX      $(SPH),4                                               3060
          PZE      FMT2                                                   3070
          AXT      1,1                                                    3080
          LDQ      BUF+1,1                                                3090
          STR                                                             3100
          TXI      *+1,1,1                                                3110
          TXL      *-3,1,12                                               3120
          TSX      $(FIL),4                                               3130
          CAL      BUF                                                    3140
          LAS      =0632142256047                                        3150
          TRA      *+2                                                    3160
          TSX      $EXIT,4            TAKE POST MORTUM                    3170
READR     AXT      **,4                                                   3180
          AXT      0,1                                                    3190
NWD       AXT      6,2                                                    3200
          LDQ      BUF,1                                                  3210
          TXI      *+1,1,1                                                3220
          NZT      DICK                                                   3230
          TRA      1,4                                                    3240
                                                                         3250
```

```
        PXA                                      3260
        LGL     6                                3270
        LAS     =054                             3280
        TRA     *+2                              3290
        TRA     READ                             3300
        STZ     DICK                             3310
        LAS     =053                             3320
        TRA     *+2                              3330
        TRA     *+3                              3340
        LGR     6                                3350
        TRA     2,4                              3360
        TIX     *+1,2,1                          3370
        TRA     1,4                              3380
BUFAD   CAL     *                                3390
        ANA     =0000000077777                   3395
        ADD     =21                              3400
        SSM                                      3410
        ADD     TOP                              3420
        ALS     18                               3430
        STO*    1,4                              3440
        TRA     4,4                              3450
        OCT     607777777777                     3460
BUF     BES     11                               3470
        PZE                                      3480
PTPLC   PZE                                      3490
BCDN    PZE                                      3500
        PZE                                      3510
SIGN    PZE                                      3520
FLOO    PZE                                      3530
ABCDN   PZE     BCDN                             3540
TEMP    PZE                                      3550
FXANS                                            3560
SAVE    PZE                                      3570
DICK    PZE                                      3580
        BCI     2,)      X,12A6                  3590
FMT2    BCI     1,(1H0,5                         3600
FMT1    BCI     1,(12A6)                         3610
BLANK   OCT     60                               3620
COMMA   OCT     73                               3630
ONE     OCT     1000000                          3640
TWO     OCT     2000000                          3650
THREE   OCT     3000000                          3660
FOUR    OCT     4000000                          3670
FIVE    OCT     5000000                          3680
SIX     OCT     6000000                          3690
FLO     OCT     233000000000                     3700
        DEC     10000000.                        3710
        DEC     1000000.                         3720
        DEC     100000.                          3730
        DEC     10000.                           3740
        DEC     1000.                            3750
        DEC     100.                             3760
        DEC     10.                              3770
DIV     PZE                                      3780
        DEC     10000000                         3790
        DEC     1000000                          3800
        DEC     100000                           3810
        DEC     10000                            3820
        DEC     1000                             3830
        DEC     100                              3840
```

```
          DEC       10                                                    3850
  NUM     DEC       1                                                     3860
  U       COMMON    119                                                   3870
  TOP     COMMON    1                                                     3880
          END                                                            3890
*         LIST 8
*         LABEL
          SUBROUTINE PRERR(J)
C         ERROR PRINTOUT SUBROUTINE
C         STRESS...STRUCTURAL ENGINEERING SYSTEM SOLVER, VERSION III
C         13 AUGUST 1963
          DIMENSION IU(36)
          COMMON IU
          GO TO (1,2,3,4,5,6,7,22,9,23,24,12,25,27,15,16,17,13,19,20,21,35,
     1 28,30,31,32,33,34,8,10,11,14,29,18,26,36),J
    1 PRINT 801
  500 RETURN
  801 FORMAT (1H1)
    2     PRINT 802,IU(112)
          GO TO 500
  802 FORMAT (37H PROBLEM TOO LARGE FOR MEMORY. ISOLV=I4)
    3 PRINT 803
          GO TO 500
  803 FORMAT (55H EXECUTION DELETED - FOLLOWING STATEMENTS SCANNED ONLY.
     1 )
    4 PRINT 804
          GO TO 500
  804 FORMAT (31H PROBLEM INCORRECTLY SPECIFIED. )
    5 PRINT 805
          GO TO 500
  805 FORMAT (33H PROBLEM FAILED DURING EXECUTION. )
    6 PRINT 806
          GO TO 500
  806 FORMAT (19H PROBLEM COMPLETED. )
    7 PRINT 807
          GO TO 500
  807 FORMAT (26H STATEMENT NOT ACCEPTABLE. )
    8 PRINT 808
          GO TO 500
  808 FORMAT (29H MODIFICATION NOT ACCEPTABLE. )
    9 PRINT 809
          GO TO 500
  809 FORMAT (35H FIXED POINT NUMBER IN WRONG PLACE. )
   10 PRINT 810
          GO TO 500
  810 FORMAT (54H JOINT NUMBER GREATER THAN NUMBER OF JOINTS SPECIFIED.)
   11 PRINT 811
          GO TO 500
  811 FORMAT (56H MEMBER NUMBER GREATER THAN NUMBER OF MEMBERS SPECIFIED
     1. )
   12 PRINT 812
          GO TO 500
  812 FORMAT (38H FLOATING POINT NUMBER IN WRONG PLACE. )
   13 PRINT 813
          GO TO 500
  813 FORMAT (34H NUMBER OF LOADINGS NOT SPECIFIED. )
   14 PRINT 814
          GO TO 500
  814 FORMAT (58H LOADING NUMBER GREATER THAN NUMBER OF LOADINGS SPECIFI
     1ED. )
```

```
   15 PRINT 815
      GO TO 500
  815 FORMAT (32H NUMBER OF JOINTS NOT SPECIFIED. )
   16 PRINT 816
      GO TO 500
  816 FORMAT (34H NUMBER OF SUPPORTS NOT SPECIFIED. )
   17 PRINT 817
      GO TO 500
  817 FORMAT (33H NUMBER OF MEMBERS NOT SPECIFIED. )
   18 PRINT 818
      GO TO 500
  818 FORMAT (60H NUMBER OF LOADINGS GIVEN NOT EQUAL TO THE NUMBER SPECI
     1FIED. )
   19 PRINT 819
      GO TO 500
  819 FORMAT (33H TYPE OF STRUCTURE NOT SPECIFIED. )
   20 PRINT 820
      GO TO 500
  820 FORMAT (34H METHOD OF SOLUTION NOT SPECIFIED. )
   21 PRINT 821
      GO TO 500
  821    FORMAT(42H A SOLUTION IS NOT AVAILABLE FOR PRINTING.)
   22 PRINT 822
      GO TO 500
  822 FORMAT (20H UNACCEPTABLE DITTO. )
   23 PRINT 823
      GO TO 500
  823 FORMAT (30H FIXED POINT NUMBER NOT FOUND. )
   24 PRINT 824
      GO TO 500
  824 FORMAT (36H TYPE OF MODIFICATION NOT SPECIFIED. )
   25 PRINT 825
      GO TO 500
  825 FORMAT (34H TYPE OF STRUCTURE NOT ACCEPTABLE. )
   26 PRINT 826
      GO TO 500
  826 FORMAT (32H METHOD SPECIFIED NOT AVAILABLE. )
   27 PRINT 827
      GO TO 500
  827 FORMAT (17H TOO MANY BLANKS. )
   28 PRINT 828
      GO TO 500
  828    FORMAT(63H NUMBER OF FREE JOINTS GIVEN NOT EQUAL TO THE NUMBER SPE
     1CIFIED. )
   29 PRINT 829
      GO TO 500
  829 FORMAT (60H NUMBER OF SUPPORTS GIVEN NOT EQUAL TO THE NUMBER SPECI
     1FIED. )
   30 PRINT 830
      GO TO 500
  830 FORMAT (59H NUMBER OF MEMBERS GIVEN NOT EQUAL TO THE NUMBER SPECIF
     1IED. )
   31 PRINT 831
      GO TO 500
  831 FORMAT (76H NUMBER OF MEMBER PROPERTIES GIVEN NOT EQUAL TO THE NUM
     1BER OF MEMBERS GIVEN. )
   32 PRINT 832
      GO TO 500
  832 FORMAT (32H UNACCEPTABLE STATEMENTS PRESENT   )
   33 PRINT 833
```

```
          GO TO 500
      833 FORMAT (26H STRUCTURAL DATA INCORRECT )
       34 PRINT 834
          GO TO 500
      834 FORMAT (23H LOADING DATA INCORRECT )
       35 PRINT 835
          GO TO 500
      835 FORMAT (19H NO LOADS SPECIFIED )
       36 PRINT 836
          GO TO 500
      836 FORMAT(55H LOADING DATA MUST BE GIVEN FOLLOWING A LOADING HEADER.)
          END
*         LIST8
*         LABEL
CRESTOR   ROUTINE FOR RESTORING IN MACHINE INTIAL DATA / LOGCHER 1-7-64
          SUBROUTINE RESTOR
          DIMENSION LABL(12),BETA(90),SYSFIL(17),PRBFIL(10),CWFIL(53),U(2),
         1IU(2) ,NTAPE(5),NXFILE(5)
          COMMON U,IU,A,IA,LABL,BETA,IB,IS,IL,INDEX,IN,
         1CHECK,NMAX,INORM,ISOLV,ISCAN,III,IMOD,JJJJ,ICONT,ISUCC,IMERG,TOP,N
         21,NL,NT,NREQ,TN,LFILE,TOLER,IPRG,IRST,IRLD,IRPR,NTAPE,NXFILE,SYSFI
         3L,NJ,NB,NDAT,ID,JF,NSQ,NCORD,IMETH,NLDS,NFJS,NSTV,NMEMV,IPSI,NMR,N
         4JR,ISODG,NDSQ,NDJ,IPDBP,IUDBP,NBB,NFJS1,JJC,JDC,JMIC,JMPC,JLD,JEXT
         5N,MEXTN,LEXTN,JLC,NLDSI,PRBFIL,
         6NAME,KXYZ,KJREL,JPLS,JMIN,MTYP,KPSI,MEMB,LOADS,MODN,KS,KMKST,KSTDB
         7,KATKA,KPPLS,KPMNS,KUV,KPPRI,KR,KSAVE,KS1,KS2,KS3,KS4,KS5,KS6,KS7,
         8KS8,KS9,KS10,KDIAG,KOFDG,IOFDG,LOADN,MEGAO,JEXT,JINT,KUDBP,KMEGA,
         9KPDBP,JTYP,MTYP1,KB,MLOAD,JLOAD,KATR,LINT,CWFIL
          EQUIVALENCE(U,IU,A,IA),(U(2),IU(2),LABL)
          LT=LFILE
          CALL FILES(3,N,JJJJ,KSAVE,TOP)
          LFILE=LT
          DO 11 I=1,4
          N =NTAPE(I)
          NXFILE(I)=1
       11 REWIND N
          RETURN
          END
*         LIST8
*         LABEL
          SUBROUTINE SELOUT(JLX,IZ,NUM)
C         STRESS PROGRAM...STRUCTURAL ENGINEERING SYSTEMS SOLVER
C         KENNETH F. REINSCHMIDT, ROOM 1-255, EXT. 2117
C         STRESS VERSION III...9 SEPTEMBER 1963
C         JLX = EXTERNAL LOADING NUMBER...IF JLX NOT EQUAL TO JLX OF
C         PREVIOUS CALL, PRINT STRUCTURE NAME, MODIFICATION NAME, AND
C         LOADING NAME ON NEW PAGE. IF EQUAL, NO PRINT.
C         IZ = PRINT CODE...IF IZ NOT EQUAL  TO IZ OF PREVIOUS CALL, PRINT
C         HEADING OF OUTPUT TYPE. IF EQUAL, NO PRINT.
C         NUM = MEMBER OR JOINT NUMBER DESIRED. IF NUM LESS THAN OR EQUAL
C         TO ZERO, PRINT ALL QUANTITIES.
          DIMENSION U(36),IU(36),Y(6,6),T(6,6),JUNK(12),V(12),
         1 SYSFIL(27),PROFIL(6), KRAY(10)
          COMMON U,IU,Y,T,K1,K2,KU,KPMIN,KPPL,KUPR,JT,JTEST,JLO,JUNK,KRR,
         1 KTEM,KSUP,V,
         4CHECK,NMAX,INORM,ISOLV,ISCAN,IIII,IMOD,JJJJ,ICONT,ISUCC,IMERG,
         5TOP,N1,NL,NT,NREQ,TN,LFILE,TOLER,IPRG,IRST,IRLD,IRPR,SYSFIL,
         8NJ,NB,NDAT,ID,JF,NSQ,NCORD,IMETH,NLDS,NFJS,NSTV,NMEMV,IPSI,NMR,
         9NJR,ISODG,NDSQ,NDJ,IPDBP,IUDBP,NBB,NFJS1,JJC,JDC,JMIC,JMPC,JLD
          COMMON JEXTN,MEXTN,LEXTN,JLC,NLDSI,IYOUNG,ISHEAR,IEXPAN,IDENS,
```

```
      1 PROFIL,
      4NAME,KXYZ,KJREL,JPLS,JMIN,MTYP,KPSI,MEMB,LOADS,MODN,KS,KMKST,KSTDB
      5,KATKA,KPPLS,KPMNS,KUV,KPPRI,KR,KSAVE,KRAY,KDIAG,KOFDG,
      6 IOFDG,LOADN,MEGAO,JEXT,JINT,KUDBP,KMEGA,KPDBP,JTYP,MTYP1,KB
      7,MLOAD,JLOAD,KATR,LEXT,KYOUNG,KSHEAR,KEXPAN,KDENS
       EQUIVALENCE (U,IU,Y)
C
       CALL ALOCAT(LEXT)
       CALL ALOCAT(LOADS,0)
       CALL ALOCAT(LOADS,JLX)
       K1 = LOADS+JLX
       JLO = IU(K1)+1
       CALL UPACW(U(JLO),K1,      JUNK(2),JUNK(3),JUNK(4),JT)
       IF (K1-1) 190,50,50
C      CONVERT TO INTERNAL LOADING NUMBERS
   50 JLI = LEXT+JLX
      JLI = IU(JLI)
      CALL RELEAS(LEXT)
      IF (JLX-JLXP) 40,108,40
   40 CALL ALOCAT(NAME)
  105 K1 = NAME+1
      K2 = NAME+12
      PRINT 300,(U(I),I=K1,K2)
      IF(MODN)106,107,106
  106 CALL ALOCAT(MODN)
      K1 = MODN+1
      K2 = MODN+12
      PRINT 302,(U(I),I=K1,K2)
      CALL RELEAS(MODN)
  107 JLO=LOADS+JLX
      JLO=IU(JLO)+1
      K1=JLO+2
      K2=JLO+13
      PRINT 301,(U(I),I=K1,K2)
      CALL RELEAS(NAME)
      JLXP = JLX
      IZP = 0
  108 GO TO(15,20,25,30),IZ
C     PRINT JOINT LOADS AND SUPPORT REACTIONS
   20 CALL ALOCAT(JINT)
      CALL ALOCAT(KR,0)
      CALL ALOCAT(KR,JLI)
      CALL ALOCAT(JTYP)
      JTP=JTYP+NUM
      JTP=3-IU(JTP)
      K1 = KR+JLI
      KRR = IU(K1)
      IF(JTP-3)102,131,102
C     PRINT MEMBER DISTORTIONS
   25 CALL ALOCAT(MTYP)
      CALL ALOCAT(KUV,0)
      CALL ALOCAT(KUV,JLI)
      K1 = KUV+JLI
      KU = IU(K1)
      GO TO 102
C     PRINT JOINT DISPLACEMENTS
   30 CALL ALOCAT(JINT)
      CALL ALOCAT(KPDBP)
      CALL ALOCAT(JTYP)
      JTP=JTYP+NUM
```

```
      JTP=3-IU(JTP)
      CALL ALOCAT(KPPRI)
   10 KUPR=KPPRI+NSTV*(JLI-1)
      IF(JTP-3)81,131,81
   81 IF (NJR) 90,90,80
C     SUPPORT DISPLACEMENTS
80    KSUP=KPDBP+(JLI-1)*NDJ
      GO TO 102
C     NO SUPPORT DISPLACEMENTS...USE KPDBP AS ZERO VECTOR
   90 KSUP = KPDBP
      GO TO(131,102),JTP
C     PRINT MEMBER FORCES
   15 CALL ALOCAT(MTYP)
      CALL ALOCAT(JPLS)
      CALL ALOCAT(JMIN)
      CALL ALOCAT(KPPLS,0)
      CALL ALOCAT(KPMNS,0)
      CALL ALOCAT(KPPLS,JLI)
      K1 = KPPLS+JLI
      KPPL = IU(K1)
      CALL ALOCAT(KPMNS,JLI)
      K1 = KPMNS+JLI
      KPMIN = IU(K1)
  102 IF (IZ-IZP) 120,125,120
  120 IZP = IZ
      IHOP = 1
      GO TO 150
  125 IHOP = 2
  150 GO TO(130,151,130,151),IZ
  151 GO TO(152,153),IHOP
152   IHOP=JTP
      JTPP=JTP
      GO TO 130
  153 IF(JTPP-JTP)155,154,155
  154 IHOP=JTP+2
      GO TO 130
  155 JTPP=JTP
      IHOP=JTP+4
  130 CALL ANSOUT (IZ,NUM,IHOP)
  131 CALL RELEAS(KXYZ)
      CALL RELEAS(KR,JLI)
      CALL RELEAS(KUV,JLI)
      CALL RELEAS(KPPLS,JLI)
      CALL RELEAS(KPMNS,JLI)
      CALL RELEAS(KR,0)
      CALL RELEAS(LOADS,0)
      CALL RELEAS(KUV,0)
      CALL RELEAS(KPPLS,0)
      CALL RELEAS(KPMNS,0)
      CALL RELEAS(JPLS)
      CALL RELEAS(JMIN)
      CALL RELEAS(JINT)
      CALL RELEAS(MTYP)
      CALL RELEAS(KPDBP,0,0)
      CALL RELEAS(KATKA,0,0)
      CALL RELEAS(KPPRI,0,0)
  190 CALL RELEAS(LOADS,JLX)
      RETURN
  300 FORMAT(1H0,//3X,12A6)
  302 FORMAT(1H0,3X,12A6)
```

286

```
  301 FORMAT (1H0,3X,12A6////)
      END
*     LIST8
*     LABEL
CANSOUT  ANSWER PRINTING ROUTINE - REINSCHMIDT - LOGCHER 1-17-64
      SUBROUTINE ANSOUT (IZ,NUM,IHOP)
C     STRESS PROGRAM...STRUCTURAL ENGINEERING SYSTEMS SOLVER
C     IZ = 1, PRINT MEMBER FORCES
C     IZ = 3, PRINT MEMBER DISTORTIONS
C     IZ = 4, PRINT JOINT DISPLACEMENTS
C     IZ = 2, PRINT JOINT LOADS AND SUPPORT REACTIONS
      DIMENSION U(36),IU(36),Y(6,6),T(6,6),JUNK(6),V(12),
     1 SYSFIL(27),PROFIL(6), KRAY(10)
      COMMON U,IU,Y,T,K1,K2,KU,KPMIN,KPPL,KUPR,JT,JTEST,JLO,NM,NP,
     1     KA,KB,KC,KD,JUNK,KRR,KTEM,KSUP,V,
     4CHECK,NMAX,INORM,ISOLV,ISCAN,IIII,IMOD,JJJJ,ICONT,ISUCC,IMERG,
     5TOP,N1,NL,NT,NREQ,TN,LFILE,TOLER,IPRG,IRST,IRLD,IRPR,SYSFIL,
     8NJ,NB,NDAT,ID,JF,NSQ,NCORD,IMETH,NLDS,NFJS,NSTV,NMEMV,IPSI,NMR,
     9NJR,KSTOP,NDSQ,NDJ,IPDBP,IUDBP,NBB,NFJS1,JJC,JDC,JMIC,JMPC,JLD
      COMMON JEXTN,MEXTN,LEXTN,JLC,NLDSI,IYOUNG,ISHEAR,IEXPAN,IDENS,
     1 PROFIL,
     4NAME,KXYZ,KJREL,JPLS,JMIN,MTYP,KPSI,MEMB,LOADS,MODN,KS,KMKST,KSTDB
     5,KATKA,KPPLS,KPMNS,KUV,KPPRI,KR,KSAVE,KRAY,KDIAG,KOFDG,
     6 IOFDG,LOADN,MEGAO,JEXT,JINT,KUDBP,KMEGA,KPDBP,JTYP,MTYP1,KBX
     7,MLOAD,JLOAD,KATR,LEXT,KYOUNG,KSHEAR,KEXPAN,KDENS
      EQUIVALENCE (U,IU,Y)
      XLOCF(NM,NP) = NM+(NP-1)*JF
      GO TO (10,5,10,7),IZ
    5     GO TO (350,350,370,370,6,6),IHOP
    6 IHOP=IHOP-4
      GO TO 370
    7     GO TO (250,250,270,270,8,8),IHOP
    8 IHOP=IHOP-4
      GO TO 270
   10 GO TO (20,30),IHOP
   30     GO TO (70,370,170),IZ
   20     GO TO (50,350,150),IZ
C     PRINT MEMBER FORCES
   50 PRINT 600
      PRINT 601
      GO TO (61,62,63,61,65),ID
   61 PRINT 611
      GO TO 70
   62 PRINT 611
      PRINT 612
      GO TO 70
   63 PRINT 613
      GO TO 70
   65 PRINT 614
      PRINT 615
   70 IF (NUM) 72,72,71
   71 LA = NUM
      LB = NUM
      GO TO 73
   72 LA = 1
      LB = MEXTN
   73 DO 100 I = LA,LB
C     TEST IF THE I TH MEMBER IS PRESENT
      K1 = MTYP+I
      IF (U(K1)) 75,100,75
```

```
   75 KA = JPLS+I
      KB = JMIN+I
      KA = IU(KA)
      KB = IU(KB)
      KC = XLOCF(KPPL,I)
      KD = XLOCF(KPMIN,I)
      K1 = KC+1
      K2 = KC+KSTOP
      PRINT 602,I,KA,(U(J),J=K1,K2)
      DO 80 J = 1,KSTOP
      K1 = KD+J
C     CHANGE SIGNS AT RIGHT END OF MEMBER
      V(J) = -U(K1)
   80 CONTINUE
      PRINT 602,I,KB,(V(J),J=1,KSTOP)
  100 CONTINUE
  101 RETURN
C     PRINT MEMBER DISTORTIONS
  150 PRINT 700
      PRINT 701
      GO TO (161,162,163,161,165),ID
  161 PRINT 711
      GO TO 170
  162 PRINT 711
      PRINT 712
      GO TO 170
  163 PRINT 713
      GO TO 170
  165 PRINT 715
      PRINT 615
  170 IF (NUM) 172,172,171
  171 LA = NUM
      LB = NUM
      GO TO 173
  172 LA = 1
      LB = MEXTN
  173 DO 200 I = LA,LB
C     TEST IF THE I TH MEMBER IS PRESENT
      K1 = MTYP+I
      IF (U(K1)) 175,200,175
  175 KC = XLOCF(KU,I)
      DO 180 J = 1,KSTOP
      K1 = KC+J
C     CHANGE SIGNS OF MEMBER DISTORTIONS
      V(J) = -U(K1)
  180 CONTINUE
      PRINT 702,I,(V(J),J=1,KSTOP)
  200 CONTINUE
      GO TO 101
C     PRINT JOINT DISPLACEMENTS
  250 PRINT 800
      PRINT 801
      GO TO (261,262,263,264,265),ID
  261 PRINT 811
      GO TO 270
  262 PRINT 811
      PRINT 812
      GO TO 270
  263 PRINT 813
      GO TO 270
```

```
      264 PRINT 811
          PRINT 814
          GO TO 270
      265 PRINT 811
          PRINT 814
          PRINT 815
      270 IF (NUM) 272,272,271
      271 LA = NUM
          LB = NUM
          GO TO(275,310,278,311),IHOP
      272 LA = 1
          LB = JEXTN
          IF(NJR)305,305,275
C         JOINT RELEASES, PRINT SUPPORT DISPLACEMENTS
      275 PRINT 817
      278 DO 300 I = LA,LB
          KA = I + JINT
          IF (IU(KA)) 300,300,280
      280 IF (IU(KA)-NFJS)300,300,285
      285 KC=XLOCF(KSUP,(IU(KA)-NFJS))
          K1 = KC+1
          K2 = KC+JF
          PRINT 702,I,(U(J),J=K1,K2)
      300 CONTINUE
          IF(NUM)101,305,101
      305 IF(NFJS)101,101,310
C         PRINT FREE JOINT DISPLACEMENTS
      310 PRINT 816
      311 DO 330 I = LA,LB
          KA = I+ JINT
          IF (IU(KA)) 330,330,315
      315 IF (IU(KA)-NFJS) 320,320,330
      320 KC=XLOCF(KUPR,IU(KA))
          K1 = KC+1
          K2 = KC+JF
          PRINT 702,I,(U(J),J=K1,K2)
      330 CONTINUE
          GO TO 101
C         PRINT JOINT LOADS AND SUPPORT REACTIONS
      350 PRINT 801
          GO TO (361,362,363,364,365),ID
      361 PRINT 911
          GO TO 370
      362 PRINT 911
          PRINT  912
          GO TO 370
      363 PRINT 913
          GO TO 370
      364 PRINT 911
          PRINT 914
          GO TO 370
      365 PRINT 911
          PRINT 914
          PRINT 915
      370 IF (NUM) 372,372,371
      371 LA = NUM
          LB = NUM
          GO TO(374,401,373,402),IHOP
      372 LA = 1
          LB = JEXTN
```

```
  374 PRINT 901
  373 DO 400 I = LA,LB
      KA = JINT + I
      IF (IU(KA)) 400,400,375
  375 IF (IU(KA)-NFJS) 400,400,380
  380 KC = XLOCF(KRR,IU(KA))
      K1 = KC+1
      K2 = KC+JF
      PRINT 702, I,(U(J),J=K1,K2)
  400 CONTINUE
      IF(NUM)405,405,101
405   IF(NFJS)101,101,401
  401 PRINT 902
  402 DO 430 I = LA,LB
      KA = JINT + I
      IF (IU(KA)) 430,430,410
  410 IF (IU(KA)-NFJS) 415,415,430
  415 KC = XLOCF(KRR,IU(KA))
      K1 = KC+1
      K2 = KC+JF
      PRINT 702,I,(U(J),J=K1,K2)
  430 CONTINUE
      GO TO 101
  600 FORMAT (1H0,3X,13HMEMBER FORCES//)
  601 FORMAT (14H MEMBER  JOINT )
  602 FORMAT(2I7,2X,6F17.7)
  611 FORMAT (1H+,21X,11HAXIAL FORCE )
  612 FORMAT (1H+,38X,28HSHEAR FORCE   BENDING MOMENT )
  613 FORMAT (1H+,21X,45HSHEAR FORCE TORSIONAL MOMENT   BENDING MOMENT )
  614 FORMAT (1H+,43X,6HFORCES,44X,7HMOMENTS )
  615 FORMAT (28X,5HAXIAL,6X,11H  SHEAR  Y ,6X,11H  SHEAR  Z ,8X,9HTORSI
     1ONAL,4X,13H  BENDING  Y ,4X,13H  BENDING  Z  )
  700 FORMAT (1H0,3X,18HMEMBER DISTORTIONS  //)
  701 FORMAT (7H MEMBER )
  702 FORMAT (I7,9X,6F17.7)
  711 FORMAT (1H+,16X,16HAXIAL DISTORTION )
  712 FORMAT (1H+,33X,33HSHEAR DISTORTION BENDING ROTATION )
  713 FORMAT (1H+,16X,50HSHEAR DISTORTION   TWIST ROTATION BENDING ROTAT
     1ION )
  715 FORMAT (1H+,38X,11HDISTORTIONS,42X,9HROTATIONS )
  800 FORMAT (1H0,3X,19HJOINT DISPLACEMENTS  //)
  801 FORMAT (1H0,//7H  JOINT )
  811 FORMAT (1H+,18X,31HX DISPLACEMENT    Y DISPLACEMENT )
  812 FORMAT (1H+,58X,8HROTATION )
  813 FORMAT (1H+,22X,10HDEFLECTION,7X,10HX ROTATION,7X,10HY ROTATION )
  814 FORMAT (1H+,52X,14HZ DISPLACEMENT )
  815 FORMAT (1H+,73X,10HX ROTATION, 7X,10HY ROTATION,7X,10HZ ROTATION )
  816 FORMAT (/1H0,37X,24HFREE JOINT DISPLACEMENTS )
  817 FORMAT (    38X,21HSUPPORT DISPLACEMENTS )
  901 FORMAT (    38X,17HSUPPORT REACTIONS )
  902 FORMAT (/1H0,37X,19HAPPLIED JOINT LOADS)
  911 FORMAT (1H+,25X,7HX FORCE,10X,7HY FORCE )
  912 FORMAT (1H+,52X,14HBENDING MOMENT )
  913 FORMAT (1H+,27X,5HSHEAR,9X,8HX MOMENT,9X,8HY MOMENT )
  914 FORMAT (1H+,59X,7HZ FORCE )
  915 FORMAT (1H+,75X,8HX MOMENT,9X,8HY MOMENT,9X,8HZ MOMENT )
      END
*     FAP
*     RELEASE COMPONENT COUNTER FUNCTION
*     R.D.LOGCHER   DECEMBER 31,1963
```

```
        COUNT    20
        LBL      COUNT
        ENTRY    ICNT
 ICNT   CAL*     1,4
        SXA      RET,4
        SXA      RET-1,1
        AXT      0,4
        AXT      12,1
        ALS      6
        PBT
        TRA      *+2
        TXI      *+1,4,1
        ALS      1
        TIX      *-4,1,1
        PXD      ,4
        AXT      **,1
 RET    AXT      **,4
        TRA      2,4
        END
 *      LIST8
 *      LABEL
CPHAS1B  SUBROUTINE PHAS1B LOGCHEK - FENVES  1-28-64
        SUBROUTINE PHAS1B
        DIMENSION LABL(12),BETA(86),SYSFIL(27),PRBFIL(4),CWFIL(49),U(2),
       1IU(2)
        COMMON U,IU,A,IA,LABL,BETA,ITABLE,J,NE,ITS,IB,IS,IL,INDEX,IN,
       1CHECK,NMAX,INORM,ISOLV,ISCAN,III,IMOD,JJJJ,ICONT,ISUCC,IMERG,TOP,N
       21,NL,NT,NREC,TN,LFILE,TOLER,IPRG,IRST,IRLD,IRPR,SYSFIL,NJ,NB,NDAT,
       3ID,JF,NSQ,NCORD,IMETH,NLDS,NFJS,NSTV,NMEMV,IPSI,NMR,NJR,ISODG,NDSQ
       4,NDJ,IPDBP,IUDBP,NBB,NFJS1,JJC,JDC,JMIC,JMPC,JLD,JEXTN,MEXTN,LEXTN
       5,JLC,NLDSI,IYOUNG,ISHEAR,IEXPAN,IDENS,NBNEW,NLDG,PRBFIL,
       6NAME,KXYZ,KJREL,JPLS,JMIN,MTYP,KPSI,MEMB,LOADS,MODN,KS,KMKST,KSIDB
       7,KATKA,KPPLS,KPMNS,KUV,KPPRI,KR,KSAVE,KS1,KS2,KS3,KS4,KS5,KS6,KS7,
       8KS8,KS9,KS10,KDIAG,KOFDG,IOFDG,LOADN,MEGAO,JEXT,JINT,KUDBP,KMEGA,
       9KPDBP,JTYP,MTYP1,KB,MLOAD,JLOAD,KATR,LINT,KYOUNG,KSHEAR,KEXPAN
        COMMON KDENS,CWFIL
        EQUIVALENCE(U,IU,A,IA),(U(2),IU(2),LABL)
C       JOINT PROCESSSING
        IT=0
 B      CHECK=CHECK*000004774377
        IF(NJ) 100,140,100
  100   NFJS=NJ-NDAT
        NFJS1=NFJS+1
        NBB=NB
        NBNEW=NB
        NDJ=JF*NDAT
        IF(JJC-NFJS) 102,101,102
 B101   CHECK=CHECK+000000000400
        GO TO 103
  102   IT=1
  103   IF(JDC-NDAT) 163,104,163
 B104   CHECK=CHECK+000001000000
        IF(IT) 163,105,163
  105   J1=1
        J2=NJ
        CALL DEFINE(JINT,JEXTN,0,0,1)
        CALL DEFINE(JEXT,NJ,0,0,1)
        CALL CLEAR(JINT)
        CALL CLEAR(JEXT)
        CALL ALOCAT(JINT)
```

```
        CALL ALOCAT(JEXT)
        CALL ALOCAT(KJREL)
        DO 130 J=1,JEXTN
        IJ=JTYP+J
        IF(IU(IJ)) 106,130,106
  106   IK=JINT+J
        IL=JLOAD+J
        IGO = IU(IJ)
        GO TO (107,117,120),IGO
  107   IU(IK)=J1
        IK=JEXT+J1
        IU(IK)=J
        J1=J1+1
        IS=1
        NDEX=1
  108   CALL UPADP(IU(IL),X,Y,NSZ)
        IF(NSZ-2)130,130,118
  118   CALL ALOCAT(JLOAD,J)
        IL=JLOAD+J
        IK=IU(IL)+2
        NBL=IU(IK)
        IK=IK+1
        IF(NBL)109,131,109
  109   DO 116 I=1,NBL
  110   CALL UPADP(IU(IK),INDEX,X,Y)
        IK=IK+IU(IK+1)
        IF(INDEX) 111,110,111
  111   IF(INDEX-NDEX) 112,116,112
  112   IF(IS) 115,1121,115
 1121   IF(NJR) 113,115,113
  113   IL=KJREL+1
        NR=5*IU(IL)+IL
        IL=IL+1
        DO 114 K=IL,NR,5
        IF(J-IU(K)) 114,116,114
  114   CONTINUE
  115   CALL PRER2(10,I,J)
 8      CHECK=CHECK*777777757777
  116   CONTINUE
        GO TO 131
  117   IU(IK)=J2
        IK=JEXT+J2
        IU(IK)=J
        J2=J2-1
        IS=0
        NDEX=5
        GO TO 108
  120   IU(IJ)=0
        CALL UPADP(IU(IL),X,Y,NSZ)
        IF(NSZ-2)130,130,119
  119   CALL ALOCAT(JLOAD,J)
        IL=JLOAD+J
        IK=IU(IL)+2
        NBL=IU(IK)
        IF(NBL) 121,129,121
  121   IXX=3
        DO 124 I=1,NBL
        IL=JLOAD+J
  122   IK=IU(IL)+IXX
        CALL UPADP(IU(IK),INDEX,JC,ILD)
```

```
            IXX=IXX+IU(IK+1)
            IF(INDEX) 123,122,123
      123   CALL ALOCAT(LOADS,JC)
            IL=LOADS+JC
            IM=IU(IL)+ILD+14
            IU(IM)=0
      124   CALL RELEAS(LOADS,JC)
            IL=JLOAD+J
            IM=IU(IL)+1
            IU(IM)=3
            IU(IM+1)=0
      129   CALL DEFINE(JLOAD,J,2,0,0,1)
      131   CALL RELEAS(JLOAD,J)
      130   CONTINUE
            CALL RELEAS(JEXT)
            IF(NJR)139,139,135
      135   IF(ID)139,139,136
      136   GO TO(1351,1352,1353,1354,1355),ID
      1351  IA=3
            GO TO 1356
      1352  IA=35
            GO TO 1356
      1353  IA=28
            GO TO 1356
      1354  IA=7
            GO TO 1356
      1355  IA=63
      1356  IL=KJREL+1
            NJR=0
            NR=5*IU(IL)+IL
            IL=IL+1
            DO 137 K=IL,NR,5
            IF(IU(K))132,137,132
      B132  X=U(K+1)*A
            NJR=NJR+ICNT(X)
            IK=JTYP+IU(K)
            IF(IU(IK)-2)133,137,133
      133   IF(IU(K+1))134,137,134
      B 134 CHECK=CHECK*777777767777
            CALL PRER2(19,IU(K),A)
      137   CONTINUE
      139   CALL RELEAS(KJREL)
      C     MEMBER PROCESSING
      140     IF(NB)1402,170,1402
      1402  CALL DEFINE(KATR,0,NJ,1,0,1)
            CALL ALOCAT(MTYP)
            NMR=0
            DO 160 I=1,MEXTN
            IJ=MTYP+I
            NMR=NMR+ICNT(IU(IJ))
            IF(IU(IJ)) 142,151,142
      142   IJ=JPLS+I
            JP=IU(IJ)
            IF(JP) 1420,1420,1419
      1419  IF(JP-JEXTN) 1421,1421,1420
      1420  CALL PRER2(16,JP,I)
      B     CHECK=CHECK*777777757777
      1421  IJ=JTYP+JP
            IF(IU(IJ)) 143,150,143
      143   IJ=JMIN+I
```

```
       JM=IU(IJ)
       IF(JM) 1430,1430,1429
 1429  IF(JM-JEXTN) 1431,1431,1430
 1430  CALL PRER2(16,JM,I)
B      CHECK=CHECK*777777757777
 1431  IJ=JTYP+JM
       IF(IU(IJ)) 160,150,160
 150   IK=MTYP+I
       NMR=NMR-ICNT(IU(IK))
       IU(IK)=0
       CALL DEFINE(MEMB,I,0,0,0,0)
       CALL PRER2(17,I,J)
       JMIC=JMIC-1
       JMPC=JMPC-1
 151   IL=MLOAD+I
       CALL UPADP(IU(IL),X,Y,NSZ)
       IF(NSZ-2)1511,160,1511
1511   CALL ALOCAT(MLOAD,I)
       IL=MLOAD+I
       IK=IU(IL)+2
       NBL=IU(IK)
       IF(NBL) 152,159,152
 152   IXX=3
       DO 155 M=1,NBL
       IL=MLOAD+I
 153   IK=IU(IL)+IXX
       CALL UPADP(IU(IK),INDEX,JC,ILD)
       IXX=IXX+IU(IK+1)
       IF(INDEX) 154,153,154
 154   CALL ALOCAT(LOADS,JC)
       IL=LOADS+JC
       IM=IU(IL)+ILD+14
       IU(IM)=0
 155   CALL RELEAS(LOADS,JC)
       IL=MLOAD+I
       IM=IU(IL)+1
       IU(IM)=3
       IU(IM+1)=0
 159   CALL DEFINE(MLOAD,I,2,0,0,1)
       CALL RELEAS(MLOAD,I)
 160   CONTINUE
       CALL RELEAS(MTYP)
       CALL RELEAS(MEMB,0)
 163   IF(JMIC-NB) 165,164,165
B164   CHECK=CHECK+000000001000
 165   IF(JMPC-NB) 170,166,170
B166   CHECK=CHECK+000000002000
C      LOAD PROCESSING
 170   IF(NLDS) 171,198,171
 171   IF(NLDG-NLDS) 200,172,200
B172   CHECK=CHECK+000002000000
       CALL DEFINE (LINT,LEXTN,0,0,1)
       CALL ALOCAT(LINT)
       NLDSI=0
       JD=NLDS
       DO 174 L=1,LEXTN
       CALL ALOCAT(LOADS,L)
       IJ=LOADS+L
       IK=IU(IJ)+1
       CALL UPADP(IU(IK),LT,X,Y)
```

```
        IJ = LINT + L
        IU(IJ) = 0
        IF(LT-1) 174,173,175
  175   IU(IJ)=JD
        JD=JD-1
        GO TO 174
  173   NLDSI=NLDSI+1
        IU(IJ)=NLDSI
  174   CALL RELEAS(LOADS,L)
        GO TO 200
 B198   CHECK=CHECK*777777777767
 C      METHOD TEST
 C200   IF(IMETH-1) 210,202,201
 C201   CHECK=CHECK*777773777777
 C      GO TO 210
 B200   CHECK=CHECK+000000000040
        NSTV=NFJS*JF
        NMEMV=MEXTN*JF
        NDSQ=NJR*NJR
 C      FINAL TEST
 B210   FIN=000007777777
        CALL ITEST(CHECK,FIN,15,ISUCC)
        IF(ISUCC) 300,211,300
  211   ISOLV=2
        ISUCC=1
        IF(ICONT-1) 220,212,220
  212   CALL SAVE
  220   IT=0
        IF(IPSI) 226,227,226
  226   IT=1
  227   CALL RELEAS(KPSI,0,IT)
        CALL RELEAS(JLOAD,0)
        CALL RELEAS(MLOAD,0)
        CALL RELEAS(JTYP)
        CALL RELEAS(KXYZ)
        CALL RELEAS(LOADS,0)
  228   CALL RELEAS(LINT)
  229   CALL DEFINE(KS,MEXTN*(NCORD+1),0,0,1)
        CALL DEFINE(KMKST,0,MEXTN,1,0,1)
        CALL DEFINE(KSTDB,0,MEXTN,1,0,1)
        CALL DEFINE(KPPLS,0,NLDS,1,0,1)
        CALL DEFINE(KPMNS,0,NLDS,1,0,1)
        CALL DEFINE(KUV  ,0,NLDS,1,0,1)
        CALL DEFINE(KPPRI,NLDS*NSTV,0,0,1)
        CALL DEFINE(KDIAG,0,NJ,1,1,1)
        CALL DEFINE(KOFDG,0,NB,1,1,1)
        CALL ALOCAT(KPPLS,0)
        CALL ALOCAT(KPMNS,0)
        CALL ALOCAT(KUV,0)
        DO 230  I=1,NLDS
        CALL DEFINE(KPPLS,I,MEXTN*JF,0,0,1)
        CALL DEFINE(KPMNS,I,MEXTN*JF,0,0,1)
  230   CALL DEFINE(KUV,I,MEXTN*JF,0,0,1)
        CALL RELEAS(KPPLS,0)
        CALL RELEAS(KPMNS,0)
        CALL RELEAS(KUV,0)
        CALL ALOCAT(KDIAG,0)
        CALL ALOCAT(KOFDG,0)
        CALL ALOCAT(KMKST,0)
        CALL ALOCAT(KSTDB,0)
```

```
      DO 250 I = 1,NJ
250   CALL DEFINE(KDIAG,I,NSQ,0,0,1)
      DO 260 I = 1,MEXTN
      CALL DEFINE(KSTDB,I,NSQ,0,0,1)
260   CALL DEFINE(KMKST,I,NSQ,0,0,1)
      DO 270  I=1,NB
270   CALL DEFINE(KOFDG,I,NSQ,0,0,1)
      CALL RELEAS(KDIAG,0)
      CALL RELEAS(KOFDG,0)
      CALL RELEAS(KSTDB,0)
      CALL RELEAS(KMKST,0)
      CALL ALOCAT(KATR,0)
      CALL ALOCAT(MTYP)
      CALL DEFINE(IOFDG,0,0,1,0,0)
      CALL ALOCAT(IOFDG,0)
      CALL DEFINE(IOFDG,0,NJ,1,1,1)
      DO 141 I=1,NJ
      CALL DEFINE(KATR,I,7,0,0,1)
      CALL ALOCAT(KATR,I)
      IK=KATR+I
      IL=IU(IK)+1
      IU(IL)=0
141   IU(IL+1)=7
      DO 404 I=1,MEXTN
      IJ=MTYP+I
      IF(IU(IJ)) 401,404,401
401   IJ=JPLS+I
      JP=IU(IJ)
      IJ=JMIN+I
      JM=IU(IJ)
144   IJ=JINT+JP
      IJJ=IU(IJ)
      IS=1
      DO 147 M=1,2
      JPI=IU(IJ)
      IK=KATR+JPI
      IL=IU(IK)+1
      NBRI=IU(IL)
      IU(IL)=NBRI+1
      NS=IU(IL+1)
      IF(NBRI-NS+2) 146,145,145
145   NS=NS+5
      IU(IL+1)=NS
      CALL DEFINE(KATR,JPI,NS,0,0,1)
      IK=KATR+JPI
146   IL=IU(IK)+NBRI+3
      IU(IL)=I*IS
      IS=-1
147   IJ=JINT+JM
      IK=IOFDG+XMAXOF(IJJ,JPI)
      IU(IK)=IU(IK)-1
404   CONTINUE
      DO 405 I=1,NJ
      IF(I-NFJS) 407,407,406
406   IJJ=0
      GO TO 408
407   IJJ=30
408   IK=IOFDG+I
      NS=-IU(IK)+1+XMINOF(I/5,(I*NB)/(NJ*10),IJJ)
      IU(IK)=0
```

```
  405   CALL DEFINE(IOFDG,I,NS,0,1,1)
        CALL RELEAS(IOFDG,0)
        CALL RELEAS(MTYP)
        CALL RELEAS(JPLS)
        CALL RELEAS(JMIN)
        DO 162 I=1,NJ
  162   CALL RELEAS(KATR,I)
        CALL RELEAS(KATR,0)
        CALL RELEAS(JINT)
        GO TO 301
  300   CALL PRERR(4)
        ISUCC = 2
  301   RETURN
        END
  *     LIST8
  *     LABEL
CSAVE     SUBROUTINE FOR SAVING INITIAL DATA SPECIFICATION - LOGCHER 1-7-64
        SUBROUTINE SAVE
        DIMENSION LABL(12),BETA(90),SYSFIL(27),PRBFIL(10),CWFIL(53),U(2),
       1IU(2)
        COMMON U,IU,A,IA,LABL,BETA,IB,IS,IL,INDEX,IN,
       1CHECK,NMAX,INORM,ISOLV,ISCAN,III,IMOD,JJJJ,ICONT,ISUCC,IMERG,TOP,N
       21,NL,NT,NREQ,TN,LFILE,TOLER,IPRG,IRST,IRLD,IRPR,SYSFIL,
       3NJ,NB,NDAT,ID,JF,NSQ,NCORD,IMETH,NLDS,NFJS,NSTV,NMEMV,IPSI,NMR,NJR
       4,ISODG,NDSQ,NDJ,IPDBP,IUDBP,NBB,NFJS1,JJC,JDC,JMIC,JMPC,JLD,JEXTN,
       5MEXTN,LEXTN,JLC,NLDSI,PRBFIL,
       6NAME,KXYZ,KJREL,JPLS,JMIN,MTYP,KPSI,MEMB,LOADS,MODN,KS,KMKST,KSTDB
       7,KATKA,KPPLS,KPMNS,KUV,KPPRI,KR,KSAVE,KS1,KS2,KS3,KS4,KS5,KS6,KS7,
       8KS8,KS9,KS10,KDIAG,KOFDG,IOFDG,LOADN,MEGAO,JEXT,JINT,KUDBP,KMEGA,
       9KPDBP,JTYP,MTYP1,KB,MLOAD,JLOAD,KATR,LINT,CWFIL
        EQUIVALENCE(U,IU,A,IA),(U(2),IU(2),LABL)
C       THIS ROUTINE REPLACES THE DUMMY ROUTINES
        CALL UPADP (NL,A,B,ML)
        KSAVE=32442-ML
        JJJJ=LFILE
        CALL FILES(2,1,JJJJ,KSAVE,TOP)
        RETURN
        END
  *     FAP
         COUNT 50
         ENTRY   ITEST
         LBL     ITEST
  *        SUBROUTINE ITEST(A,B,L,I)
  *STRESS PROGRAMMING SYSTEM ... VERSION IIA
  *S.J.FENVES 3/21/63
  * I = 0 IF B LOGICALLY CONTAINED IN A, 1 OTHERWISE
  *IF L=0, NO DIAGNOSTIC PRINTING
  *IF L NOT 0, PRINT L+BIT POSITION OF B NOT CONT. IN A
  *
  ITEST SXA     END,4           PLANT LINK
        CAL*    1,4             A
        ANA*    2,4             AND B
        ERA*    2,4             EXCL. OR B
        TZE     OK                  IF ZERO, OK
        SLW     TEMP                NOT ZERO, SAVE
        CAL     ONE
        SLW*    4,4             I = 1
        NZT*    3,4             TEST L
        TRA     END             ZERO-NO PRINT
        CAL*    3,4
```

```
        SLW     L                       GET L
        CAL     TEMP            GET DIFFERENCE
        AXT     0,4             SET TO COUNT SHIFTS
 BACK   LBT                     TEST LOW BIT
        TRA     NEXT            ZERO OK
        SLW     TEMP                    SAVE
        PXD     ,4              GET INDEX
        ADD     L
        SLW     IPR                     PRINT PARAMETER
        SXA     SAVE,4          SAVE INDEX
        CALL    PRERR,IPR
        CAL     TEMP            RESTORE
        LXA     SAVE,4
 NEXT   ARS     1               SHIFT
        TXL     INC,4,34        TEST
        TRA     END
 INC    TXI     BACK,4,1
 OK     STZ*    4,4
 END    AXT     0,4
        TRA     5,4
 ONE    OCT     000001000000
 TEMP   PZE
 L      PZE
 IPR    PZE
 SAVE   PZE
        END
*       LIST8
*       LABEL
CMAIN2/6
        DIMENSION  Y(6,6),T(6,6),Q(6,6),U(36),IU(36),SYPA(40),FILL(6)
        COMMON U,T,Q,CHECK,NMAX,INORM,ISOLV,ISCAN,IIII,IMOD,ILINK,ICONT,
       1ISUCC,SYPA,NJ,NB,NDAT,ID,JF,NSQ,NCORD,IMETH,NLDS,NFJS,NSTV
       2,NMEMV,IPSI,NMR,NJR,ISODG,NDSQ,NDJ,SDJ,NPR,NBB,NFJS1,JJC,JDC,
       3JMIC,JMPC,JLD,JEXTN,MEXTN,LEXTN,JLC,NLDSI,IYOUNG,ISHER,IEXPAN,
       4IDENS,FILL,
       5NAME,KXYZ,KJREL,JPLS,JMIN,MTYP,KPSI,MEMB,
       6LOADS, INPUT,KS,KMKST,KSTOB,KATKA,KPPLS,KPMIN,KUV,KPPRI,KR,KMK,
       7KV,K33,LA2R,LA2RT,NV   ,NTP  ,NSCR7,NSCR8,NSCR9,NSCR10,KDIAG,KOFD
       8G,IOFDG,LDNM,MEGAO,JEXT,JINT,KUDBP,KMEGA,KPDBP,JTYP,MTYP1,KB,MLOAD
       9,JLOAD,KATR,LEXT,KYOUNG,KSHER,KEXPAN,KDENS
        EQUIVALENCE(U(1),IU(1),Y(1))
        CALL MEMBER
        ISUCC=ISUCC
        GO TO (1,20),ISUCC
 1      ISOLV=3
        IF(NMR)2,3,2
 2      CALL MRELES
        ISUCC=ISUCC
        GO TO (3,20),ISUCC
 3      ISOLV=4
        CALL CHAIN(3,A4)
 20     CALL CHAIN(1,A4)
        END
*       LABEL
*       LIST8
*       SYMBOL TABLE
        SUBROUTINE MEMBER
C       LEON R.L.WANG
C       STRESS PROGRAM...STRUCTURAL ENGINEERING SYSTEMS SOLVER
C       VERSION III...27 AUGUST 1963 BY KENNETH REINSCHMIDT
```

```
C      MAIN SUBROUTINE PERFORMS ALL MEMBER COMPUTATIONS
C      DESCRIPTION OF NOTATIONS
C      JM=NAME OF A MEMBER
C      NB  =NUMBERS OF MEMBERS
C      NP=+NODE
C      NM=-NODE
C      MTYP=TYPE OF MEMBER GIVEN
C      MTYP(1)=SECTION PROPERTIES GIVEN  -    PRISMATIC MEMBER
C      MTYP(2)=LOCAL STIFFNESS OF MEMBER GIVEN
C      MTYP(3)=LOCAL FLEXIBILITY OF MEMBER GIVEN
C      MTYP(4)=MEMBER OF STRAIGHT VARIABLE SECTION
C      MTYP(5)=COMPUTE STEEL SECTION PROPERTIES
C       JF=DEGREE OF FREEDOM
C      ID=TYPE OF STRUCTURE
C      ID(1)=PLANE TRUSS
C      ID(2)=PLANE FRAME
C      ID(3)=PLANE GRID
C      ID(4)=SPACE TRUSS
C      ID(5)=SPACE FRAME
C      Y=FLEXIBILITY OR STIFFNESS MATRICES
       DIMENSION U(36),IU(36),Y(6,6),T(6,6),V(6),W(6),FILL( 8),
      1 SYSFIL(27),PROFIL(6), KRAY(10),BL(4)
       COMMON  U,IU,Y,T,V,W,JLD,NLS,H,JT,J,K,JM,N,NS,L,BL,FILL,EGO,S,
      4CHECK,NMAX,INORM,ISOLV,ISCAN,IIII,IMOD,JJJJ,ICONT,ISUCC,IMERG,
      5TOP,N1,NL,NT,NREQ,TN,LFILE,TOLER,IPRG,IRST,IRLD,IRPR,SYSFIL,
      8NJ,NB,NDAT,ID,JF,NSQ,NCORD,IMETH,NLDS,NFJS,NSTV,NMEMV,IPSI,NMR,
      9NJR,ISODG,NDSQ,NDJ,IPDBP,IUDBP,NBB,NFJS1,JJC,JDC,JMIC,JMPC,JXX
       COMMON JEXTN,MEXTN,LEXTN,JLC,NLDSI,IYOUNG,ISHEAR,IEXPAN,IDENS,
      1 PROFIL,
      4NAME,KXYZ,KJREL,JPLS,JMIN,MTYP,KPSI,MEMB,LOADS,MODN,KS,KMKST,KSTDB
      5,KATKA,KPPLS,KPMNS,KUV,KPPRI,KR,KSAVE,KRAY,KDIAG,KOFDG,
      6 IOFDG,LOADN,MEGAO,JEXT,JINT,KUDBP,KMEGA,KPDBP,JTYP,MTYP1,KB
      7,MLOAD,JLOAD,KATR,LEXT,KYOUNG,KSHEAR,KEXPAN,KDENS
       EQUIVALENCE (U,IU,Y)
       CALL ALOCAT (KXYZ)
       CALL ALOCAT (JPLS)
       CALL ALOCAT (JMIN)
       CALL ALOCAT (MTYP)
       CALL ALOCAT (KS)
       CALL ALOCAT(KMKST,0)
       H=0.0
       IF (IYOUNG-1) 709,400,703
  709 EGO = 1.0
       GO TO 410
  400 CALL ALOCAT(KYOUNG)
C
C TEST FOR MOD OF EL OR SHEAR MOD NOT SPECIFIED
C
       JS=0
       L=KYOUNG
  413 NS=0
  406 CONTINUE
       DO 401  JM=1,MEXTN
       I=L+JM
       IF(U(I))403,403,404
  403 IF(H)401,401,405
  404 H=U(I)
       IF(NS)408,408,401
  408 NS=1
       GO TO 406
```

```
 405   U(I)=H
       CALL PRER2(14,JM,H)
 401   CONTINUE
       IF(JS)410,410,430
 703 EGO = U(191)
 410 IF (ISHEAR-1) 430,420,430
 420 CALL ALOCAT(KSHEAR)
     H=0.0
     JS=1
     L=KSHEAR
     GO TO 413
 430 CALL ALOCAT (MEMB,0)
 1   DO 3000 JM=1,MEXTN
C       TEST IF JM TH MEMBER IS PRESENT IN THIS MODIFICATION
C       UNPACK MTYP AND NS
     IA=MTYP+JM
     CALL UPACW(U(IA),IX,IX,L,NS,IX)
     IF (L) 100,3000,100
 100 CALL ALOCAT (MEMB,JM)
     CALL ALOCAT(KMKST,JM)
     NS = NS/7
     JU=KMKST+JM
     JU=IU(JU)
     IK=MEMB+JM
     JT=IU(IK)
C       FIND JOINT (EXTERNAL) NUMBERS
     NP = JPLS+JM
     NM = JMIN+JM
     NP = IU(NP)
     NM = IU(NM)
C       COMPUTE LENGTH
C       STORE S
     KM = KXYZ + (NM-1)*3 + 1
     KP = KXYZ + (NP-1)*3 + 1
     JKS=KS+(JM-1)*(NCORD+1)+1
     U(JKS)=0.0
     DO 5  I=1,NCORD
     JKSS=JKS+I
     U(JKSS)=U(KM)-U(KP)
     KM = KM + 1
     KP = KP + 1
     U(JKS)=U(JKS)+U(JKSS)**2
 5     CONTINUE
     U(JKS)=SQRTF(U(JKS))
     S=U(JKS)
     DO 8   I=1,JF
     DO 8   J=1,JF
 8   Y(I,J)=0.
     KA = JT+1
     IF (IYOUNG-1) 720,702,720
 702 I = KYOUNG+JM
     EGO = U(I)
 720 IF (ISHEAR-1) 701,721,722
 701 H = 0.4
     GO TO 780
 721 I = KSHEAR+JM
     H = U(I)/EGO
     GO TO 780
 722 H = U(192)/EGO
 780 GO TO (10,20,21,15,900),L
```

300

```
C      PRISMATIC MEMBER
   10 NS = 1
C      SEPARATE TRUSSES FOR SHORT PROCESSING
      GO TO (38,15,15,38,15),ID
   15 STOP=S
      CALL MEMFOD(STOP,U(KA),1)
      GO TO 600
C      SEPARATE TRUSSES
   20 GO TO (40,24,24,40,24),ID
   21 GO TO (50,24,24,50,24),ID
   24 DO 25 I=1, JF
      DO 25 J=1, JF
      JS=JT+ JF*(I-1)+J
   25 Y(I,J)=U(JS)
      GO TO (900,710,650),L
C      TRUSSES WITH PRISMATIC BARS ONLY
C      FIND STIFFNESS DIRECTLY
   38 Y(1,1) = U(JT+1)*EGO/U(JKS)
      GO TO 710
   40 Y(1,1)=U(JT+1)
      GO TO 710
   50 Y(1,1)=1.0/U(JT+1)
      GO TO 710
C      INVERT THE FLEXIBILITY MATRIX
  600 GO TO (620,650,650,620,650),ID
  620 Y(1,1)=1./Y(1,1)
      GO TO 710
  650 DO 700 I=1, JF
      DO 700 J=1, JF
      T(I,J)=0.
  700 T(I,I)=1.
      A=1.
      M1=XSIMEQF(6,JF,JF,Y,T,A,V)
      GO TO (710,740,730),M1
C      INVERSION OF FLEXIBILITY/STIFFNESS IS SINGULAR
  730 CALL PRER2(1,JM,0.0)
      GO TO 745
C      INVERSION OF FLEXIBILITY/STIFFNESS IS UNDER/OVER FLOW
  740 CALL PRER2(2,JM,0.0)
  745 ISUCC=2
      GO TO 900
C      STORE K*
  710 DO 715  I=1,JF
      DO 715   J=1,JF
      JX=JU+JF*(I-1)+J
  715 U(JX)=Y(I,J)
  900 CALL RELEAS (MEMB,JM)
      CALL RELEAS(KMKST,JM)
 3000 CONTINUE
      CALL RELEAS(KYOUNG)
      CALL RELEAS(KSHEAR)
      CALL RELEAS (KXYZ)
      CALL RELEAS (JPLS)
      CALL RELEAS (JMIN)
      CALL RELEAS (MTYP)
      CALL RELEAS (KS)
      CALL RELEAS(KMKST,0)
      CALL RELEAS (MEMB,0)
      RETURN
      END
```

```
*      LIST8
*      SYMBOL TABLE
*      LABEL
       SUBROUTINE MRELES
C      S.MAUCH        STUCT. ANALYSIS PROG.  RELEASES
C      THIS ROUTINE MODIFIES KSTAR TO ACCOUNT FOR FORCE CONSTRAINTS.
C   NMR=NUMBER OF MEMBER RELEASES SPECIFIED
C   ASSUMES THAT CARRY OVER MATRIX OF THE MEMBER INVOLVES ONLY THE LENGTH
C   IUR=CALLING ORIGIN(MEMBER OR JOINT RELEASE)
       DIMENSION  Y(6,6),T(6,6),Q(6,6),U(36),IU(36),SYPA(40),FILL(6)
       COMMON U,T,Q,CHECK,NMAX,INORM,ISOLVE,ISCAN,IIII,IMOD,JJJJ,ICONT,
      1ISUCC,SYPA,NJ,NB,NDAT,ID,JF,NSQ,NCORD,IMETH,NLDS,NFJS,NSTV
      2,NMEMV,IPSI,NMR,NJR,ISODG,NDSQ,NDJ,SDJ,NPR,NBB,NFJS1,JJC,JDC,
      3JMIC,JMPC,JLD,JEXTN,MEXTN,LEXTN,JLC,NLDSI,IYOUNG,ISHER,IEXPAN,
      4IDENS,FILL,
      5NAME,KXYZ,KJREL,JPLS,JMIN,MTYP,KPSI,MEMB,
      6LOADS, INPUT,KS,KMKST,KSTDB,KATKA,KPPLS,KPMIN,KUV,KPPRI,KR,KMK,
      7KV,K33,LA2R,LA2RT,NV   ,NTP  ,NSCR7,NSCR8,NSCR9,NSCR10,KDIAG,KOFD
      8G,IOFDG,LDNM,MEGAO,JEXT,JINT,KUDBP,KMEGA,KPDBP,JTYP,MTYP1,KB,MLOAD
      9,JLOAD,KATR,LEXT,KYOUNG,KSHER,KEXPAN,KDENS
       EQUIVALENCE(U(1),IU(1),Y(1))
       CALL ALOCAT(MTYP1)
       CALL DEFINE(NTP,36,0,1,1)
       CALL DEFINE(NV,6,0,1,1)
       CALL ALOCAT(NTP)
       CALL ALOCAT(NV)
       CALL ALOCAT(KS)
       CALL ALOCAT(MTYP)
       NPR=0
       CALL DEFINE(KMEGA,0,MEXTN,1,C,1)
       CALL ALOCAT(KMEGA,0)
       NMRR=NMR
       DO 6000 JM=1,MEXTN
C CLEAR DECREMENT OF MTYP1 (KMEGA REFERENCE FOR + MRELES)
       LMREL=MTYP1+JM
B      U(LMREL)=U(LMREL)*700000777777
       LMREL=MTYP +JM
       MREL=IU(LMREL)
B 95   BREL=(BREL*077777000000)
       MM=MREL
       EQUIVALENCE(BREL,MREL)
       IF(MREL)6000,6000,93
   93  MRELSV=MREL
   92  KMKS=KS+(JM-1)*(NCORD+1)+1
  700  SDJ=U(KMKS)
       SS=SDJ
  701  CALL ALOCAT(KMKST,JM)
       IE=KMKST+JM
       KMSTP=IU(IE)
       CALL COPY(Y,Y,JF,1,KMSTP)
  410  ITR=0
B 55   BBL1=BREL*000077000000
   39  IF(BBL1) 38,262,38
C
C     BRANCH TO 262 MEANS RELEASE IS ONLY AT + END
C   NRE IS USED TO TEST FOR TOTAL RELEASE
C   NRE=1 MEANS ONLY PLUS END REL.
C   NRE=-1MEANS ONLY MINUS END REL.
C   NRE=0 MEANS PLUS AND MINUS END REL.
C
```

```
 262   NRE=1
       GO TO 31
  38   IE=0
B      IF(BREL*0077C0000000)24,23,24
  23   NRE=(-1)
C  23 MEANS RELEASE IS ONLY AT - END
       GO TO 31
  24   NRE=0
  31   IF(ID-4)22, 8,22
  22   IF(ID-1)56, 8,56
C IF A TRUSS MEMBER IS OVER RELEASED STIFFNESS IS JUST SET ZERO NO FATAL MISTAKE
   8   GO TO 11
C
C   IF TRUSS IS RELEASED STIFFNESS IS SET TO ZERO  KMEGAO IS NOT SET UP
C THE CORRESPONDING CHECK IS MADE IN MREC
C IF BOTH ENDS OF A TRUSS ARE RELEASED MEESAGE IS PRINTED
C
  56   IF(NRE)3,3,260
  11   CALL COPY(Y,Y,JF,0,0)
C KSTAR IS ZERO
       GO TO 510
   4   CALL CARRY(ID,JF,SS)
  10   CALL MAMUL(Y,T,T,JF,JF,0)
C     KCSTAR IS NOW IN Y
 500   IF(ITR)3,3,510
C ITR=1 MEANS + END IS CONSIDERED AND BOTH ENDS ARE REL.
C     SET UP PERMUTATION MATRIX IN T
   3   IC=MREL
       IDD=ID
       CALL PERMUT(IC,JF,IDD)
 450   IC=IC
       IF(IDD)501,50,501
C  IF IDD IS NON-ZERO LAMBDA IS UNITY
C  IC IS RETURNED FROM PERMUT
  50   CALL MAMUL(Y,T,T,JF,JF,0)
C     NEXT STEP  PARTITIONING AND IMPOSE CONSTRAINT.
C     KCPSTAR IS NOW IN Y
 501   JC=JF-IC
       NMRR=NMRR-JC
       IF(JF-JC)281,281,280
C
C     JC=NUMBER OF RELEASES
C  281 MEANS THERE IS A TOTAL RELEASE
C  11 MEANS TOTAL RELEASE IS AT PLUS OR AT MINUS END.BUT NO RELEASE AT OTHER END
C
 281   IF(NRE)11,230,111
 111   CALL FIXM(0,0,JM)
C  NON-TRUSS MEMBER WITH TOTAL RELEAS AT START
       GO TO 11
 230   ISU=5
       GO TO 623
B614   BOL=466525512643
 630   PRINT 620,KOUT,BOL
 620   FORMAT(I10,A8)
       GO TO 600
 623   ISUCC=2
       CALL PRER2(ISU,JM,MRELSV)
       GO TO 600
 280   IF(JC-1)200,200,201
 201   DO 211 I=1,JC
```

```
         IP=I+IC
         DO 211 J=1,JC
         JP=J+IC
         IN=NTP+I+(J-1)*6
  211    U(IN)=Y(IP,JP)
         CALL COPY(Q,Q,JF,-2,0)
 2100    C=1.0
         JQ=JC
 2111    CALL BUGER(6,JQ,JQ,NTP,72,C,NV)
C        PULL KCPSTAR22 OUT OF KCPSTAR AND INVERT IT.
         KOUT=2111
         GO TO (203,614,230),JQ
  200    YJFA=Y(JF,JF)
         IF(YJFA)232,230,232
  232    IN=NTP+1
         U(IN)=1./YJFA
C        FORM KCPST12*KCPST22(-1)*KCPST21
C    KCPST22(-1) IS IN TP
  203    DO 2160 L=1,JF
         DO 2160J=1,JF
         IF(IC-L)219,218,218
  218    IF(IC-J)219,217,217
  219    Q(L,J)=0.0
         GO TO 2160
  217    Q(L,J)=Y(L,J)
         DO 216 I=1,JC
         DO 216 K=1,JC
         IIC=I+IC
         KIC=K+IC
         IN=NTP+I+(K-1)*6
  216    Q(L,J)=Q(L,J)-Y(L,IIC)*U(IN)*Y(KIC,J)
 2160    CONTINUE
C        REPLACE RESULT INTO Y
         IF(NRE+1)240,220,240
  240    IF(IE-1)220,241,220
  241    CALL FIXM(IC,MREL,JM)
  220    CALL COPY(Y,Q,JF,1,0)
C  220 COPIES RESULT OF 2160 INTO Q
  222    S=S
         IDD=ID
         IN =MREL
         CALL PERMUT(IN,JF,IDD)
  229    IF(IDD)506,221,506
C        KCPPSTTILDE IS NOW IN Y. UNPERMUTATE.
  221    CALL MAMUL(Y,T,T,JF,JF,1)
C
C    END OF SEPAR
C
  506    IF(IE)260,260,250
  260    IE=1
         MREL=MREL/64
         IF(MREL)4,510,4
  250    SS=-SDJ
         ITR=1
C  ITR=1 SIGNIFIES THAT BACK TRANSFORMATION  TO PLUS START IS TO BE MADE
C        250 INVERTS TCSTAR FOR BACKTRANS FORMATION.
         GO TO 4
  510    ISU=KMKST+JM
         KMSTV=IU(ISU)
  515    CALL COPY(Y,Y,JF,2,KMSTV)
```

```
          CALL RELEAS(KMKST,JM)
    600   IF(NMRR)6001,6001,6000
    6000  CONTINUE
    6001  CALL RELEAS(NTP,0,0)
          CALL RELEAS(NV,0,0)
          IF(NPR)6003,6003,6002
    6002  CALL RELEAS(KMEGA,0)
          CALL RELEAS(MTYP1)
    6003  RETURN
          END
*         LIST8
*         LABEL
*         SYMBOL TABLE
          SUBROUTINE MEMFOD (STOP,SP,IFOD)
C         LEON R. WANG   RM 1-255,  EXT. 2117
C         STRESS PROGRAM...STRUCTURAL ENGINEERING SYSTEMS SOLVER
C         APRIL 16, 1963
C         MAY 7, 1963
C         VERSION III...27 AUGUST 1963...KEN REINSCHMIDT
C         CORRECTION FOR REARRANGING DATA STORAGE ALSO INCLUDING SHEAR
C         DEFORMATION
C         THIS SUBROUTINE COMPUTES F* OF A NONPRISMATIC STRAIGHT MEMBER
C         AS WELL AS PRISMATIC MEMBER
C         STOP=LENGTH OF MEMBER
C         THIS SUBROUTINE CAN ALSO COMPUTE  CANTILEVER DEFLECTION AT RIGHT
C         END DUE TO A CONCENTRATED LOAD
C         STOP=DISTANCE OF APPLIED LOAD FROM LEFT END
C         DESCRIPTION OF NOTATIONS
C         JM=NAME OF A MEMBER
C         Y=FLEXIBILITY MATRICES
C         ID=TYPE OF STRUCTURE
C         ID(1)=PLANE TRUSS
C         ID(2)=PLANE FRAME
C         ID(3)=PLANE GRID
C         ID(4)=SPACE TRUSS
C         ID(5)=SPACE FRAME
C         SECTION PROPERTIES ARE IN SP,NOT COMPACTED.
C         MTYP=TYPE OF MEMBER GIVEN
C         MTYP(1)=SECTION PROPERTIES GIVEN
C         MTYP(2)=LOCAL STIFFNESS OF MEMBER GIVEN
C         MTYP(3)=LOCAL FLEXIBILITY OF MEMBER GIVEN
C         MTYP(4)=MEMBER OF STRAIGHT VARIABLE SECTION
C         MTYP(5)=COMPUTE STEEL SECTION PROPERTIES
C          JF=DEGREE OF FREEDOM
C         IFOD=1  COMPUTES FLEXIBILITY COEFF.
C         IFOD=2  COMPUTES DEFLECTION COEFF. UNDER CONC. LOAD
C         H = G/E, THE RATIO OF THE SHEAR MODULUS TO YOUNG,S MODULUS
          DIMENSION U(36),IU(36),Y(6,6),T(6,6),V(6),W(6),FILL( 8),SP(7,9),
         1 SYSFIL(27),PROFIL(6),  KRAY(10),BL(4)
          COMMON  U,IU,Y,T,V,W,JLD,NLS,H,JT,J,K,JM,N,NS,L,BL,FILL,EGO,S,
         4CHECK,NMAX,INORM,ISOLV,ISCAN,IIII,IMOD,JJJJ,ICONT,ISUCC,IMERG,
         5TOP,N1,NL,NT,NREQ,TN,LFILE,TOLER,IPRG,IRST,IRLD,IRPR,SYSFIL,
         8NJ,NB,NDAT,ID,JF,NSQ,NCORD,IMETH,NLDS,NFJS,NSTV,NMEMV,IPSI,NMR,
         9NJR,ISODG,NDSQ,NDJ,IPDBP,IUDBP,NBB,NFJS1,JJC,JDC,JMIC,JMPC,JXX
          COMMON JEXTN,MEXTN,LEXTN,JLC,NLDSI,IYOUNG,ISHEAR,IEXPAN,IDENS,
         1 PROFIL,
         4NAME,KXYZ,KJREL,JPLS,JMIN,MTYP,KPSI,MEMB,LOADS,MODN,KS,KMKST,KSTDB
         5,KATKA,KPPLS,KPMNS,KUV,KPPRI,KR,KSAVE,KRAY,KDIAG,KOFDG,
         6 IOFDG,LOADN,MEGAO,JEXT,JINT,KUDBP,KMEGA,KPDBP,JTYP,MTYP1,KB
         7,MLOAD,JLOAD,KATR,LEXT,KYOUNG,KSHEAR,KEXPAN,KDENS
```

```
      EQUIVALENCE (U, IU, Y)
      TOLER=0.02
 780  GO TO (10,30),IFOD
 10   SL=0.
      GO TO (30,600,600,18,600),L
 18   DO 22  I=1,NS
  22  SL=SL+SP(7,I)
      DI=S-SL
      IF (ABSF(DI)-TOLER*S)30,30,25
  25  CALL PRER2(3,JM,NS)
      SP(7,NS)=SP(7,NS)+DI
      CALL PRER2(11,NS,SP(7,NS))
 30   SLX=S
      DO 110  I=1,NS
      GO TO (31,600,600,32,600),L
 31   STEM=S
      GO TO 33
 32   STEM=SP(7,I)
  33  STSL=STOP-STEM
      IF (STSL )  35,38,38
 35   SLS=STOP
      GO TO 40
C     DEFLECTION COEFF. FOR ONE SEGMENT
 38   SLS=STEM
 40   S2=SLS**2
      S3=SLS*S2
      DO 41  I=1,JF
      DO  41  M=1,JF
 41   T(I,M)=0.
      GO TO (42,45,50,42,55),ID
C     PLANE FRAMES
  45  IF (SP(2,I))  46,46,48
  46  IF (SP(3,I)) 150,150,146
 146  SP(2,I) = SP(3,I)
      GO TO 48
 150  T(2,2)=S3/(3.*SP(6,I))
      GO TO 49
 48   T(2,2)=S3/(3.*SP(6,I)) +SLS/(H*SP(2,I))
 49   T(3,3)=SLS/SP(6,I)
      T(2,3) = S2/(2.*SP(6,I))
      T(3,2)=T(2,3)
C     TRUSSES PLANE OR SPACE
 42   T(1,1)=SLS/SP(1,I)
      GO TO 65
C     PLANE GRID
 50   IF (SP(3,I))  51,51,53
  51  IF (SP(2,I)) 153,153,152
 152  SP(3,I) = SP(2,I)
      GO TO 53
 153  IF (SP(1,I)) 160,160,155
 155  SP(3,I) = SP(1,I)
      GO TO  53
 160  T(1,1)=S3/(3.*SP(5,I))
      GO TO 54
 53   T(1,1)=S3/(3.*SP(5,I)) +SLS/(H*SP(3,I))
  54  IF (SP(4,I)) 162,162,180
 162  IF (SP(6,I)) 165,165,164
 164  SP(4,I) = SP(6,I)
      GO TO 180
 165  SP(4,I) = 0.0001*SP(5,I)
```

```
      180  T(2,2)=SLS/(H*SP(4,I))
           T(3,3)=SLS/SP(5,I)
           T(1,3) = -S2/(2.*SP(5,I))
           T(3,1)=T(1,3)
           GO TO 65
C          SPACE FRAME
       55  IF(SP(2,I))  56,56,58
       56  T(2,2)=S3/(3.*SP(6,I))
           GO TO 59
       58  T(2,2)=S3/(3.*SP(6,I)) +SLS/(H*SP(2,I))
       59  IF(SP(3,I))  60, 60, 62
       60  T(3,3)=S3/(3.*SP(5,I))
           GO TO 63
       62  T(3,3)=S3/(3.*SP(5,I)) +SLS/(H*SP(3,I))
       63  T(4,4)=SLS/(H*SP(4,I))
           T(5,5)= SLS/SP(5,I)
           T(6,6)=SLS/SP(6,I)
           T(2,6) = S2/(2.*SP(6,I))
           T(3,5) = -S2/(2.*SP(5,I))
           T(6,2) = T(2,6)
           T(5,3) = T(3,5)
           GO TO 42
       65  DO 66  I=1,JF
           DO 66  M=1,JF
       66  Y(I,M)=Y(I,M)+T(I,M)
           GO TO (67,68),IFOD
       67  IF (L-4) 400,68,400
C          TRANSFORMATION OF DEFLECTIONS TO RIGHT END
       68  DS=SLX-SLS
           IF(DS) 400,400,70
       70  GO TO (100,75,80,100,85),ID
       75  Y(2,2)=Y(2,2)+T(3,2)*DS
           Y(2,3)=Y(2,3)+T(3,3)*DS
           GO TO 90
       80  Y(1,1)=Y(1,1)-T(3,1)*DS
           Y(1,3)=Y(1,3)-T(3,3)*DS
           GO TO 90
       85  Y(2,2)=Y(2,2)+T(6,2)*DS
           Y(3,3)=Y(3,3)-T(5,3)*DS
           Y(2,6)=Y(2,6)+T(6,6)*DS
           Y(3,5)=Y(3,5)-T(5,5)*DS
C           DEFLECTION AT RIGHT END UNDER TRANSFORMED LOAD
       90  IF(STSL)  400,400,95
       95  GO TO (100,96,97,100,98),ID
       96  Y(2,2)=Y(2,2)+(T(2,3)+T(3,3)*DS)*STSL
           Y(3,2)=Y(3,2)+T(3,3)*STSL
           GO TO 100
       97  Y(1,1)=Y(1,1)-(T(1,3)-T(3,3)*DS)*STSL
           Y(3,1)=Y(3,1)-T(3,3)*STSL
           GO TO 100
       98  Y(2,2)=Y(2,2)+(T(2,6)+T(6,6)*DS)*STSL
           Y(3,3)=Y(3,3)-(T(3,5)-T(5,5)*DS)*STSL
           Y(6,2)=Y(6,2)+T(6,6)*STSL
           Y(5,3)=Y(5,3)-T(5,5)*STSL
      100  STOP=STSL
           SLX=DS
      110  CONTINUE
      400  DO 500 I = 1,JF
           DO 500 M = 1,JF
      500  Y(I,M) = Y(I,M)/EGO
```

```
 600 RETURN
     END
*    FAP
     LBL      BUGER
     COUNT    25
*   BUGER THE CALLING SEQUENCE FOR XSIMEQF
*      BUGGERS ARGUMENTS 4, 5, AND 7
     ENTRY    BUGER
BUGER SXA     BACK,1
      SXA     BACK+1,2
      SXA     BACK+2,4
      CAL     ARG3+1
      SUB     =204
      STA     TOP
      CLA     3,4
      STA     ARG
      CLS*    4,4
      ANA     =0077777000000
      ARS     18
      SSM
      ADD     TOP
      STA     ARG4
      CLS*    7,4
      ANA     =0077777000000
      ARS     18
      SSM
      ADD     TOP
      STA     ARG7
      CLS*    5,4
      ANA     =0077777000000
      ARS     18
      SSM
      ADD     TOP
      STA     ARG5
      CLA     1,4
      STA     ARG1
      CLA     3,4
      STA     ARG3
      CLA     6,4
      STA     ARG6
      CLA     2,4
      STA     ARG2
ARG1  CLA     *
ARG3  LDQ     *
      STQ     IN+204      INTO (TOP-2)
ARG4  LDQ     *
      STQ     IN+203      INTO TOP-3
ARG5  LDQ     *
      STQ     IN+202
ARG6  LDQ     *
      STQ     IN+201
ARG7  LDQ     *
      STQ     IN+200
ARG2  LDQ     *
      TSX     $XSIMEQ,4
ARG   STO     **      RETURN RESULT CODE IN THIRD ARG
BACK  AXT     *,1
      AXT     *,2
      AXT     *,4
      TRA     8,4
```

```
    TOP    PZE
     IN    COMMON  1
           END
  *      LABEL
  *      SYMBOL TABLE
         SUBROUTINE FIXM(IC,MREL,JM)
         DIMENSION  Y(6,6),T(6,6),Q(6,6),U(36),IU(36),SYPA(40),FILL(20)
         COMMON U,T,Q,CHECK,NMAX,INORM,ISOLVE,ISCAN,IIII,IMOD,JJJJ,ICONT,
        1ISUCC,SYPA,NJ,NB,NDAT,ID,JF,NSQ,NCORD,IMETH,NLDS,NFJS,NSTV
        2,NMEMV,IPSI,NMR,NJR,ISODG,NDSQ,NDJ,SDJ,NPR,NBB,NFJS1,FILL,
        3NAME,KXYZ,KJREL,JPLS,JMIN,MTYP,KPSI,MEMB,
        4LOADS, INPUT,KS,KMKST,KSTDB,KATKA,KPPLS,KPMIN,KUV,KPPRI,KR,KMK,
        5KV,K33,LA2R ,LA2RT,NV    ,NTP  ,NSCR7,NSCR8,NSCR9,NSCR10,KDIAG,KOFD
        6G,IOFDG,LDNM,MEGAO,JEXT,JINT,KUDBP,KMEGA,KPDBP,JTYP,MTYP1
         EQUIVALENCE(U(1),IU(1),Y(1))
C  FIRST MULTIPLY KP12*KP22(-1)
  241    NPR=NPR+1
         CALL DEFINE(NSCR7,36,0,1,1)
         CALL ALOCAT(NSCR7)
         CALL DEFINE( KMEGA,NPR,NSQ,0,0,1)
         CALL ALOCAT( KMEGA,NPR)
         JP=MTYP1+JM
         IU(JP)=NPR
C  ALTERATION OF MTYP1 IF A PLUS RELEAS IS DELETED IS NOT DONE IN MRELES
         IX=KMEGA+NPR
         CALL CLEAR(U(IX))
         IX=IU(IX)
         SDJM = -SDJ
  3      IF(IC)1,1,2
  1      CALL CARRY (ID,JF,SDJM)
         CALL COPY(T,T,JF,2,IX)
         GO TO 246
C  TOTAL RELEAS AT START OF NON TRUSS MEMBER
  2      JC=JF-IC
         DO 10  L=1,IC
         DO 10  J=1,JC
         IA=NSCR7+(L-1)*6+J
         U(IA)=0.0
  217 DO 216  I=1,JC
         IIC=IC+I
         IN=NTP+J+(I-1)*6
  216    U(IA)=U(IA)+Y(L,IIC)*U(IN)
  10     CONTINUE
C  KP12*KP22(-1) FOR JM IS NOW IN NSCR7
C  NEXT LAM1(T)*NSCR7*LAM2
         DO 20  I=1,JF
         DO 20  N=1,JF
         Y(I,N)=0.0
         DO 15  M=1,JC
         L=IC+M
         DO 15  K=1,IC
         IA=NSCR7+(K-1)*6 +M
  15     Y(I,N)=Y(I,N)+T(K,I)*U(IA)*T(L,N)
  20     CONTINUE
C  ADD LAM1*LAM1T=IPRIME
         DO 30  I=1,JF
         I=I
         MM=MREL
         CALL UNPCK(MM,I,ID)
 1000 IF(MM)30,30,29
```

```
  29   Y(I,I)=Y(I,I)+1.0
  30   CONTINUE
C  OMEGA IS NOW IN Y MUST BE MULTIPLIED BY CARRY OVER
       CALL CARRY (ID,JF,SDJM)
       CALL MAMUL(Y,T,T,JF,JF,-1)
C  T*OMEGA IS NOW IN Y
 240   CALL COPY(Y,Y,JF,2,IX)
C  CAN NOW MODIFY CANTILEVER FORCES
 245 CALL RELEAS(NSCR7)
 246 CALL RELEAS(KMEGA,NPR)
     RETURN
     END
*    LIST8
*    LABEL
     SUBROUTINE CARRY(ID,JF,SS)
     DIMENSION Y(36),T(36)
     COMMON Y,T
     DO 100 I=2,18
 100 T(I)=0.
     T(1)=1.
     T(8)=1.
     T(15)=1.
     GO TO (10,5,6,10,7),ID
   5 T(14)=SS
     GO TO 10
   6 T(13)=-SS
     GO TO 10
   7 DO 20 I=19,35
  20 T(I)=0.
     T(22)=1.
     T(29)=1.
     T(36)=1.
     T(32)=SS
     T(27)=-SS
  10 RETURN
     END
*    LIST8
*    LABEL
*    SYMBOL TABLE
     SUBROUTINE COPY(A,B,JF,IX,ICM)
     DIMENSION A(6,6),B(6,6),U(200),IU(200)
     COMMON U,IU
     EQUIVALENCE(U,IU)
C    THIS SUBROUTINE COPIES B OR U INTO A. OR PUTS ZEROS INTO A.
C    ICM POS MEANS B IS TO BE FOUND IN U. ICM IS STARTING LOCATION.
C    IX=-2 FILL A WITH UNIT MATRIX.
C    IX=-1 ADD A TO U
C    IX=0 FILL A WITH ZEROS
C  IX=1  FILL A WITH U OR B
C  IX=2  FILL U WITH A
     XLOCF(N1,N2,N3)=N1+N2+(N3-1)*JF
     IXX=IX+3
     GO TO (1,2,1,15,5),IXX
   1 DO 40 I=1,JF
     DO 3 J=1,JF
   3 A(I,J)=0.0
     IF(IX)31,40,40
  31 A(I,I)=1.0
  40 CONTINUE
     GO TO 20
```

```
15      IF(ICM)16,16,17
17      ITES=1
        GO TO 7
16      ITES=0
        GO TO 7
 5      DO 6   I=1,JF
        DO 6   J=1,JF
        IJK=XLOCF(ICM,I,J)
 6      U(IJK)=A(I,J)
        GO TO 20
 2      ITES=-1
 7      DO 30 I=1,JF
        DO 30 J=1,JF
        IF(ITES)11,9,12
11      IJK=XLOCF(ICM,I,J)
        U(IJK)=A(I,J)+U(IJK)
        GO TO 30
 9      CKK=B(I,J)
        GO TO 10
12      IJK=XLOCF(ICM,I,J)
        CKK=U(IJK)
10      A(I,J)=CKK
30      CONTINUE
20      RETURN
        END
*       LIST8
*       LABEL
*       SYMBOL TABLE
        SUBROUTINE  MAMUL(Y,T,A,JS,JT,JJ)
C DOES MULTIPLICATION  T*Y*A
C   JS IS SIZE OF SQUARE ARRAY
C   JJ=0  TRANSPOSE A. TRIPLE PRODUCT
C   JJ=1  TRANSPOSE T. TRIPLE PRODUCT
C   JJ=-1 ONLY DOUBLE PROD.T IS NOT TRANSPOSED. Y IS POST MULT.
C   A IS POST MULTIPLIER
C   T IS PRE MULTIPLIER
        DIMENSION  Q(6,6),T(6,6),Y(6,6),A(6,6)
        IF(JJ)4,9,9
 9      DO 1   I=1,JT
        DO 1   J=1,JT
        Q(I,J)=0.0
        DO 1   N=1,JS
        DO 1   K=1,JS
        IF(JJ)3,3,6
 3      CF=T(I,K)
        CB=A(J,N)
        GO TO 1
 6      CF=T(K,I)
        CB=A(N,J)
 1      Q(I,J)=Q(I,J)+CF*Y(K,N)*CB
        GO TO 100
 4      DO 40  I=1,JT
        DO 40  N=1,JS
        Q(I,N)=0.0
        DO 40  K=1,JS
40      Q(I,N)=Q(I,N)+T(I,K)*Y(K,N)
100     CALL COPY(Y,Q,JT,1,0)
15      RETURN
        END
*       LIST8
```

```
*       LABEL
*       SYMBOL TABLE
        SUBROUTINE PERMUT(MREL,JF,ID)
C       THIS SUBROUTINE SETS UP PERMUTATION MATRIX IN T
        DIMENSION Y(6,6),T(6,6),U(36),IU(36)
        COMMON U,T
        EQUIVALENCE(U(1),IU(1),Y(1))
C       THIS SUBROUTINE SETS UP PERMUTATION MATRIX IN T
        CALL COPY(T,T,JF,0,0)
  1     IC=JF
        DO 101 I=1,JF
        I=I
  2     MM=MREL
        CALL UNPCK(MM,I,ID)
        IF(MM)30,30,20
C       MM POS. MEANS RELEASE I IS PRESENT.
 20     T(IC,I)=1.0
        IF(IC-I)4,3,4
    4   ID=0
  3     IC=IC-1
        GO TO 101
 30     II=IC-JF+I
C  21 CHECKS IF LAMBDA IS UNITY
 21     IF(II-I)24,25,24
 24     ID=0
 25     T(II,I)=1.0
101     CONTINUE
 28     MREL=IC
        RETURN
        END
*       FAP
        COUNT   60
        LBL     UNPCK
        ENTRY   UNPCK
UNPCK   CLA     1,4
        STA     MREL
        CLA     2,4
        STA     INDEX
        STA     INDX
        CLA     3,4
        STA     JFR
INDX    CLA     **
        ARS     18
        SUB     =1
        TZE     JFR
        TPL     INDEX           NO CONDENSING
JFR     CLA     **
        ARS     18
        SUB     =3
        TZE     GRID
        ADD     =1
        TZE     PLNFR           PLANE FRAME
        CLA*    1,4             SPACE FRAME
        ARS     18
        STO     MRCON
        TRA     INDEX
GRID    CLA     MSKG1           MASK1 GRID
        ANA*    1,4
        ARS     20
        STO     MRCON           MREL CONDENSED
```

312

```
        TRA     INDEX
PLNFR   CLA     MSKF1           MASK1 PLANE FRAME
        ANA*    1,4
        ARS     3
        STO     TMPFR
        CLA     MSKF2
        ANA*    1,4
        ADD     TMPFR
        ARS     18
        STO     MRCCN
INDEX   CLA     **              ACTUAL DECODING OF MREL COND.
        ARS     18
        SUB     =1
        STA     *+2
        CLA     =1
        ALS     **
        ANA     MRCCN
MREL    STO     **
        TRA     4,4
TMPFR   PZE
TEMPG   PZE
MSKG1   OCT     000034000000
MSKG2   OCT     003400000000
MSKF1   OCT     000040000000
MSKF2   OCT     000003000000
MRCON   PZE
* THIS SUBROUTINE CONDENSES THE
* CODE WORD SPECIFYING THE RELEASE
* (ACCORDING TO TYPE OF STRUCTURE)
* AND DECODES THE CONDENSED WORD.
* IT IS CALLED BY SUBROUTINE
* MRELES.
        END
*       LIST8
*       LABEL
CMAIN3/6
        DIMENSION Y(6,6),T(6,6),Q(6,6),U(36),IU(36),SYPA(40),FILL(6)
        COMMON U,T,Q,CHECK,NMAX,INORM,ISOLV,ISCAN,IIII,IMOD,ILINK,ICCNT,
     1ISUCC,SYPA,NJ,NB,NDAT,ID,JF,NSQ,NCORD,IMETH,NLDS,NFJS,NSTV
     2,NMEMV,IPSI,NMR,NJR,ISODG,NDSQ,NDJ,SCJ,NPR,NBB,NFJS1,JJC,JDC,
     3JMIC,JMPC,JLD,JEXTN,MEXTN,LEXTN,JLC,NLDSI,IYOUNG,ISHER,IEXPAN,
     4IDENS,FILL,
     5NAME,KXYZ,KJREL,JPLS,JMIN,MTYP,KPSI,MEMB,
     6LOADS, INPUT,KS,KMKST,KSTDB,KATKA,KPPLS,KPMIN,KUV,KPPRI,KR,KMK,
     7KV,K33,LA2R ,LA2RT,NV    ,NTP  ,NSCR7,NSCR8,NSCR9,NSCR10,KDIAG,KOFD
     8G,IOFDG,LDNM,MEGAO,JEXT,JINT,KUDBP,KMEGA,KPDBP,JTYP,MTYP1,KB,MLOAD
     9,JLOAD,KATR,LEXT,KYOUNG,KSHER,KEXPAN,KDENS
        EQUIVALENCE(U(1),IU(1),Y(1))
        CALL LOADPC
        ISUCC=ISUCC
        GO TO (1,2),ISUCC
1       ISOLV=5
        CALL CHAIN(4,A4)
2       CALL CHAIN(1,A4)
        END
*       LIST8
*       LABEL
        SUBROUTINE FOMOD
C  THIS ROUTINE MODIFIES JOINT LOAD VECTORS IF THERE ARE FREE JOINTS
C  THIS ROUTINE SOLVES FOR JOINT DISPLACEMENTS IF THERE ARE NO FREE JOINTS
```

```
      DIMENSION  Y(6,6),T(6,6),Q(6,6),U(36),IU(36),SYPA(40),FILL(6)
      COMMON U,T,Q,CHECK,NMAX,INORM,ISOLVE,ISCAN,IIII,IMOD,JJJJ,ICONT,
     1ISUCC,SYPA,NJ,NB,NDAT,ID,JF,NSQ,NCORD,IMETH,NLDS,NFJS,NSTV
     2,NMEMV,IPSI,NMR,NJR,ISODG,NDSQ,NDJ,IPB,IUPB,NBB,NFJS1
     2,JJC,JDC,JMIC,JMPC,JLD,JEXTN,MEXTN,LEXTN,JLC,NLDSI,IYOUNG,ISHEAR,
     2IEXPAN,IDENS,FILL,
     3NAME,KXYZ,KJREL,JPLS,JMIN,MTYP,KPSI,MEMB,
     4LOADS, INPUT,KS,KMKST,KSTDB,KATKA,KPPLS,KPMIN,KUV,KPPRI,KR,KMK,
     5KV,K33,LA2R ,LA2RT,NSCR5,NSCR6,NSCR7,NSCR8,NSCR9,NSCR10,KDIAG,KOFD
     6G,IOFDG,LDNM,MEGAO,JINT,JEXT,IFDT,KMEGA,KPDBP,JTYP,MTYP1,IOFC
      EQUIVALENCE(U(1),IU(1),Y(1))
      CALL ALOCAT(KPDBP)
      CALL ALOCAT(KPPRI)
      CALL DEFINE(NSCR5,NDJ,0,1,0)
      CALL CLEAR(NSCR5)
    9 DO 200 L=1,NLDSI
      JX=(L-1)*NDJ
      IB=KPDBP+JX
      DO 8 I=1,NDJ
      IA=IB+I
      IF(IU(IA))160,8,160
    8 CONTINUE
C THIS LOADING CONDITION HAS NO FIXED END FORVES ON SUPPORTS
      GO TO 200
  160 CALL ALOCAT(NSCR5)
C
C     L=INDEX FOR LOAD CONDITION
C  OMEGAO*KPDBP FOR L TH LOAD COND  INTO NSCR5
C
      DO 171  I=1,NDJ
      II=(I-1)/JF+1
C WILL ALOCAT EACH MEGAO JF TIMES   SO WHAT
      N3=MEGAO+II
C  TEST IF MEGAO(II) IS NOT ALL ZERO (NOT DEFINED)
      IF(IU(N3))10,171,10
   10 CALL ALOCAT(MEGAO,II)
      N3=MEGAO+II
      IB=NSCR5+I
      U(IB)=0.0
      NIN=IU(N3)+(I-1-(II-1)*JF)*NDJ
      IAP=KPDBP+JX
      DO 170  K=1,NDJ
      IC=NIN+K
      IA=IAP+K
  170 U(IB)=U(IB)-U(IC)*U(IA)
      CALL RELEAS(MEGAO,II)
  171 CONTINUE
C
C OMEGA*KPDBP INTO KPDBP IN ALL CASES (=DISPLACEMENTS IF NFJS=0)
C
  174 DO 172 I=1,NDJ
      IA=IAP+I
      IB=NSCR5+I
  172 U(IA)=-U(IB)
  175 IF(NFJS)173,200,173
C FREE JOINT FORCE MODIFICATION DUE TO SUPPORT RELEASE  (ATKA//*NSCR5)
C SUBSCRIPTS OF PROD ARE    (I,J)*(J)
  173 JX=(L-1)*NSTV
      DO 110 J=NFJS1,NJ
      IA=IOFDG+J
```

```
      IA=IU(IA)+1
      CALL UPADP(IU(IA),D,LCU,LDE)
C LCU IS NUMBER OF ENTRIES FOR KOFDG IN ROW J ( 1 LESS THAN  'CUURENT LENGTH OF)
C   I IS COLUMN ORDER
      IF(LCU)1,110,1
1     CALL ALOCAT(IOFDG,J)
      DO 100 IM=1,LCU
      IA=IOFDG+J
      IB=IU(IA)+1+IM
      CALL UPADP(IU(IB),D,I,IA)
      IF(I-NFJS)6,6,100
6     CALL ALOCAT(KOFDG,IA)
      IASV=KOFDG+IA
      N2=IU(IASV)
5     N3=NSCR5+(J-NFJS1)*JF
      NRES=JX+(I-1)*JF+KPPRI
      CALL MAPROD(0,N2,N3,NRES,1,JF,4)
100   CALL RELEAS(KOFDG,IA)
110   CALL RELEAS(IOFDG,J)
200   CONTINUE
201   I=I
      CALL RELEAS(NSCR5,0,0)
555   CALL RELEAS(KPDBP)
      RETURN
      END
*     LIST8
*     LABEL
*     SYMBOL TABLE
      SUBROUTINE LOADPC
C     STRESS PROGRAM...STRUCTURAL ENGINEERING SYSTEMS SOLVER
C     REVISION TO CHANGE TO VERSION 3      AUG 28 63    S M
C     LOAD PROCESSING BY JOINT AND MEMBER
C     JLD=CURRENT LOADING CONTION
C     INDEX FOR LOADING TYPE
C     NDEX(1)=JOINT LOAD
C     NDEX(2)=MEMBER LOAD K=1 CONCENTRATED K=2 UNIFORM K=3 LINEAR
C     NDEX(3)=MEMBER END LOAD J=0 FOR RIGHT END J=1 FOR LEFT END
C     NDEX(4)=MEMBER DISTORTION
C     NDEX(5)=JOINT DISPLACEMENT
C     UPON CHANGE OF STRUCTURE TYPE ALL UNUSED FIELDS IN LOAD BLOCKS
C     ARE ASSUMED 0
C     H IS=G/E
      DIMENSION Y(6,6),T(6,6),PL(6),PR(6),FILL(11),
     1SYSFIL(32),PROFIL(6),U(36),IU(36),KSRTCH(9)
      COMMON U,IU,Y,T,PL,PR,JLD,NBL,H,JT,J,K,JM,N,FILL,IKI,K9,IM,EGO,S
     1,CHECK,NMAX,INORM,ISOLV,ISCAN,IIII,IMOD,JJJJ,ICONT,ISUCC,IMERG,
     2TOP,N1,NL,NT,NREC,TN,LFILE,SYSFIL,NJ,NB,NDAT,ID,JF,NSQ,NCORD,
     3IMETH,NLDS,NFJS,NSTV,NMEMV,IPSI,NMR,NJR,ISODG,NDSQ,NDJ,IXX,
     4IUDBP,NBB,NFJS1,JJC,JDC,JMIC,JMPC,JLT,JEXTN,MEXTN,LEXTN,JLC,NLDSI
     5,IYOUNG,ISHEAR,IEXPAN,IDENS,PROFIL,NAME,KXYZ,KJREL,
     6JPLS,JMIN,MTYP,KPSI,MEMB,LOADS,MODN,KS,KMKSI,KSTDB,KATKA,KPPLS,
     7KPMNS,KUV,KPPRI,KR,KSAVE,KWORK,KSRTCH,KDIAG,KOFDG,IOFDG,
     8LDNM,MEGAO,JEXT,JINT,KUDBP,KMEGA,KPDBP,JTYP,MTYP1,KB,MLOAD,JLOAD
     9,KATR,LINT,KYOUNG,KSHEAR,KEXPAN,KDENS
      EQUIVALENCE(U,IU,Y)
      IF(IYOUNG-1)5,3,5
3     CALL ALOCAT(KYOUNG)
5     IF(ISHEAR-1)4,1,4
1     CALL ALOCAT(KSHEAR)
4     CALL ALOCAT(KPPRI)
```

```
      CALL ALOCAT(KPPLS,0)
      CALL ALOCAT(KPMNS,0)
      CALL ALOCAT(MEMB,0)
      CALL ALOCAT(KUV,0)
      IF(IPSI)311,7,311
 311  CALL ALOCAT(KPSI)
 7    CALL ALOCAT(KMKST,0)
      CALL ALOCAT(KS)
      CALL ALOCAT(MTYP)
      CALL ALOCAT(MLOAD,0)
      IF(NMR)550,550,551
 551  CALL ALOCAT(KMEGA,0)
      CALL ALOCAT(MTYP1)
C     MAX 10 SEGMENTS PER MEMBER SP(7,10)
 550  CALL DEFINE(KWORK,36,0,0,1)
      CALL ALOCAT(KWORK)
      CALL ALOCAT(LINT)
      IF(NJR)8,8,6
 6    CALL DEFINE(KPDBP,NDJ*NLDS ,0,0,1)
      CALL ALOCAT(KPDBP)
C     CLEAR ARRAYS IF MODIFICATIONS
 8    IF(MODN)11,2,11
 11   DO 12 IM=1,NLDS
      JM=KPPLS+IM
      JT=KPMNS+IM
      NLS=KUV+IM
      CALL CLEAR(U(JM))
      CALL CLEAR(U(JT))
 12   CALL CLEAR(U(NLS))
      CALL CLEAR(KPPRI)
      CALL CLEAR(KPDBP)
 2    DO 101 IM=1,MEXTN
      NLS=MTYP+IM
      JM=IM
C     CHECK IF MEMBER DELETED
      IF(U(NLS))102,101,102
 102  IXX=MLOAD+IM
      CALL UPADP(IU(IXX),JT,JT,IXX)
      IF(IXX-2)105,101,105
B105  IF(U(NLS)*077700000000)104,103,104
C     CHECK IF MEMBER IS RELEASED AT START END.
 104  IXX=MTYP1+JM
      IXX=IU(IXX)
      CALL ALOCAT(KMEGA,IXX)
C     ALOCAT T*OMEGA FOR RELEASED MEMBER
 103  CALL ALOCAT(MLOAD,IM)
C     GET LENGTH
      NLS=KS+(JM-1)*(NCORD+1)+1
      S=U(NLS)
      NBL=MLOAD+IM
      NBL=IU(NBL)+2
      NBL=IU(NBL)
      IF(NBL)9,106,9
 9    JT=0
      CALL ALOCAT(KMKST,IM)
      ISAV=IM
      CALL LOADPS
      IM=ISAV
      CALL RELEAS(KMKST,IM)
 106  CALL RELEAS(MLOAD,IM)
```

```
101   CONTINUE
      CALL RELEAS(MTYP)
      CALL RELEAS(MLOAD,0)
      CALL RELEAS(MEMB,0)
      CALL RELEAS(KYOUNG)
      CALL RELEAS(KSHEAR)
      CALL ALOCAT(JTYP)
      CALL ALOCAT(JINT)
      CALL ALOCAT(JLOAD,0)
C     JM=EXT JOINT NO.
      DO 400 JM=1,JEXTN
C     GET INTERNAL JOINT NUMBER IM
      IM=JINT+JM
      IM=IU(IM)
C     CHECK IF JOINT IS DELETED
      JLD=JTYP+JM
      IF(IU(JLD))211,400,211
211   IXX=JLOAD+JM
      CALL UPADP(IU(IXX),JT,JT,IXX)
      IF(IXX-2)210,400,210
210   CALL ALOCAT(JLOAD,JM)
      NBL=JLOAD+JM
      NBL=IU(NBL)+2
      NBL=IU(NBL)
      IF(NBL)390,399,390
390   JT=1
      ISAV=JM
      CALL LOADPS
      JM=ISAV
399   CALL RELEAS(JLOAD,JM)
400   CONTINUE
      CALL RELEAS(KWORK,0,0)
      CALL RELEAS(KPPRI)
      CALL RELEAS(KPPLS,0)
      CALL RELEAS(KPMNS,0)
      CALL RELEAS(KUV,0)
      CALL RELEAS(KPSI)
      CALL RELEAS(KMKST,0)
      CALL RELEAS(KPDBP)
      CALL RELEAS(KS)
      CALL RELEAS(JLOAD,0)
      CALL RELEAS(JTYP)
      CALL RELEAS(LINT)
      CALL RELEAS(JINT)
589   IF(NMR)421,420,421
421   CALL RELEAS(KMEGA,0)
      CALL RELEAS(MTYP1)
420   RETURN
      END
*     LIST8
*     LABEL
*     SYMBOL TABLE
      SUBROUTINE LOADPS
C  THIS SUBROUTINE BRANCHES ON NDEX AND COMPACTS LOAD DATA FOR ONE LOAD
C     INTO PL (PL AND PR IF NDEX =3)  ALSO TRANSMITS NDEX TO LSTOR
      DIMENSION Y(6,6),T(6,6),PL(6),PR(6),FILL(14) ,
     1 SYSFIL(32), PROFIL(10),              U(36), IU(36)  ,KSRTCH(9)
      COMMON U,IU,Y,T,PL,PR,JLD,NLS,H   ,JT,J,K,JM,N,FILL,NDEX,S,
     1 CHECK,NMAX,INORM,ISOLV,ISCAN,IIII,IMOD,JJJJ,ICONT,ISUCC,IMERG,
     2 TOP,N1,NL,NT,NREQ,TN,LFILE,SYSFIL,NJ,NB,NDAT,ID,JF,NSQ,NCORD,
```

```
      3 IMETH, NLDS,NFJS,NSTV,NMEMV,IPSI,NMR,NJR,ISODG,NDSQ,NDJ,IPDBP,
      4 IUDBP,NBB,NFJS1,JJC,JDC,JMIC,JMPC,JLT,JEXTN,MEXTN,LEXTN,JLC,NLDSI
      5,PROFIL,NAME,KXYZ,KJREL,
      6 JPLS,JMIN,MTYP,KPSI,MEMB,LOADS,MODN,KS,KMKST,KSTDB,KATKA,KPPLS,
      7 KPMNS,KUV,KPPRI,KR,KSAVE,KWORK,KSRTCH,KDIAG,KOFDG,IOFDG,
      8LDNM,MEGAO,JEXT,JINT,KUDBP,KMEGA,KPDBP,JTYP,MTYP1,KB,MLOAD,JLOAD
      9,KATR,LINT,KYOUNG,KSHEAR,KEXPAN,KDENS
        EQUIVALENCE (U, IU, Y)
C       NBL FROM LOADPC SAVED, NAME HAS CHANGED
C       NEW NLS USED IN SUBSEQUENT ROUTINES.
        NBL=NLS
  144 IPL=3
C LOOP ON LOAD BLOCKS
  147 DO 150 I=1,NBL
        IF(JT)1,1,2
    1 IX=MLOAD+JM
        GO TO 3
    2 IX=JLOAD+JM
    3 NLS=IU(IX)+IPL+1
C CHECK IF LOAD COND DELETED
        IF(U(NLS-1))4,148,4
  148 IPL=IPL+IU(NLS)
        GO TO 3
    4 CALL UPADP(U(NLS-1),NDEX,JC,ILD)
C  UNPACK LOAD TYPE K
        CALL UPACW(U(NLS),A,N,J,K,B)
C  JC =LOAD COND  EXT.
   12   JLD=LINT+JC
        JLD=IU(JLD)
C  JLD = INERNAL LOADING NUMBER
C   NDEX=1, JOINT LOAD, DONT ALOCAT ALL THIS JUNK
        IF(NDEX-2)7,400,400
  400   CALL ALOCAT(KPPLS,JLD)
        CALL ALOCAT(KPMNS,JLD)
        CALL ALOCAT(KUV ,JLD)
        IF(JT)6,6,7
    6 IX=MLOAD+JM
        GO TO 8
    7 IX=JLOAD+JM
    8 NLS=IU(IX)+IPL+1
C  FOR MEMBER LOADS NDEX =2 LEAVE LOAD DATA IN MLOAD(IT DOES NOT DEPEND
C  ON STRUCTURE TYPE
        IF(NDEX-2)206,205,206
  205 CALL ALOCAT(MEMB,JM)
        NLS=MLOAD+JM
        NLS=IU(NLS)+IPL+1
        JC=MEMB+JM
        JC=IU(JC)
        GO TO 30
  206   NDEX3=NDEX-3
        GO TO (210,220,230,210,210),ID
C       TRUSSES AND SPACE FRAMES
  210   DO 211  I=1,JF
        IX=NLS+I
  211   PL(I)=U(IX)
  215   IF(NDEX3)213,212,213
  212   DO 214  I=1,JF
        IX=NLS+I+6
  214   PR(I)=-U(IX)
        GO TO 61
```

318

```
C   PLANE FRAME
 220   PL(1)=U(NLS+1)
       PL(2)=U(NLS+2)
       PL(3)=U(NLS+6)
       IF(NDEX3)213,222,213
 222   PR(1)=-U(NLS+7)
       PR(2)=-U(NLS+8)
       PR(3)=-U(NLS+12)
       GO TO 61
C   PLANE GRID
 230   DO 231  I=1,3
       IX=NLS+I+2
 231   PL(I)=U(IX)
       IF(NDEX3)213,232,213
 232   DO 233 I=1,3
       IX=NLS+I+8
 233   PR(I)=-U(IX)
       GO TO 61
 213   GO TO (10, 30, 61, 40, 50), NDEX
 10    CALL JTLOAD
       GO TO 149
 30    CALL MEMBLD(U(JC+1))
       CALL RELEAS(MEMB,JM)
       NDEX=2
C   NDEX(77307) RESTORED FOR LSTOR BECAUSE MEMBLD WIPES IT WITH EMD.
       GO TO 60
 40    CALL MDISTN
       GO TO 60
 50    CALL JDISPL
C      RESTORE NLS FOR LOADPS
       NLS=JLOAD+JM
       NLS=IU(NLS)+IPL+1
       GO TO 149
   61 DO 62 I= 1,JF
       SAVE=-PL(I)
       PL(I)=PR(I)
 62    PR(I)=SAVE
       CALL CASE2(72,78)
       GO TO 149
 60    CALL LSTOR
 149   IPL=IPL+N
       CALL RELEAS(KUV,JLD)
       CALL RELEAS(KPMNS,JLD)
 150   CALL RELEAS(KPPLS,JLD)
 300   RETURN
       END
 *     LIST8
 *     LABEL
       SUBROUTINE MDISTN
C      STRESS PROGRAM...STRUCTURAL ENGINEERING SYSTEMS SOLVER
C      VERSION III...28 AUGUST 1963
C      KENNETH F. REINSCHMIDT, ROOM 1-255, EXT. 2117
C      GETS COMPACTED MEMBER DISTORTIONS IN PL
C      STORES IN KUV
C      COMPUTES FIXED END FORCES IN PL AND PR FOR LSTOR AND CASE2
       DIMENSION Y(6,6),T(6,6), PL(6), PR(6),FILL(8),U(36),IU(36),
      1 SYSFIL(27),PROFIL(6), KRAY(10)
       COMMON U,IU,Y,T,PL,PR,JLD,NLS,H,JT,J,K,JM,N,NS,L,JX,JU,KUS,KUI,
      1 FILL,EGO,S,
      4CHECK,NMAX,INORM,ISOLV,ISCAN,IIII,IMOD,JJJJ,ICONT,ISUCC,IMERG,
```

```
      5TOP,N1,NL,NT,NREQ,TN,LFILE,TOLER,IPRG,IRST,IRLD,IRPR,SYSFIL,
      8NJ,NB,NDAT,ID,JF,NSQ,NCORD,IMETH,NLDS,NFJS,NSTV,NMEMV,IPSI,NMR,
      9NJR,ISODG,NDSQ,NDJ,IPDBP,IUDBP,NBB,NFJS1,JJC,JDC,JMIC,JMPC,JXX
       COMMON JEXTN,MEXTN,LEXTN,JLC,NLDSI,IYOUNG,ISHEAR,IEXPAN,IDENS,
      1 PROFIL,
      4NAME,KXYZ,KJREL,JPLS,JMIN,MTYP,KPSI,MEMB,LOADS,MODN,KS,KMKST,KSTDB
      5,KATKA,KPPLS,KPMNS,KUV,KPPRI,KR,KSAVE,KRAY,KDIAG,KOFDG,
      6 IOFDG,LOADN,MEGAO,JEXT,JINI,KUDBP,KMEGA,KPDBP,JTYP,MTYP1,KB
      7,MLOAD,JLOAD,KATR,LEXT,KYOUNG,KSHEAR,KEXPAN,KDENS
       EQUIVALENCE (U, IU, Y)
C     STORE MEMBER DISTORTION IN KUV
      KUS = KUV+JLD
      KUS = IU(KUS)+JF*(JM-1)
      DO 4  I=1,JF
      KUI=KUS+I
    4 U(KUI) = U(KUI)+PL(I)
C     MULTIPLY MEMBER DISTORTION BY STIFFNESS (K*)
      JU=KMKST+JM
      JU=IU(JU)
      DO 8  I=1,JF
      PR(I) = 0.0
      DO 8  M=1,JF
      JX = JU+(M-1)*JF+I
      PR(I) = PR(I)+U(JX)*PL(M)
    8 CONTINUE
C     PR NOW CONTAINS MINUS (RIGHT) END FORCE
C     TRANSLATE  MINUS (RIGHT) END FORCE TO PLUS (LEFT) END, CHANGING
C     SIGN FOR COMPATIBILITY WITH CASE2
      DO 20  I=1,JF
   20 PL(I)=-PR(I)
C     MEMBER LENGTH IS IN S...STORED BY LOADPC
      GO TO (300,270,280,300,290),ID
C     STRUCTURE IS A PLANE FRAME
  270 PL(3) = PL(3)+      S*PL(2)
      GO TO 300
C     STRUCTURE IS A PLANE GRID
  280 PL(3) = PL(3)-      S*PL(1)
      GO TO 300
C     STRUCTURE IS A SPACE FRAME
  290 PL(5) = PL(5)-      S*PL(3)
      PL(6) = PL(6)+      S*PL(2)
  300 RETURN
      END
*     LIST8
*     LABEL
*     SYMBOL TABLE
      SUBROUTINE MEMBLD(SP)
C  CALLED BY LOADPS IF NDEX=2
C     STRESS PROGRAM...STRUCTURAL ENGINEERING SYSTEMS SOLVER
C  VERSION III REVISION AUG 28        S   M
      DIMENSION Y(6,6), T(6,6), PL(5), PR(7), BL(4)  ,KSRTCH(9),
      1 SYSFIL(32), PROFIL(6),                U(36), IU(36),SP(7,10 )
      COMMON U,IU,Y,T,PL,PR,JLD,NLS,H,JT,J,K,JM,N,NS,L,BL,AA,I,BB,SB
      COMMON  A,B,C,D,EMD,S,
      1 CHECK,NMAX,INORM,ISOLV,ISCAN,IIII,IMOD,JJJJ,ICONT,ISUCC,IMERG,
      2 TOP,N1,NL,NT,NREQ,TN,LFILE,SYSFIL,NJ,NB,NDAT,ID,JF,NSQ,NCORD,
      3 IMETH, NLDS,NFJS,NSTV,NMEMV,IPSI,NMR,NJR,ISODG,NDSQ,NDJ,IPDBP,
      4 IUDBP,NBB,NFJS1,JJC,JDC,JMIC,JMPC,JLT,JEXTN,MEXTN,LEXTN,JLC,NLDSI
      5,IYOUNG,ISHEAR,IEXPAN,IDENS,PROFIL,NAME,KXYZ,KJREL,
      6 JPLS,JMIN,MTYP,KPSI,MEMB,LOADS,MODN,KS,KMKST,KSTDB,KATKA,KPPLS,
```

```
      7 KPMNS,KUV,KPPRI,KR,KSAVE,KWORK,KSRTCH,KDIAG,KOFDG,IOFDG,
      8LDNM,MEGAO,JEXT,JINT,KUDBP,KMEGA,KPDBP,JTYP,MTYP1,IOFC,MLOAD,JLOAD
      9,KATR,LINT,KYOUNG,KSHEAR,KEXPAN,KDENS
        EQUIVALENCE (U, IU, Y)
C       THIS SUBROUTINE COMPUTES THE DEFLECTION UNDER APPLIED LOAD
C       AND TRANSFERS TO THE END OF THE MEMBER AND SUMS DEFLECTIONS AND
C       LOADS
C       IFOD=2   COMPUTES DEFLECTION COEFF. UNDER CONC. LOAD
C       ARRANGEMENT OF KWORK
C       KWORK+1-6      STORES CANTILEVER FORCES ON LEFT END
C       KWORK+31-36 6STORES CANTILEVER DEFLECTIONS ON RIGHT END
C       UNPACK MTYP,RELEASE AND NS
        IA=MTYP+JM
        CALL UPACW(U(IA),IX,REL,L,NS,IX)
        NS=NS/7
        DO 3001   I=1,36
        KBI=KWORK+I
 3001 U(KBI)=0.
        GO TO (3101,3051,3051,3121,3101),L
C                       STIFFNESS OR FLEX GIVEN
 3051 CALL PRER2(12,JM,L)
        GO TO 8000
C  PRISMATIC MEMBER CR STEEL
 3101 NS=1
C  VARIABLE EI
C       PROCESS ONE LOAD
C  LOAD DATA STILL IN MLOAD(JM)     MOVE IT TO BL(4)
 3121 DO 3151 I=1,4
        JS=NLS+I
 3151 BL(I)=U(JS)
        KB=KWORK
        KF=KWORK+30
C       GET H, THE RATIO OF G TO E
        IF(IYOUNG-1)5,4,6
5       EMD=1.0
        GO TO 7
4       IK=KYOUNG+JM
        EMD=U(IK)
        GO TO 7
6       EMD=U(191)
7       IF(ISHEAR-1)8,9,10
8       H=0.4
        GO TO 2
9       IK=KSHEAR+JM
        H=U(IK)/EMD
        GO TO 2
10      H=U(192)/EMD
C   SECTION PROPERTIES ARE IN MEMB(JM)   =U(MEMB+JM)=SP
 2      DO 1  I=1,6
        DO 1  M=1,6
 1      Y(I,M)=0.
C       TRUNCATES LOAD DIRECTION J
        GO TO (26,16,20,26,26),ID
16      IF(J-6)  26,18,26
18      J=3
        GO TO 26
20      J=J-2
26      KBJ=KB+J
        GO TO (28,55,90),K
C  K=1  CONC LOAD, K=2 UNIFORM ,K=3  LINEAR
```

```
 28    STOP=BL(3)
       A=BL(1)
       B=BL(3)
        CALL STICLD
 30    CALL MEMFOD(STOP,SP,2)
       DO 35   I=1,JF
       KFI=KF+I
 35    U(KFI)=U(KFI)+Y(I,J)*A
       GO TO 5000
 55    A=BL(1)
       B=BL(1)
       C=BL(3)
       D=BL(4)
       GO TO 93
 90    A=BL(1)
       B=BL(2)
       C=BL(3)
       D=BL(4)
 93    AA=A
       CC=ABSF(C)
 96    IF(D)   97,97,98
 97    D=S
 98    CALL STICLD
C      COMPUTATION OF DEFLECTIONS AT RIGHT END OF A SEGMENT UNDER A
C      UNIFORMLY OR LINEARLY DISTRIBUTED LOAD
 100   SL=0.
       SBS=0.
       SAV2=U(KBJ)
       DO 2000 I=1,NS
       I=I
C MOVE SECTION PROPERTIES OF CURRENT SEGMENT FROM SP TO PR
       DO 101   IK=1,7
 101   PR(IK)=SP(IK,I)
       IF(NS-1)105,106,105
 106   PR(7)=S
 105   SL=SL+PR(7)
       CALL LINEAR(SL,CC)
C      COMPUTES DEFLECTION BY TRANSFORMED CONCENTRATED LOADS
 205   IF (SB)   1000,1000,214
 214   DDC=D-CC
       TEMP=0.5*AA*DDC
       TEMQ=0.5*B*DDC
       U(KBJ)=TEMP+TEMQ
       SBS=SBS+SB
       SBB=S-SBS
       SB2=SB**2
       SB3=SB2*SB
       C1=1.
       KG=2
       KFM=KF+2
       KBM=KB+2
       TAA=SB/PR(1)
       KBI=KB+3
       KBH=KBI
       KFH=KF+3
       SAV=U(KBI)
       SAV1=U(KBH)
       GO TO (310,300,400,310,300),ID
 300   TBB=SB3/(3.*PR(6))
       TCC=SB/PR(6)
```

```
       TBC=SB2/(2.*PR(6))
       IF(ID-3)510,510,500
C  PLANE FRAME
C  TRUSSES
  510 GO TO (310,316,330),J
  310 U(KF+1)=U(KF+1)+U(KB+1)*TAA
       GO TO 1000
  316 U(KBH)=(TEMP*(DDC/3.+CC-SBS)+TEMQ*(2.*DDC/3.+CC-SBS))*C1
  318 IF(PR(KG))321,325,321
  321 TBB=TBB+SB/(H*PR(KG))
  325 U(KFM)=U(KFM)+TBB*U(KBM)+TBC*U(KBH)+(TCC*U(KBH)+TBC*U(KBM))*SBB*C1
       U(KFH)=U(KFH)+TCC*U(KBH)+TBC*U(KBM)
       GO TO 998
  330 U(KFM)=U(KFM)+TBC*U(KBH)+TCC*U(KBH)*SBB*C1
       U(KFH)=U(KFH)+TCC*U(KBH)
       GO TO 1000
C  PLANE GRID
  400 TBB=SB3/(3.*PR(5))
       TAA=SB/(H*PR(4))
       TCC=SB/PR(5)
       TBC=-SB2/(2.*PR(5))
       C1=-1.
       KBM=KB+1
       KFM=KF+1
       KG=3
       GO TO (316,440,330),J
  440 U(KF+2)=U(KF+2)+TBB*U(KB+2)
       GO TO 1000
C  SPACE FRAME
  500 TFF=SB3/(3.*PR(5))
       TDD=SB/(H*PR(4))
       TEE=SB/PR(5)
       TCE=-SB2/(2.*PR(5))
       KBI=KB+5
       KBH=KB+6
       KFH=KF+6
       SAV=U(KBI)
       SAV1=U(KBH)
       GO TO (310,316,532,540,550,330),J
  532 U(KB+5)=-TEMP*(DDC/3.+CC-SBS)-TEMQ*(2.*DDC/3.+CC-SBS)
  534 IF (PR(3))    535,538,535
  535 TCC=TCC+SB/(H*SP(3,I))
  538 U(KF+3)=U(KF+3)+TFF*U(KB+3)+TCE*U(KB+5)-(TEE*U(KB+5)+TCE*U(KB+3))*
      1SBB
       U(KF+5)=U(KF+5)+TEE*U(KB+5)+TCE*U(KB+3)
       GO TO 998
  540 U(KF+4)=U(KF+4)+TDD*U(KB+4)
       GO TO 1000
  550 U(KF+3)=U(KF+3)+TCE*U(KB+5)-TEE*U(KB+5)*SBB
       U(KF+5)=U(KF+5)+TEE*U(KB+5)
       GO TO 1000
  998 U(KBH)=SAV1
       U(KBI)=SAV
 1000 IF(SL-CC) 2000,2000,1300
 1300 IF(D-SL) 4000,4000,1400
 1400 SB=SL-CC
       CC=SL
       AA=BB
       GO TO 205
 2000 CONTINUE
```

```
 4000 U(KBJ)=SAV2
      K1 = KF+1
      K2 = KF+JF
      DO 420 I = K1,K2
  420 U(I) = U(I)/EMD
 5000 CALL EFVDTL
 8000 RETURN
      END
*     LABEL
      SUBROUTINE CASE2(J1,J2)
C     STRESS PROGRAM...STRUCTURAL ENGINEERING SYSTEMS SOLVER
C     VERSION III...28 AUGUST 1963
C     KEN REINSCHMIDT, ROOM 1-255, EXT. 2117
C     JM = MEMBER NUMBER
C     J2 = LOCATION OF MINUS END LOAD VECTOR = PR
C     J1 = LOCATION OF  PLUS END LOAD VECTOR = PL
C     DIMENSION STATEMENT FOR ARRAYS IN COMMON
      DIMENSION U(36),IU(36),Y(6,6),T(6,6),V(6,2),    JUNK(5),MJNK(4),
     1 SYSFIL(27),PROFIL(5), KRAY(10)
      COMMON U,IU,Y,T,V,JLD,JUNK,JM,MMM,JA,JB,MJNK,NM,KSTOP,K1,K2,K3,K4,
     1K5,I6,EGO,S,
     4CHECK,NMAX,INORM,ISOLV,ISCAN,IIII,IMOD,JJJJ,ICONT,ISUCC,IMERG,
     5TOP,N1,NL,NT,NREQ,TN,LFILE,TOLER,IPRG,IRST,IRLD,IRPR,SYSFIL,
     8NJ,NB,NDAT,ID,JF,NSQ,NCORD,IMETH,NLDS,NFJS,NSTV,NMEMV,IPSI,NMR,
     9NJR,ISODG,NDSQ,NDJ,IPDBP,IUDBP,NBB,NFJS1,JJC,JDC,JMIC,JMPC,JXX
      COMMON JEXTN,MEXTN,LEXTN,JLC,NLDSI,IYOUNG,ISHEAR,IEXPAN,IDENS,
     1ISUPLD,PROFIL ,
     4NAME,KXYZ,KJREL ,JPLS,JMIN,MTYP,KPSI,MEMB,LOADS,MODN,KS,KMKST,KSTDB
     5,KATKA,KPPLS,KPMNS,KUV,KPPRI,KR,KSAVE,KRAY,KDIAG,KOFDG,
     6 IOFDG,LOADN,MEGAO,JEXT,JINT,KUDBP,KMEGA,KPDBP,JTYP,MTYP1,KB
     7,MLOAD,JLOAD,KATR,LEXT,KYOUNG,KSHEAR,KEXPAN,KDENS
      EQUIVALENCE (U,IU,Y)
      NM = JMIN+JM
      NM = IU(NM)+JINT
      CALL TRAMAT(JM,2)
      KSTOP = 2
C     ADD MINUS END LOAD VECTOR TO PMINUS AND TO PPRIME
   10 K1 = KPMNS+JLD
      K1 = IU(K1)+(JM-1)*JF
      K2 = IU(NM)
      SIGN = 1.0
      J3 = J2
      K6 = KPPRI+(JLD-1)*NSTV
      K7 = KPDBP+(JLD-1)*NDJ
   45 K5 =      K6 +(K2-1)*JF
   50 DO 200 I = 1,JF
      K3 = K1+I
      J4=J3+I
      U(K3)=U(K3) + SIGN*U(J4)
C     IF THE NODE IS IN THE DATUM, ADD TO P PRIME ONLY IF THERE ARE
C     JOINT RELEASES
      IF (K2-NFJS) 150,150,130
  130 IF (NJR) 140,200,140
  140 K4 = K7+(K2-NFJS1)*JF+I
      GO TO 160
  150 K4 = K5+I
  160 TEMP = 0.0
      DO 100 J = 1,JF
C     INVERT THE GIVEN ROTATION MATRIX TO TRANSFORM FROM MEMBER
C     LOCAL TO JOINT GLOBAL COORDINATES
```

```
      J4 = J3+J
      TEMP = T(J,I)*U(J4)        +TEMP
  100 CONTINUE
      U(K4) = U(K4) + TEMP
  200 CONTINUE
      GO TO (300,210), KSTOP
  210 KSTOP = 1
C     SUBTRACT PLUS END LOAD VECTOR FROM P+ AND ADD TO PPRIME
   30 K1 = KPPLS+JLD
      K1 = IU(K1)+(JM-1)*JF
      J3 = J1
      NM = JPLS+JM
      NM = IU(NM)+JINT
      K2 = IU(NM)
      SIGN = -1.0
      GO TO 45
  300 RETURN
      END
*     LIST8
*     SYMBOL TABLE
*     LABEL
      SUBROUTINE JDISPL
C     STRESS PROGRAM...STRUCTURAL ENGINEERING SYSTEMS SOLVER
C     VERSION III...28 AUGUST 1963
C     KENNETH F. REINSCHMIDT, ROOM 1-255, EXT. 2117
C     OBTAINS COMPACTED SUPPORT JOINT DISPLACEMENTS IN PL
C     COMPUTES RESULTING MEMBER DISTORTIONS AND STORES IN KUV
C     COMPUTES RESULTING MEMBER FORCES AND SAVES IN PL AND PR FOR LSTOR
      DIMENSION Y(6,6),T(6,6), PL(6), PR(6),FILL(9),  U(36),IU(36),
     1 SYSFIL(27),PROFIL(6), KRAY(10) ,TEMP(6)
      COMMON U,IU,Y,T,PL,PR,JLD,NLS,H,JT,J,K,NM,N,NS,L,FILL, IKI,
     1K9,JM,EGO,S,
     4CHECK,NMAX,INORM,ISOLV,ISCAN,IIII,IMOD,JJJJ,ICONT,ISUCC,IMERG,
     5TOP,N1,NL,NT,NREQ,TN,LFILE,TOLER,IPRG,IRST,IRLD,IRPR,SYSFIL,
     8NJ,NB,NDAT,ID,JF,NSQ,NCORD,IMETH,NLDS,NFJS,NSTV,NMEMV,IPSI,NMR,
     9NJR,ISODG,NDSQ,NDJ,IPDBP,IUDBP,NBB,NFJS1,JJC,JDC,JMIC,JMPC,JXX
      COMMON JEXTN,MEXTN,LEXTN,JLC,NLDSI,IYOUNG,ISHEAR,IEXPAN,IDENS,
     1 PROFIL,
     4NAME,KXYZ,KJREL,JPLS,JMIN,MTYP,KPSI,MEMB,LOADS,MODN,KS,KMKST,KSTDB
     5,KATKA,KPPLS,KPMNS,KUV,KPPRI,KR,KSAVE,KRAY,KDIAG,KOFDG,
     6 IOFDG,LOADN,MEGAO,JEXT,JINT,KUDBP,KMEGA,KPDBP,JTYP,MTYP1,KB
     7,MLOAD,JLOAD,KATR,LEXT,KYOUNG,KSHEAR,KEXPAN,KDENS
      EQUIVALENCE (U, IU, Y)
C     IM (77310) IS INTERNAL JOINT NUMBER FROM LOADPS  -  IT IS JM IN JDISPL
C     JDISPL
C     GET A TRANSPOSE MATRIX FOR THE JM TH NODE
      CALL ALOCAT(KATR,0)
      CALL ALOCAT(KATR,JM)
      K1 = KATR+JM
C     FIND NUMBER OF BRANCHES INCIDENT ON THIS NODE
      K1=IU(K1)+1
    1 NBRI=IU(K1)
C YOU HAVE TO SAVE THE RAW DATA FOR THE LATER JOINTS ........
      DO 3 I=1,JF
    3 TEMP(I)=PL(I)
C     LOOP FOR ALL INCIDENT BRANCHES
      DO 400 I = 1,NBRI
C     GET I TH BRANCH NUMBER
C     NM (77327) IS TEMP MEMBER NUMBER FOR TRAMAT AND LSTOR
      K1=KATR+JM
```

```
        K1=IU(K1)+2
        NM = K1+I
        NM = IU(NM)
C     NM IS NOW SIGNED MEMBER NUMBER
C RESTORE PL FROM TEMP
        DO 4 JJ=1,JF
   4    PL(JJ)=TEMP(JJ)
     2 IF(NM) 130,400,30
C     I TH MEMBER IS POSITIVELY INCIDENT ON THE JM TH NODE
C        TRANSLATE THE + END DEFLECTION GIVEN TO THE - END
C        USE TRANSLATION MATRIX MINUS TO PLUS, TRANSPOSED
    30 CALL TRAMAT(NM,1)
C        PL CONTAINS THE RAW DATA , HERE THE JOINT DISPLACEMENTS
        DO 50 JJ = 1,JF
        PR(JJ) = 0.0
        DO 50 KK = 1,JF
        PR(JJ) = PR(JJ)+T(KK,JJ)*PL(KK)
    50 CONTINUE
        GO TO 200
C        I TH MEMBER IS NEGATIVELY INCIDENT ON THE JM TH NODE
C        SUBTRACT MINUS END APPLIED JOINT DEFLECTION
   130 DO 150 JJ = 1,JF
        PR(JJ) = -PL(JJ)
   150 CONTINUE
C        NOW ROTATE DISTORTION INTO LOCAL (MEMBER) COORDINATES
   200 NM=XABSF(NM)
        CALL ALOCAT(KMKST,NM)
        CALL TRAMAT(NM,2)
        DO 210 JJ = 1,JF
        PL(JJ) = 0.0
        DO 210 KK = 1,JF
        PL(JJ) = PL(JJ)+T(JJ,KK)*PR(KK)
   210 CONTINUE
C        NOW ADD EFFECTIVE MINUS END DISTORTION TO V IN POOL
   211 K3=KUV+JLD
        K3 = IU(K3)+JF*(NM-1)
        DO 220 JJ = 1,JF
        IKI = K3+JJ
        U(IKI) = U(IKI)+PL(JJ)
   220 CONTINUE
C        NOW MULTIPLY THE DISPLACEMENT BY THE I TH MEMBER STIFFNESS
        K3=KMKST+NM
        K3=IU(K3)
        DO 250 JJ = 1,JF
        PR(JJ) = 0.0
        DO 250 KK = 1,JF
        IKI = K3+(KK-1)*JF+JJ
        PR(JJ) = PR(JJ)+U(IKI)*PL(KK)
   250 CONTINUE
        CALL RELEAS(KMKST,NM)
C        NOW TRANSLATE FORCE TO PLUS (LEFT) END, CHANGING THE SIGN FOR
C        COMPATIBILITY WITH CASE2
        DO 260 JJ = 1,JF
        PL(JJ) = -PR(JJ)
   260 CONTINUE
C        GET MEMBER LENGTH
        IKI = KS+(NM-1)*(NCORD+1)+1
        GO TO (300,270,280,300,290), ID
C        PLANE FRAME
   270 PL(3) = PL(3)+U(IKI)*PL(2)
```

```
      GO TO 300
C     PLANE GRID
  280 PL(3) = PL(3)-U(IKI)*PL(1)
      GO TO 300
C     SPACE FRAME
  290 PL(5) = PL(5)-U(IKI)*PL(3)
      PL(6) = PL(6)+U(IKI)*PL(2)
C     CALL LSTOR TO PERFORM P = KU AND P(PRIME) = -A(TRANSPOSE)KU
C     MEMBER NUMBER IS IN NM
  300 CALL LSTOR
  400 CONTINUE
  410 CALL RELEAS(KATR,JM)
      CALL RELEAS(KATR,0)
      RETURN
      END
*     LIST8
*     LABEL
*     SYMBOL TABLE
      SUBROUTINE JTLOAD
C     STRESS PROGRAM...STRUCTURAL ENGINEERING SYSTEMS SOLVER
C     THIS SUBROUTINE UNPACKS LOADS AND STORES IN KPPRI
C     JM NOW IS JOINT NUMBER
C     LEON RU-LIANG WANG    RM. 1-255    EXT.  2117
      DIMENSION Y(6,6),T(6,6),PL(6),PR(6),FILL(13),KSRTCH(9 ) ,
     1 SYSFIL(32), PROFIL(22),                U(36), IU(36)
      COMMON U,IU,Y,T,PL,PR,JLD,NLS,JLC,LJ,J,K,NM,N,FILL,JM,EGO,S,
     1 CHECK,NMAX,INORM,ISOLV,ISCAN,IIII,IMOD,JJJJ,ICONT,ISUCC,IMERG,
     2 TOP,N1,NL,NT,NREQ,TN,LFILE,SYSFIL,NJ,NB,NDAT,ID,JF,NSQ,NCORD,
     3 IMETH, NLDS,NFJS,NSTV,NMEMV,IPSI,NMR,NJR,ISODG,NDSQ,NDJ,IPDBP,
     4 IUDBP,PROFIL,
     5 NAME,KXYZ,KJREL,
     6 JPLS,JMIN,MTYP,KPSI,MEMB,LOADS,MODN,KS,KMKST,KSTDB,KATKA,KPPLS,
     7 KPMNS,KUV,KPPRI,KR,KSAVE,KWORK,KSRTCH,KDIAG,KOFDG,IOFDG,LOADN,
     8 MEGAO,JEXT,JINT,KUDBP,KMEGA,KPDBP
      EQUIVALENCE (U, IU, Y)
C     UNPACK  LOADS
C     STORES JTLOAD IN KPPRI
C  JM IS INTERNAL JOINT NUMBER
      IF(JM-NFJS)1,1,5
    5 KP=KPDBP+(JM-NFJS-1)*JF+(JLD-1)*NDJ
      GO TO 8
    1 KP=KPPRI+    (JM-1)*JF+(JLD-1)*NSTV
    8 DO 10 I=1,JF
      KPI=KP+I
   10 U(KPI)=U(KPI)+PL(I)
   11 KPI=KPI
      RETURN
      END
*     LIST8
*     LABEL
*     SYMBOL TABLE
      SUBROUTINE LSTOR
C     STRESS PROGRAM...STRUCTURAL ENGINEERING SYSTEMS SOLVER
C     THIS SUBROUTINE CHECKS MEMBER RELEASES AND MODIFIES PL AND PR,
C     CALL CASE2 STORES END LOADS IN KPPRI
C     MENDL ENDM AND MDISTN WILL ENTER THIS SUBROUTINE AFTER THEIR
C     COMPUTATION FOR W AND V
C     JUNE 17, 1963
C     JULY 23 63
C     LEON RU-LIANG WANG RM. 1-255 EXT. 2117
```

```
C      AUG 02 63 S MAUCH
C      LSTOR CONTAINS MRELC (FROM 650 TO 60) TO MODIFY FIXED END FORCES
C      FOR RELEASES
C      U(KWORK) CONTAINS FCA=START CANTILEVER FORCES
C      CHANGE TO VERSION III AUG 30 S M
C      PL IS IN V, PR=RBBAR IS IN W
       DIMENSION Y(6,6),T(6,6),V(6),W(6),FILL(12),
      1SYSFIL(32),PROFIL(22),U(36),IU(36)
       COMMON U,IU,Y,T,V,W,JLO,NLS,H,JT,J,K,JM,N,NS,L,FILL,NDEX,S,
      1CHECK,NMAX,INORM,ISOLV,ISCAN,IIII,IMOD,JJJJ,ICONT,ISUCC,IMERG,
      2TOP,N1,NL,NT,NREQ,TN,LFILE,SYSFIL,NJ,NB,NDAT,ID,JF,NSQ,NCORD,
      3IMETH,NLDS,NFJS,NSTV,NMEMV,IPSI,NMR,NJR,ISODG,NDSQ,NDJ,IXX,
      4IUDBP,PROFIL,
      5NAME,KXYZ,KJREL,
      6JPLS,JMIN,MTYP,KPSI,MEMB,LOADS,MODN,KS,KMKST,KSTOB,KATKA,KPPLS,
      7KPMNS,KUV,KPPRI,KR,KSAVE,KWORK,ARRFIL,SCR3,SCR4,SCR5,SCR6,SCR7,
      8SCR8,SCR9,SCR10,KDIAG,KOFDG,IOFDG,LDNM,MEGAO,JEXT,JINI,KUDBP,
      9KMEGA,KPDBP,JTYP,MTYP1
       EQUIVALENCE(U,IU,Y)
C      UNPACK MTYP,RELEASE AND NS
       IA=MTYP+JM
       CALL UPACW(U(IA),IX,REL,L,NS,IX)
C      CHECK RELEASES
B600   TEST=REL*007700000000
       IF(TEST)650,800,650
 650   IF(ID-1)1,2,1
  1    IF(ID-4)3,2,3
C
C      START-RELEASED TRUSS MEMBER
C
  2    DO 9 I=1,JF
       IBB=KWORK+I
       W(I)=U(IBB)
  9    V(I)=0.0
       GO TO 800
C
C      START RELEASED NON-TRUSS MEMBER
C
  3    IF(NDEX-4)5,800,800
C      IF MDISTN OR JDISPL FCA(CANT LEVER FORCE AT A)=0.0 NO MODIFICATION
  5    IXX=MTYP1+JM
       IXX=IU(IXX)
       CALL ALOCAT(KMEGA,IXX)
       JP=KMEGA+IXX
       JP=IU(JP)
C      MODIFY MINUS END FORCE FOR START RELEASE
       DO 10 I=1,JF
       IA=JP+I
       DO 10 NX=1,JF
       IAA=IA+(NX-1)*JF
       IBB=KWORK+NX
 10    W(I)=W(I)+U(IAA)*U(IBB)
       CALL CARRY(ID,JF,S)
C      NOW TRANSLATE MODIFIED FIXED END FORCE TO PLUS END
       DO 60 I=1,JF
       IBB=KWORK+I
       V(I)=U(IBB)
       DO 60 JP=1,JF
 60    V(I)=V(I)-T(I,JP)*W(JP)
       CALL RELEAS(KMEGA,IXX)
```

```
800   CALL CASE2(72,78)
      RETURN
      END
*     LIST8
*     LABEL
*     SYMBOL TABLE
      SUBROUTINE TRAMAT (JM,JT)
C     THIS SUBROUTINE COMPUTES THE FORCE TRANSLATION OR ROTATION MATRIX
C     STRESS PROGRAM...STRUCTURAL ENGINEERING SYSTEMS SOLVER
C     VERSION III...22 AUGUST 1963
C     KEN REINSCHMIDT, ROOM 1-255, EXT 2117
C     JM = MEMBER NUMBER
C     JT = 1, FORM TRANSLATION MATRIX ONLY, MINUS NODE TO PLUS NODE
C     JT = 2, FORM ROTATION MATRIX ONLY...GLOBAL TO LOCAL
C     ID = 1, PLANE TRUSS, COORDINATES1,2
C     ID = 2, PLANE FRAME, COORDINATES 1, 2, 6
C     ID = 3, PLANE GRID, COORDINATES 3, 4, 5
C     ID = 4, SPACE TRUSS, COORDINATES 1, 2, 3
C     ID = 5, SPACE FRAME, COORDINATES 1, 2, 3, 4, 5, 6
C     IPSI = FLAG FOR EXISTENCE OF TWIST TABLE
C     KPSI = FIRST LOCATION OF TWIST ANGLE TABLE
C     DIMENSION STATEMENT FOR ARRAYS IN COMMON
      DIMENSION Y(6,6),T(6,6),Q(6,6), U(36),IU(36),
     1 SYSFIL(27),PROFIL(6), KRAY(10)
      COMMON U,IU,Y,T,Q,
     4CHECK,NMAX,INORM,ISOLV,ISCAN,IIII,IMOD,JJJJ,ICONT,ISUCC,IMERG,
     5TOP,N1,NL,NT,NREQ,TN,LFILE,TCLER,IPRG,IRST,IRLD,IRPR,SYSFIL,
     5TOP,N1,NL,NT,NREQ,TN,LFILE,SYSFIL,
     8NJ,NB,NDAT,ID,JF,NSQ,NCORD,IMETH,NLDS,NFJS,NSTV,NMEMV,IPSI,NMR,
     9NJR,ISODG,NDSQ,NDJ,IPDBP,IUDBP,NBB,NFJS1,JJC,JDC,JMIC,JMPC,JLD
      COMMON JEXTN,MEXTN,LEXTN,JLC,NLDSI,IYOUNG,ISHEAR,IEXPAN,IDENS,
     1 PROFIL,
     4NAME,KXYZ,KJREL,JPLS,JMIN,MTYP,KPSI,MEMB,LOADS,MODN,KS,KMKST,KSTDB
     5,KATKA,KPPLS,KPMNS,KUV,KPPRI,KR,KSAVE,KRAY,KDIAG,KOFDG,
     6 IOFDG,LOADN,MEGAO,JEXT,JINT,KUDBP,KMEGA,KPDBP,JTYP,MTYP1,KB
     7,MLOAD,JLOAD,KATR,LEXT,KYOUNG,KSHEAR,KEXPAN,KDENS
      EQUIVALENCE (U,IU,Y)
      KM = KS+(JM-1)*(NCORD+1)+1
      DO 12 I = 1,6
      DO 10 J = 1,6
      T(I,J) = 0.0
   10 CONTINUE
      T(I,I) = 1.0
   12 CONTINUE
      GO TO (50,100),JT
C     FORM THE SKEW-SYMMETRIC TRANSLATION MATRIX
   50 T(5,3) = -U(KM+1)
      T(6,2) = U(KM+1)
      T(4,3) = U(KM+2)
      T(6,1) = -U(KM+2)
      IF (NCORD-3)  420,60,60
   60 T(4,2) = -U(KM+3)
      T(5,1) = U(KM+3)
   65 GO TO 420
C     ROTATION MATRIX...GLOBAL TO LOCAL
C     FORM THE DIRECTION COSINES
  100 IF (IPSI) 105,105,110
  105 SPSI = 0.0
      CPSI = 1.0
      GO TO 115
```

```
  110 KK = KPSI+JM
       PIMOD=U(KK)/57.2957795
       SPSI=SINF(PIMOD)
       CPSI=COSF(PIMOD)
C      DIRECTION COSINES OF X1 LOCAL AXIS
C      REVISED PORTION OF TRAMAT   1/13/64
  115  SL=U(KM)
       DELX=U(KM+1)
       DELY=U(KM+2)
       DELZ=U(KM+3)
       IF(NCORD-3)130,140,140
  130  DELZ=0.0
  140  D=SQRTF(DELX*DELX+DELZ*DELZ)
C      D IS UNSIGNED PROJECTION IN GLOBAL X-Z PLANE
C      IS MEMBER PERPENDICULAR TO THAT PLANE
       IF(D/SL-1.0E-06)200,160,160
C      MEMBER IS NOT PERPENDICULAR TO GLOBAL X-Z PLANE
  160  S1=DELZ/D
       S2=DELY/SL
       IF(ID-4)164,165,165
C PLANE STRUCTURE SET PSI1=0.0
  164  C1=1.0
       C2=DELX/SL
       GO TO 166
C      SPACE STRUCTURE
  165  C1=DELX/D
       C2=D/SL
  166  T(1,1)=C2*C1
       T(1,2)=S2
        T(1,3)=C2*S1
       T(2,1)=-S1*SPSI-C1*S2*CPSI
       T(2,2)=-C2*CPSI
       T(2,3)= C1*SPSI-S1*S2*CPSI
       T(3,1)= C1*S2*SPSI-S1*CPSI
       T(3,2)=-C2*SPSI
       T(3,3)=C1*CPSI+S1*S2*SPSI
       GO TO 220
C      MEMBER IS PERPENDICULAR TO THE X1-X3 PLANE
  200  SIGN=ABSF(DELY)/DELY
       T(1,1)=0.0
       T(1,2)=DELY/SL
       T(1,3)=0.0
       T(2,1)=-CPSI*SIGN
       T(2,2) = 0.0
       T(2,3) = SPSI
       T(3,1) = SPSI*SIGN
       T(3,2) = 0.0
       T(3,3) = CPSI
C      FILL OUT MATRIX LOWER RIGHT
  220  DO 230 I = 1,3
       KK = I+3
       DO 230 J = 1,3
       JK = J+3
       T(KK,JK) = T(I,J)
  230 CONTINUE
C     DELETE AS REQUIRED
  420 GO TO (600,440,500,600,600),ID
C     PLANE FRAME
  440 DO 460 I = 1,2
       T(I,3) = T(I,6)
```

```
             T(3,I) = T(6,I)
     460 CONTINUE
             T(3,3) = T(6,6)
             GO TO 600
C        PLANAR GRID
     500 DO 520 I = 1,3
             KK = I+2
             DO 520 J = 1,3
             JK = J+2
             T(I,J) = T(KK,JK)
     520 CONTINUE
     600 CONTINUE
             RETURN
             END
*        LIST8
*        LABEL
*        SYMBOL TABLE
         SUBROUTINE LINEAR(SL,CC)
C        STRESS PROGRAM...STRUCTURAL ENGINEERING SYSTEMS SOLVER
C        MAY 9, 1963
C        LEON RU-LIANG WANG    RM. 1-255    EXT.   2117
C        REVISED FOR VERSION III ON 30 AUGUST 1963 BY KEN REINSCHMIDT
C        THIS SUBROUTINE COMPUTES DEFLECTIONS AT RIGHT END OF A SEGMENT
C        UNDER A LINEARLY DISTRIBUTED LOAD,TRANSFORMS TO RIGHT END OF A
C        MEMBER AND SUMS   OCT 9     S M   STORED IN KWORK+30 COMPACTED
         DIMENSION U(36),IU(36),Y(6,6),T(6,6),PL(5),PR(7),BL(4),
        1 SYSFIL(27),PROFIL(6), KRAY(9)
         COMMON U,IU,Y,T,PL,PR,JLD,NLS,H, JT, J,K,JM,N,NS,L,BL,AA, I,BB,SB,
        1 A,B,C,D,EGO,S,
        4CHECK,NMAX,INORM,ISOLV,ISCAN,IIII,IMOD,JJJJ,ICONT,ISUCC,IMERG,
        5TOP,N1,NL,NT,NREQ,TN,LFILE,TOLER,IPRG,IRST,IRLD,IRPR,SYSFIL,
        8NJ,NB,NDAT,ID,JF,NSQ,NCORD,IMETH,NLDS,NFJS,NSTV,NMEMV,IPSI,NMR,
        9NJR,ISODG,NDSQ,NDJ,IPDBP,IUDBP,NBB,NFJS1,JJC,JDC,JMIC,JMPC,JXX
         COMMON JEXTN,MEXTN,LEXTN,JLC,NLDSI,IYOUNG,ISHEAR,IEXPAN,IDENS,
        1 PROFIL,
        4NAME,KXYZ,KJREL,JPLS,JMIN,MTYP,KPSI,MEMB,LOADS,MODN,KS,KMKST,KSTDB
        5,KATKA,KPPLS,KPMNS,KUV,KPPRI,KR,KSAVE,KWORK,KRAY,KDIAG,KOFDG,
        6 IOFDG,LOADN,MEGAO,JEXT,JINT,KUDBP,KMEGA,KPDBP,JTYP,MTYP1,KB
        7,MLOAD,JLOAD,KATR,LEXT,KYOUNG,KSHEAR,KEXPAN,KDENS
         EQUIVALENCE (U, IU, Y)
C        PR IS LOADED  WITH SECTION PROPERTIES BY MEMBLD       LENGTH IN PR(7)
C        USEFUL FUNCTION DEFINITIONS
         FIRSTF(X) = (AA/2.0 + BB) * C2/(3.0 * X)
         SECONF(X) = (AA/3.0 + BB) * C3/(8.0 * X)
         THIRDF(X) = (AA + 2.75 * BB) * C4/(30.0 * X)
         FORTHF(X) = (AA + 1.6666667 * BB) * C3/ (8.0 * X)
         KF=KWORK+ 30
C        COMPUTATION FOR A LINEAR DISTRIBUTED LOAD
     24  IF(SL-CC) 40, 40, 200
     40  SB=PR(7)
         GO TO 400
     200 SB= PR(7) -(SL-CC)
     205 IF(D-SL)  208,208,210
     208 BB=B
         DD=0
         GO TO 212
     210 BB=AA+(B-AA)*(SL-CC)/(D-CC)
         DD=SL
     212 CC1=DD-CC
         C2=CC1**2
```

```
      C3=C2*CC1
      C4=C3*CC1
      SS=S-DD
C     DEFLECTIONS UNDER LINEAR LOAD ,  TRANSFORMS TO RIGHT END AND SUMS
      GO TO (213,220,245,213,270),ID
C     TRUSS, PLANAR OR SPATIAL
  213 IF(J-1)275,275,395
C     PLANE FRAME
  220 I=KF+3
      GO TO (275,280,240,395,395,395),J
  240 TEMQ= FIRSTF(PR(6))
      TEMP=FORTHF(PR(6))
      GO TO 285
C     PLANE GRID
C  J IS ALREADY COMPACTED TO TYPE
  245 IL=KF+1
       GO TO (248,300,305,395,395,395),J
  248 TEMQ=-SECONF(PR(5))
      GO TO 252
C     SPACE FRAME
  270 I=KF+6
      IL=KF+3
      GO TO (275,280,290,300,305,240),J
  275 U(KF+1)=U(KF+1)+FIRSTF(PR(1))
      GO TO 400
  280 TEMQ= SECONF(PR(6))
      TEMP = THIRDF(PR(6))
      IF (PR(2)) 284,285,284
  284 TEMP = TEMP + FIRSTF(PR(2))/H
  285 U(KF+2)=U(KF+2)+TEMP+TEMQ*SS
      U(I)=U(I)+TEMQ
      GO TO 400
  290 TEMQ=-SECONF(PR(5))
  252 TEMP = THIRDF(PR(5))
      IF (PR(3)) 294,295,294
  294 TEMP = TEMP + FIRSTF(PR(3))/H
  295 U(IL)=U(IL)+TEMP-TEMQ*SS
      U(IL+2)=U(IL+2)+TEMQ
      GO TO 400
  300 U(IL+1)=U(IL+1)+FIRSTF(PR(4))/H
      GO TO 400
  305 TEMQ = FIRSTF(PR(5))
      TEMP = -FORTHF(PR(5))
      GO TO 295
C     WRONG FORCE DIRECTION FOR STRUCTURE
  395 CALL PRER2(13,JM,JLD)
  400 RETURN
      END
*     LIST8
*     SYMBOL TABLE
*     LABEL
      SUBROUTINE EFVDTL
C     STRESS PROGRAM...STRUCTURAL ENGINEERING SYSTEMS SOLVER
C     VERSION 2
C     APRIL 4, 1963
C     LEON R.  WANG ,RM  1-255  EXT. 2117
C     THIS SUBROUTINE COMPUTES PR=-K*U AND PL BY STATIC
C     Y=STIFFNESS MATRICES
C  VERSION III    S M  AUG 30 63
      DIMENSION Y(6,6),T(6,6), PL(6), PR(6),BL(4) ,
```

```
      1 SYSFIL(32), PROFIL(10), U(36),IU(36)
        COMMON U,IU,Y,T,PL,PR,JLD,NLS,H,JT,J,K,JM,N,NS,L,BL,AA,I,BB,SB
        COMMON  A,B,C,D,EMD,S,
      1 CHECK,NMAX,INORM,ISOLV,ISCAN,IIII,IMOD,JJJJ,ICONT,ISUCC,IMERG,
      2 TOP,N1,NL,NT,NREQ,TN,LFILE,SYSFIL,NJ,NB,NDAT,ID,JF,NSQ,NCORD,
      3 IMETH, NLDS,NFJS,NSTV,NMEMV,IPSI,NMR,NJR,ISODG,NDSQ,NDJ,IPDBP,
      4 IUDBP,NBB,NFJS1,JJC,JDC,JMIC,JMPC,JLT,JEXTN,MEXTN,LEXTN,JLC,NLDSI
      5,PROFIL,NAME,KXYZ,KJREL,
      6 JPLS,JMIN,MTYP,KPSI,MEMB,LOADS,MODN,KS,KMKST,KSTDB,KATKA,KPPLS,
      7 KPMNS,KUV,KPPRI,KR,KSAVE,KWORK
        EQUIVALENCE (U, IU, Y)
        KA=KWORK
        KD=KWORK+ 18
        KF=KWORK+ 30
C KF REFERS TO CANTILEVER DEFLECTION
C     GET K* INTO Y.
C  NOV 21 63  CHANGE TO KMKST SECOND LEVEL      S MAUCH
        JU=KMKST+JM
        JU=IU(JU)
        DO 10  I=1,JF
        PR(I)=0.0
        DO 10   J=1,JF
        JX=JU+JF*(I-1)+J
   10   Y(I,J)=U(JX)
C      RIGHT END FORCE VECTOR =K* TIMES CANTILEVER DEFLECTIONS WHICH
   50   DO 60   I=1,JF
        DO 60   M=1,JF
         KFM=KF+M
   60    PR(I)=PR(I)+Y(I,M)*U(KFM)
C       TRANSFORMS RIGHT END FORCE VECTOR TO LEFT END
        DO 65   I=1,JF
         KDI=KD+I
   65   U(KDI)=PR(I)
        GO TO (100,70,80,100,90),ID
   70   U(KD+3)=U(KD+3)+PR(2)*S
        GO TO 100
   80   U(KD+3)=U(KD+3)-PR(1)*S
        GO TO 100
   90   U(KD+5)=U(KD+5)-PR(3)*S
        U(KD+6)=U(KD+6)+PR(2)*S
C      LEFT END FORCE VECTOR BY STATIC =APPLIED FORCE-PR
  100   DO 110 I=1,JF
        KAI=KA+I
        KDI=KD+I
  110   PL(I)=U(KAI)-U(KDI)
        RETURN
        END
   *    LIST8
   *    LABEL
   *    SYMBOL TABLE
        SUBROUTINE STICLD
C       MAY 7, 1963
C       STRESS PROGRAM...STRUCTURAL ENGINEERING SYSTEMS SOLVER
C       LEON RU-LIANG WANG   RM. 1-255   EXT.  2117
C  VERSION III   S M  AUG 30 63
C       THIS SUBROUTINE COMPUTES RIGHT END AND LEFT END CANTILEVER FORCES
C       BY STATICS IN COMPACTED FORM
C       LEFT END CANTILEVER FORCES STORE IN 1ST COLUMN OF Q
C       RIGHT END CANTILEVER FORCES STORE IN 3RD COLUMN OF Q
        DIMENSION Y(6,6),T(6,6), PL(6), PR(6),BL(4)   ,
```

```
      1 SYSFIL(32), PROFIL(10), U(36),IU(36)
       COMMON U,IU,Y,T,PL,PR,JLD,NLS,H,JT,J,K,JM,N,NS,L,BL,AA,I,BB,SB
       COMMON   A,B,C,D,EMD,S,
      1 CHECK,NMAX,INORM,ISOLV,ISCAN,IIII,IMOD,JJJJ,ICONT,ISUCC,IMERG,
      2 TOP,N1,NL,NT,NREQ,IN,LFILE,SYSFIL,NJ,NB,NDAT,ID,JF,NSQ,NCORD,
      3 IMETH, NLDS,NFJS,NSTV,NMEMV,IPSI,NMR,NJR,ISODG,NDSQ,NDJ,IPDBP,
      4 IUDBP,NBB,NFJS1,JJC,JDC,JMIC,JMPC,JLT,JEXTN,MEXTN,LEXTN,JLC,NLDSI
      5,PROFIL,NAME,KXYZ,KJREL,
      6 JPLS,JMIN,MTYP,KPSI,MEMB,LOADS,MODN,KS,KMKST,KSTDB,KATKA,KPPLS,
      7 KPMNS,KUV,KPPRI,KR,KSAVE,KWORK
       EQUIVALENCE (U, IU, Y)
       KA=KWORK
       KAJ=KA+J
C      CHECK TYPE OF LOADING DEPENDING ON K
  9    GO TO (10,90,90),K
C
C      TRANSFORMS CONCENTRATED LOAD TO LEFT END AND SUMS
C      SAME PROCEDURES TO RIGHT END
  10   U(KAJ)=U(KAJ)+A
       GO TO (200,15,20,200,25),ID
  15   GO TO (200,18,200),J
  18   U(KA+3)=U(KA+3)+A*B
       GO TO 200
  20   GO TO (22,200,200),J
  22   U(KA+3)=U(KA+3)-A*B
       GO TO 200
  25   GO TO (200,30,40,200,200,200),J
  30   U(KA+6)=U(KA+6)+A*B
       GO TO 200
  40   U(KA+5)=U(KA+5)-A*B
       GO TO 200
C
C      TRANSFORMS LINEAR DISTRIBUTED LOAD TO LEFT END END AND SUMS
C      SAME PROCEDURES TO RIGHT END
  90   TEMP=0.5*A*(D-C)
       TEMQ=0.5*B*(D-C)
       AVX=2.*C/3.+D/3.
       AVY=C/3.+2.*D/3.
       U(KAJ)=U(KAJ)+TEMP +TEMQ
       GO TO (200,95,100,200,120),ID
  95   GO TO (200,98,200),J
  98   U(KA+3)=U(KA+3)+TEMP*AVX +TEMQ*AVY
       GO TO 200
  100  GO TO (110,200,200),J
  110   U(KA+3)=U(KA+3)-TEMP*AVX -TEMQ*AVY
       GO TO 200
  120  GO TO (200,130,140,200,200,200,200),J
  130  U(KA+6)=U(KA+6)+TEMP*AVX +TEMQ*AVY
       GO TO 200
  140  U(KA+5)=U(KA+5)-TEMP*AVX -TEMQ*AVY
  200    RETURN
       END
*      FAP
*      CLEARS AN ARRAY WHOSE CODEWORD IS THE ONLY ARGUEMENT
       COUNT   16
       LBL     CLEAR
       ENTRY   CLEAR
CLEAR  CLA*    1,4
       ANA     =077777
       TZE     BACK+1
```

```
        CLA*      1,4
        TMI       DFILE
        SXA       BACK,1
        PAX       ,1
*       LENGTH OF ARRAY INTO XR1
        ARS       18
        ANA       =0000000077777
        SSM
        ADD       TOP
        STA       *+1
        STZ       **,1
        TIX       *-1,1,1
BACK    AXT       **,1
        TRA       2,4
DFILE   ANA       =0000000777777
        SSM
        STO*      1,4
        TRA       BACK+1
  TOP   TSX       U+1
  U     COMMON    1
        END
*       LIST8
*       LABEL
CMAIN4/6
        DIMENSION  Y(6,6),T(6,6),Q(6,6),U(36),IU(36),SYPA(40),FILL(6)
        COMMON U,T,Q,CHECK,NMAX,INORM,ISOLV,ISCAN,IIII,IMOD,ILINK,ICONT,
       1ISUCC,SYPA,NJ,NB,NDAT,ID,JF,NSQ,NCORD,IMETH,NLDS,NFJS,NSTV
       2,NMEMV,IPSI,NMR,NJR,ISODG,NDSQ,NDJ,SDJ,NPR,NBB,NFJS1,JJC,JDC,
       3JMIC,JMPC,JLD,JEXTN,MEXTN,LEXTN,JLC,NLDSI,IYOUNG,ISHER,IEXPAN,
       4IDENS,FILL,
       5NAME,KXYZ,KJREL,JPLS,JMIN,MTYP,KPSI,MEMB,
       6LOADS, INPUT,KS,KMKST,KSTDB,KATKA,KPPLS,KPMIN,KUV,KPPRI,KR,KMK,
       7KV,K33,LA2R ,LA2RT,NV    ,NTP   ,NSCR7,NSCR8,NSCR9,NSCR10,KDIAG,KOFD
       8G,IOFDG,LDNM,MEGAO,JEXT,JINT,KUDBP,KMEGA,KPDBP,JTYP,MTYP1,KB,MLOAD
       9,JLOAD,KATR,LEXT,KYOUNG,KSHER,KEXPAN,KDENS
        EQUIVALENCE(U(1),IU(1),Y(1))
        CALL TRANS
        ISUCC=ISUCC
        GO TO (1,20),ISUCC
1       ISOLV=6
        CALL ATKA
        ISUCC=ISUCC
        GO TO (2,20),ISUCC
2       ISOLV=7
        IF(NJR)3,8,3
3       CALL JRELES
        ISUCC=ISUCC
        GO TO(4,20),ISUCC
4       ISOLV=8
        CALL FOMOD
        ISUCC=ISUCC
        GO TO(8,20),ISUCC
8       ISOLV=9
        IF(NFJS)5,6,5
5       CALL CHAIN(5,A4)
6       ISOLV=10
        CALL CHAIN(6,A4)
20      CALL CHAIN(1,A4)
        END
*       LIST8
```

```
*       LABEL
*       SYMBOL TABLE
        SUBROUTINE TRANS
C       THIS SUBROUTINE ROTATES THE MEMBER STIFFNESS MATRIX(KMKST) INTO
C       GLOBAL COORDINATES(KSTDB)
C NOV 19 CHANGING KMKST ,KSTDB TO 2 ND LEVEL   NEEDED RECOMPILATION OF
C PHAS1B MEMBER MRELES TRANS ATKA LOADPC LOADPS MDISTN JDISPL BAKSUB AVECT
C  NEWADR     NOV 17 TO 19 Y 1963
C       STRESS PROGRAMMING SYSTEM
C       THE PRIMITIVE STIFFNESS MATRICES ARE IN KMKST
C       THE ROTATED STIFFNESS MATRICES ARE PLACED IN KSTDB
        DIMENSION U(36),IU(36),Y(6,6),T(6,6),JUNK(20),
       1SYSFIL(27),PROFIL(6),KRAY(10)
        COMMON U,IU,Y,T,K3,K4,IDOWN,JDOWN,LDOWN,IUP,JUP,LUP,JBOTH,LBOTH,
       1JBIG,JSML,LPLS,LMIN,NUP,MUP,JUNK,
       4CHECK,NMAX,INORM,ISOLV,ISCAN,IIII,IMOD,JJJJ,ICONT,ISUCC,IMERG,
       5TOP,N1,NL,NT,NREQ,TN,LFILE,TOLER,IPRG,IRST,IRLD,IRPR,SYSFIL,
       8NJ,NB,NDAT,ID,JF,NSQ,NCORD,IMETH,NLDS,NFJS,NSTV,NMEMV,IPSI,NMR,
       9NJR,ISODG,NDSQ,NDJ,IPDBP,IUDBP,NBB,NFJS1,JJC,JDC,JMIC,JMPC,JLD
        COMMON JEXTN,MEXTN,LEXTN,JLC,NLDSI,IYOUNG,ISHEAR,IEXPAN,IDENS,
       1PROFIL,
       4NAME,KXYZ,KJREL,JPLS,JMIN,MTYP,KPSI,MEMB,LOADS,MODN,KS,KMKST,KSTDB
       5,KATKA,KPPLS,KPMNS,KUV,KPPRI,KR,KSAVE,KRAY,KDIAG,KOFDG,
       6IOFDG,LOADN,MEGAO,JEXT,JINT,KUDBP,KMEGA,KPDBP,JTYP,MTYP1,KB
       7,MLOAD,JLOAD,KATR,LEXT,KYOUNG,KSHEAR,KEXPAN,KDENS
        EQUIVALENCE(U,IU,Y)
C       GET NECESSARY ARRAYS
        CALL ALOCAT(KS)
  458   IF(IPSI)50,60,50
   50   CALL ALOCAT(KPSI)
C       ZERO KSTDB
   60   CALL ALOCAT(KSTDB,0)
        CALL ALOCAT(MTYP)
C       LOOP FOR ALL MEMBERS
        J=0
  200   J=J+1
        I=J
C       TEST FOR I TH MEMBER PRESENT
        K1=MTYP+I
        IF(U(K1))70,201,70
   70   CALL ALOCAT(KMKST,I)
        CALL ALOCAT(KSTDB,I)
        K2=KSTDB+I
        CALL CLEAR(IU(K2))
        K2=IU(K2)
        K1=KMKST+I
        K1=IU(K1)
C       GET THE ROTATION MATRIX, JOINT GLOBAL TO MEMBER LOCAL
        CALL TRAMAT(I,2)
C       MULTIPLY AND STORE
        CALL MATRIP(K1,K2,1)
        CALL RELEAS(KMKST,I)
        CALL RELEAS(KSTDB,I)
  201   IF(J-MEXTN)200,220,220
  220   CALL RELEAS(KMKST,0)
        CALL RELEAS(KPSI)
        RETURN
        END
*       LIST8
*       LABEL
```

```
*       SYMBOL TABLE
        SUBROUTINE ATKA
C       THIS SUBROUTINE COMPUTES THE STIFFNESS TABLES
C       KENNETH F. REINSCHMIDT, ROOM 1-255, EXT.2117
C       STRESS PROGRAMMING SYSTEM
C       VERSION III...22 AUGUST 1963
C       KDIAG...TABLE OF STIFFNESS SUBMATRICES ALONG MAIN DIAGONAL
C       KOFDG...OFF DIAGONAL STIFFNESS MATRICES
C       IOFDG...INDEX TABLE FOR OFF DIAGONAL STIFFNESSES
        DIMENSION U(36),IU(36),Y(6,6),T(6,6),JUNK(27),
       1SYSFIL(27),PROFIL(3),KRAY(10)
        COMMON U,IU,Y,T,K3,K4,K1,K2,NT,
       1JBIG,JSML,LPLS,LMIN,JUNK,
       4CHECK,NMAX,INORM,ISOLV,ISCAN,IIII,IMOD,JJJJ,ICONT,ISUCC,IMERG,
       5TOP,N1,NL,NZ,NREQ,TN,LFILE,TOLER,IPRG,IRST,IRLD,IRPR,SYSFIL,
       8NJ,NB,NDAT,ID,JF,NSQ,NCORD,IMETH,NLDS,NFJS,NSTV,NMEMV,IPSI,NMR,
       9NJR,ISODG,NDSQ,NDJ,IPDBP,IUDBP,NBB,NFJS1,JJC,JDC,JMIC,JMPC,JLD
        COMMON JEXTN,MEXTN,LEXTN,JLC,NLDSI,IYOUNG,ISHEAR,IEXPAN,IDENS,
       1NBNEW,NLDG,JTSTAB,PROFIL ,
       4NAME,KXYZ,KJREL,JPLS,JMIN,MTYP,KPSI,MEMB,LOADS,MODN,KS,KMKST,KSTDB
       5,KATKA,KPPLS,KPMNS,KUV,KPPRI,KR,KSAVE,KRAY,KDIAG,KOFDG,
       6 IOFDG,LOADN,MEGAO,JEXT,JINT,IFDT ,KMEGA,KPDBP,JTYP,MTYP1,IOFC
       7,MLOAD,JLOAD,KATR,LEXT,KYOUNG,KSHEAR,KEXPAN,KDENS
        EQUIVALENCE (U,IU,Y)
C
C DEFINE BIT REPRESENTATION OF ATKA (IOFDG TRANSPOSE)
C
        JBIG=(NJ*(NJ-1))/14400+1
        CALL DEFINE(IFDT,0,JBIG,1,1,1)
        CALL ALOCAT(IFDT,0)
        DO 400   I=1,JBIG
        CALL DEFINE(IFDT,I,200,0,1,1)
        IF(MODN)435,400,435
  435   LE=IFDT+I
        CALL CLEAR(IU(LE))
  400   CONTINUE
C       GET IN THE NECESSARY ARRAYS
C       KS AND KSTDB ALLOCATED BY TRANS
  437   K1=K1
        CALL ALOCAT(JPLS)
        CALL ALOCAT(JMIN)
        CALL ALOCAT (JINT)
        CALL ALOCAT(KDIAG,0)
        CALL ALOCAT(KOFDG,0)
        CALL ALOCAT(IOFDG,0)
C
C IF MODIFICATION FIRST  PART CLEAR ALL IOFDG
C     AND KDIAG
C
        IF(MODN)2,3,2
  2     DO 1 L=1,NJ
        LD=KDIAG+L
  1     CALL CLEAR(IU(LD))
C LM IS THE COUNT FOR THE KOFDG ARRAYS
  3     LM=0
        DO 700 L = 1,MEXTN
C       TEST IF L TH MEMBER IS PRESENT
        K1 = MTYP+L
        IF (U(K1)) 60,700,60
  60    LPLS=JPLS+L
```

```
       LMIN = JMIN+L
       LM=LM+1
C      CONVERT TO INTERNAL JOINT NUMBERS
       LPLS = IU(LPLS)+JINT
       LMIN = IU(LMIN)+JINT
       LPLS=IU(LPLS)
       LMIN=IU(LMIN)
       JBIG=XMAXOF(LPLS,LMIN)
       JSML=XMINOF(LPLS,LMIN)
C      GET TRANSLATION MATRIX, MINUS NODE TO PLUS NODE
       CALL TRAMAT(L,1)
C      ADD IN SUBMATRICES ON DIAGONAL
C      AT MINUS NODE, ADD IN DIRECTLY
       CALL ALOCAT(IOFDG,JBIG)
       CALL ALOCAT(KDIAG,LMIN)
       CALL ALOCAT(KDIAG,LPLS)
       CALL ALOCAT(KOFDG,LM)
C KMKST AND KSTDB ARE MEXTN LONG BUT KOFDG IS NB LONG I E ONLY FOR NON DELETED M
C MEMBERS
       K2=KOFDG+LM
       CALL CLEAR(IU(K2))
       CALL ALOCAT(KSTDB,L)
       K1=KSTDB+L
       K1=IU(K1)
       K2=KDIAG+LMIN
       K2 = IU(K2)
       DO 80  I = 1,NSQ
       K3 = K1+I
       K4 = K2+I
       U(K4) = U(K4)+U(K3)
    80 CONTINUE
C      AT PLUS NODE, PRE AND POST MULTIPLY BY THE TRANSLATION MATRIX
       K2=KDIAG+LPLS
       K2 = IU(K2)
       CALL MATRIP(K1,K2,2)
C      NOW GET OFF DIAGONAL SUBMATRICES IN LOWER TRIANGLE MATRIX ONLY
C      ASSUME PLUS AND MINUS NODES DISTINCT
       K2=KOFDG+LM
       K2 = IU(K2)
320    IF(LPLS-LMIN)340,700,350
  340 NT = 1
       GO TO 355
  350 NT = 2
  355 DO 500 I = 1,JF
       DO 500 K = 1,JF
       K4 = K2+(K-1)*JF+I
       DO 500 J = 1,JF
       GO TO (100,300),NT
C       T IS THE POSTMULTIPLIER
  100 K3 = K1+(J-1)*JF+I
  120 U(K4) = U(K4)-U(K3)*T(K,J)
       GO TO 500
C       T IS THE PREMULTIPLIER
  300 K3 = K1+(K-1)*JF+J
  330 U(K4) = U(K4)-T(I,J)*U(K3)
  500 CONTINUE
       CALL RELEAS(KDIAG,LPLS)
       CALL RELEAS(KDIAG,LMIN)
       CALL ADRESS(JBIG,JSML,0,0)
  360  JB=IOFDG+JBIG
```

```
C
C  FILL IOFDG CURRENT LENGTH IS FIRST WORD OF DATA ARRAY
C  ADRESS PUTS THE BIT FOR THIS ARRAY INTO IFDT
C
C  UNPACK LENGTH OF IOFDG(JB) INTO IB
       CALL UPADP(IU(JB  ),A,B,IB)
       JB  =IU(JB  )
C    INCREASE NUMBER OF ELEMENTS (IA) BY 1
       IA=IU(JB  +1)+1
       CALL PADP(IU(JB  +1),0,IA,IB)
       JB=JB+IA+1
       CALL PADP(IU(JB  ),0,JSML,LM)
       CALL RELEAS(IOFDG,JBIG)
       CALL RELEAS(KOFDG,LM)
       CALL RELEAS(KSTDB,L)
  700 CONTINUE
       CALL RELEAS(KS)
  438 K1=K1
       CALL RELEAS(KSTDB,0)
       CALL RELEAS(KOFDG,0)
       CALL RELEAS(IOFDG,0)
       CALL RELEAS(IFDT,0)
       CALL RELEAS(JPLS)
       CALL RELEAS(JMIN)
       CALL RELEAS(JINT)
       CALL RELEAS(MTYP)
       IF(JTSTAB)800,850,800
  800 IF(NFJS)801,850,801
  801 CALL ALOCAT(JEXT)
       DO 950 I=1,NFJS
       CALL ALOCAT(KDIAG,I)
       K2=JEXT+I
       K2=IU(K2)
       D=1.00E-6
       K3 = KDIAG+I
       K3=IU(K3)
       K4=XDETRMF(JF,JF,U(K3),D)
       GO TO (950,860,870),K4
  860 CALL PRER2(21,K2,0.0)
       GO TO 952
  870 CALL PRER2(20,K2,0.0)
  952 ISUCC=2
  950 CALL RELEAS(KDIAG,I)
       CALL RELEAS(JEXT)
  850 CALL RELEAS(KDIAG,0)
       RETURN
       END
*      LIST8
*      LABEL
*      SYMBOL TABLE
       SUBROUTINE STEP5
C   THIS SUBROUTINE IS ONLY CALLED IF THERE ARE FREE JOINTS NFJS IS POS
C   NOV 12 RECOMP TO ELIMINATE CARRY-OVER OF ADDRESSES PAST CALL ALOCATS
C   SUBSCRIPTS ARE M,I)*(I,J)*(J,K)
C   MULTUPLY ATKA//*(MEGAO*A//TKA
C   IOFDG=MATRIX OF INDICES  FOR AT/KA/ ONE INTEGER PER SUBMATRIX IN KOFDG
CTHIS MATRIX HAS NJ   ROWS  FIRST IS OF LENGTH ZERO
C   MATRIX IS LOWER HALF OF A/TKA/
C REVISION DEC 31 63  TO NOT REDEFINE KOFDG(0) EVERY TIME THERE IS A NEW ARRAY
       DIMENSION  Y(6,6),T(6,6),Q(6,6),U(36),IU(36),SYPA(40),FILL(14),FIL
```

```
          1L1(5)
          COMMON U,T,Q,CHECK,NMAX,INORM,ISOLVE,ISCAN,IIII,IMOD,JJJJ,ICONT,
         1ISUCC,SYPA,NJ,NB,NDAT,ID,JF,NSQ,NCORD,IMETH,NLDS,NFJS,NSTV
         2,NMEMV,IPSI,NMR,NJR,ISODG,NDSQ,NDJ,IPB,IUPB,NBB,NFJS1 ,FILL,NBNEW,
         9FILL1,
         3NAME,KXYZ,KJREL,JPLS,JMIN,MTYP,KPSI,MEMB,
         4LOADS, INPUT,KS,KMKST,KSTDB,KATKA,KPPLS,KPMIN,KUV,KPPRI,KR,KMK,
         5KV,K33,LA2R ,LA2RT,NSCR5,NSCR6,NSCR7,NSCR8,NSCR9,NSCR10,KDIAG,KOFD
         6G,IOFDG,LDNM,MEGAO,JEXT,JINT,IFDT ,KMEGA,KPDBP,JTYP,MTYP1
          EQUIVALENCE(U(1),IU(1),Y(1))
C        TEST IF A//TKA// IS HYPERDIAGONAL.
C NBNEW IS PRESENT LENGHT OF KOFDG(0)
2010   IF(NJ-NFJS1)2,2,550
C TEST FOR HYPERDIAG
 550   IP=NJ-1
       DO 480  K=NFJS1,IP
       CALL ADRESS(K+1,K,IS,4)
 461   IF(IS)470,480,470
 480   CONTINUE
 2     IHYPD=0
       GO TO 390
 470   IHYPD=1
C  470 MEANS A//TKA// IS NOT HYPERDIAGONAL
 390   IA=IA
C
C START K LOOP
C
 3     DO 500 K=1,NFJS
C   SEARCH FOR NONZERO COLUMN OF A//TKA
C
C START J LOOP
C
 400   J=NFJS1
C TEST IF THIS MEGAO IS NOT ALL ZERO AND NOT ALLOCATED OR DEFINED
C   ADDRESS WITH LAST ARG =4 SEARCHES DOWN TO NJ
 477   CALL ADRESS(J,K,IA3,4)
 4     IF(IA3)404,500,404
 404   CALL ALOCAT(KDIAG,K)
       JMS=J-NFJS
       CALL RELEAS(IOFDG,J)
       I22=MEGAO+JMS
       IF(IU(I22))250,391,250
C   NEXT FIND NON ZERO ROW IN ATKA//
C SINCE ATKA IS STORED BY LOWER HALF ONLY M IS LARGER THAN K
C
C   START M LOOP
C START IP LOOP
C   WRITE DO LOOP ON IP IN LONG HAND BECAUSE IP IS REDEFINED INSIDE LOOP
C
 250   CALL ALOCAT(KOFDG,IA3)
       CALL ALOCAT(MEGAO,JMS)
C NEED JMS COLUMN OF MEGAO BUT SINCE SYMM JMS TH ROW IS OK   (MUST SWITCH INDICES
       DO 421  M=K,NFJS
       IP=NFJS
 408   IP=IP+1
       IF(IHYPD)423,424,423
 424   IP=J
       IA1=IA3
C MUST ONLY CHECK 1 ELEMENT IN IOFDG(IP) SINCE A//TKA// IS HYPERDIAG  SO LA=2
 423 CALL ADRESS(IP,M,IA1,4)
```

```
      406 IF(IA1)428,472,428
C     472   RELEAS IOFDG IP WHICH WAS ALLOCATED BY ADDRESS
C IF IA1=0 NO MORE ELEMENTS IN COLUMN IP NEXT M
  428 I=IP-NFJS
      I22=MEGAO+I
C THIS CHECKS WETHER ELEMENT MEGAO(I,JMS) IS A PRIORI ZERO
      IF(IU(I22))429,472,429
  429 IMEGZ=MEGAO+JMS
      IMEGZ=IU(IMEGZ)+(I-1)*JF
      DO 4000 NN=1,JF
      NIN=IMEGZ+(NN-1)*NDJ
      DO 4000  MIM=1,JF
      MXM=MIM+NIN
      IF(U(MXM))427,4000,427
 4000 CONTINUE
      GO TO 472
  427 CALL ALOCAT(KOFDG,IA1)
      IF(K-M)450,440,450
  440 IRES=KDIAG+K
      IRES=IU(IRES)
      GO TO 443
  450 CALL ADRESS(M,K,IAR,2)
  441 IF (IAR)444,447,444
C 447 CREATE A NEW KOFDG
  447 NBB=NBB+1
      IAR=NBB
      LA=IOFDG+M
      LA=IU(LA)+1
      CALL UPADP(IU(LA),DUM,LCU,LDE)
      LCU=LCU+1
      IF(LCU-LDE)430,431,431
  431 LDE=LDE+XMINOF((M-LCU)/2+2,LCU+1,30)
      CALL DEFINE(IOFDG,M,LDE,0,1,1)
      LA=IOFDG+M
      LA=IU(LA)+1
  430 CALL PADP(IU(LA),0,LCU,LDE)
      LA=LA+LCU
      CALL PADP(IU(LA),0,K,NBB)
      CALL ADRESS(M,K,0,0)
C TEST IF NEED REDEFINE KOFDG(0)
      IF(NBB-NBNEW)449,449,448
  448 NBNEW=NBNEW+NB/4+5
      CALL DEFINE(KOFDG,0,NBNEW,1,1,1)
  449 CALL DEFINE(KOFDG,NBB,NSQ,0,0,1)
  444 CALL RELEAS(IOFDG,M)
      CALL ALOCAT(KOFDG,IAR)
      IRES=KOFDG+IAR
      IRES=IU(IRES)
C SUBSCRIPTS ARE   (LA,NA)*(NA,NX)*(NX,LB)
C NOW COMPUTE ADRESSES FOR USED ARRAYS
  443 I1=KOFDG+IA1
      I1=IU(I1)
      I3=KOFDG+IA3
      I3=IU(I3)
      I22=MEGAO+JMS
      I22=IU(I22)+(I-1)*JF
      DO 460 LA=1,JF
      DO 460 LB=1,JF
      NADR=IRES+(LB-1)*JF+LA
      DO 460 NA=1,JF
```

```
       I2=I22+NA
       NAD1=I1+(LA-1)*JF+NA
       DO 460  NX=1,JF
       NAD2=I2+(NX-1)*NDJ
       NAD3=I3+(LB-1)*JF+NX
 460   U(NADR)=U(NADR)-U(NAD1)*U(NAD2)*U(NAD3)
       IF(K-M)417,420,417
 417   CALL RELEAS(KOFDG,IAR)
 420   CALL RELEAS(KOFDG,IA1)
 472   CALL RELEAS(IOFDG,IP)
C TEST OF IP LOOP
       IF(IHYPD)471,421,471
 471   IF(IP-NJ)408,421,421
 421   CONTINUE
 402   CALL RELEAS(MEGAO,JMS)
 391   CALL RELEAS(KOFDG,IA3)
 401   J=J+1
       IF(J-NJ)477,477,500
 500   CALL RELEAS(KDIAG,K)
C      END STEP 5
       RETURN
       END
*      LIST8
*      SYMBOL TABLE
*      LABEL
       SUBROUTINE JRELES
C  AUG 05 1963    VERSION III
C  REVISION  NOVEMBER 14 63    S MAUCH
C  NJR IS THE TOTAL NUMBER OF RELEASED DIRECTIONS SPECIFIED AT JOINTS
       DIMENSION  Y(6,6),T(6,6),Q(6,6),U(36),IU(36),SYPA(40),FILL(20)
       COMMON U,T,Q,CHECK,NMAX,INORM,ISOLVE,ISCAN,IIII,IMOD,JJJJ,ICONT,
      1ISUCC,SYPA,NJ,NB,NDAT,ID,JF,NSQ,NCORD,IMETH,NLDS,NFJS,NSTV
      2,NMEMV,IPSI,NMR,NJR,ISODG,NDSQ,NDJ,IPB,IUPB,NBB,NFJS1 ,FILL,
      3NAME,KXYZ,KJREL,JPLS,JMIN,MTYP,KPSI,MEMB,
      4LOADS,  INPUT,KS,KMKST,KSTDB,KATKA,KPPLS,KPMIN,KUV,KPPRI,KR,KMK,
      5KV,K33,LA2R ,LA2RT,NSCR5,NSCR6,NSCR7,NSCR8,NSCR9,NSCR10,KDIAG,KOFD
      6G,IOFDG,LDNM,MEGAO,JEXT,JINT,KUDBP,KMEGA,KPDBP,JTYP,MTYP1
       EQUIVALENCE(U(1),IU(1),Y(1))
       CALL ALOCAT(KDIAG,0)
       CALL ALOCAT(KOFDG,0)
       CALL ALOCAT(IOFDG,0)
       CALL ALOCAT(KS)
       CALL ALOCAT(KJREL)
       CALL ALOCAT(JMIN)
       CALL ALOCAT(JPLS)
       CALL ALOCAT(KXYZ)
       CALL DEFINE(KV,NJR,0,0,1)
       CALL ALOCAT(KV)
       CALL ALOCAT(JINT)
 50    NDSQ=NJR*NJR
C
C    STEP 1
C     GENERATE LA2R
C  IOFDG IS SECOND LEVEL ARRAY
C
       LEN=NJR*JF
       CALL DEFINE(LA2R,LEN,0,1,1)
       CALL ALOCAT(LA2R)
       IA=NJR+KV+1
       NDU=0
```

```
         LEN=LA2R+LEN
         NBOT=NJR
         I1=KJREL+1
         KST=IU(I1)
         DO 100  NPI=1,KST
         I2=I1+(NPI-1)*5+1
         JNR=IU(I2)
         N=JINT+JNR
         N=IU(N)
C   N IS NOW THE INTERNAL JOINT NUMBER
         JREL=IU(I2+1)
C  JREL IS NOW THE RELEAS CODE
         JD=N-NFJS
         IF(JREL)100,100,2
  2      CALL TTHETA(I2+1,ID,JF)
C                     TRANSFORMATION TCO// FOR JOINT JD (INTO Q) ,DONE BY TTHETA
         DO 1 IDIR=1,JF
  20     MM=JREL
         CALL UNPCK(MM,IDIR,ID)
         IF(MM)1,1,4
  4      NDU=NDU+1
         NLAA=LEN-JF*NDU
         NBOT=NBOT-1
         DO 3 IR=1,JF
         NLA=NLAA+IR
C        COPY ROW IDIR OF TCO// INTO LA2R
  3      U(NLA)=Q(IDIR,IR)
C        STORE LOGIC OF LA2R IN KV.
         IKV=IA-NDU
         IU(IKV)=JD
         IF(NBOT)6,1,1
  1      CONTINUE
 100     CONTINUE
C        LA2R IS NOW AVAILABLE, STORED AS TABLE ONLY.
C
C     END STEP 1
C     STEP 2
C        NEXT STEP SET UP K33 AND INVERT. M
C        LOGICAL TRIPLE PRODUCT (LA2R) (SQARE) (LA2RTRANSP)      =K33
C
  6      CALL STEP2
C     STEP 3
C   INVERT K33      VIA BUGER
C
 21      IA=IA
         IF(NJR-1)62,62,67
 62      I=K33+1
         IF(U(I))66,622,66
 66      U(I)=1./U(I)
         GO TO 61
C ALOCAT NSCR6 AND USE IT AS UNIT MATRIX
 67      CALL DEFINE(NSCR6,NJR*(NJR+1),0,0,1)
         CALL ALOCAT(NSCR6)
C  CLEAR NSCR6
         CALL CLEAR(NSCR6)
C        ESTABLISH UNIT MATRIX TO INVERT K33
         DO 60 I=1,NJR
         IA=NSCR6+(I-1)*NJR+I
 60      U(IA)=1.0
         NDSS=NJR
```

```
      IAVEC=NSCR6+NDSQ
      C=0.0000
      CALL BUGER   (NDSS,NDSS,NDSS,K33  ,NSCR6 ,C,IAVEC)
      CALL RELEAS(NSCR6,0,0)
 65    GO TO (61,622,622),NDSS
  622 ISUCC=2
      CALL PRER2(7,2,NJR)
 8999 GO TO 9000
C      K33(-1) IS NOW IN K33
C   END STEP 3
C   STEP 4
C DEFINE MEGAO AND CLEAR ITS DATA ARRAYS
C MEGAO WILL HAVE NDAT ARRAYS  EACH CONTAINS JF ROWS OF MEGAO
C CLEAR CODE WORD ARRAY BEFORE DEFINING DATA.NOT ALL DATA MAY BE ALLOCATED
CAND CODEWORDS ARE TESTED FOR ZERO IN FOMOD STEP5 (MODS)
 61    CALL DEFINE(MEGAO,0,0,1,0,1)
      CALL ALOCAT(MEGAO,0)
      CALL DEFINE(MEGAO,0,NDAT,1,0,1)
C      TRIPLEPROD LA2RT*K33(-1)*LA2R.JRELES
C SUBSCRIPTS ARE ( I,L)*(L,M)*(M,J)
      NDU=0
      DO 2000 L=1,NJR
      I=KV+L
      I=IU(I)
      IF(I-NDU)1602,1601,1602
 1602 NDU=I
C STATEMENT 1600 IS ONLY REACHED 1 TIME     (SPECIAL PROPERTY OF LA2R
 1600   CALL DEFINE(MEGAO,I,NDJ*JF,0,0,1)
      CALL ALOCAT(MEGAO,I)
      IRES=MEGAO+I
      CALL CLEAR(IU(IRES))
 1601 DO 1900 II=1,JF
      I1=LA2R+(L-1)*JF+II
      DO 1800 M=1,NJR
      I2=K33+(M-1)*NJR+L
      IF(U(I2))1701,1800,1701
 1701 J=KV+M
      J=IU(J)
      DO 1750 JJ=1,JF
      I3=LA2R+(M-1)*JF+JJ
      IRESU=IU(IRES)+JJ+(II-1)*NDJ+(J-1)*JF
 1750 U(IRESU)=U(IRESU)+U(I1)*U(I2)*U(I3)
 1800 CONTINUE
 1900 CONTINUE
      CALL RELEAS(MEGAO,I)
 2000 CONTINUE
C
C   MEGAO NOW CONTAINS OMEGA ZERO  SIZE = (NDAT*JF)*(NDAT*JF)
      CALL RELEAS(K33,0,0)
C
 1604 NDU=0
C   END STEP 4
C   STEP 5
C   MULTIPLY ATKA//*(K33(-1))*A//TKA
C
      IF(NFJS)9000,9000,8001
 8001 CALL STEP5
 9000 CALL RELEAS(LA2R,0,0)
      CALL RELEAS(KV,0,0)
 9001  RETURN
```

```
          END
*         LABEL
*         LIST8
          SUBROUTINE MATRIP(K1,K2,NT)
C         SUBROUTINE MATRIP COMPUTES THE MATRIX TRIPLE PRODUCT
C          T(TRANSPOSE)*Y*T OR T*Y*T(TRANSPOSE)
C         AND ADDS THE RESULT INTO POOL AT K2
C         KEN REINSCHMIDT, ROOM 1-255, EXT. 2117
C         STRESS PROGRAM...STRUCTURAL ENGINEERING SYSTEMS SOLVER
C         VERSION III...22 AUGUST 1963
C         K1 = PCOL LOCATION OF THE MIDDLE MULTIPLIER
C         K2 = PCOL LOCATION WHERE PRODUCT IS TO BE STORED
C         NT = 1 IF THE PREMULTIPLIER IS TO BE TRANSPOSED
C         NT = 2 IF THE POSTMULTIPLIER IS TO BE TRANSPOSED
C         T IS ASSUMED FILLED BY TRAMAT
C         DIMENSION STATEMENT FOR ARRAYS IN COMMON
          DIMENSION U(36),IU(36),Y(6,6),T(6,6),JUNK(20),
         1 SYSFIL(27),PROFIL(6), KRAY(10)
          COMMON U,IU,Y,T,K3,K4,IDOWN,JDOWN,LDOWN,IUP,JUP,LUP,JBOTH,LBOTH,
         1JBIG,JSML,LPLS,LMIN,NUP,MUP,JUNK,
         4CHECK,NMAX,INORM,ISOLV,ISCAN,IIII,IMOD,JJJJ,ICONT,ISUCC,IMERG,
         5TOP,N1,NL,NT,NREQ,TN,LFILE,TCLER,IPRG,IRST,IRLD,IRPR,SYSFIL,
         8NJ,NB,NDAT,ID,JF,NSQ,NCORD,IMETH,NLDS,NFJS,NSTV,NMEMV,IPSI,NMR,
         9NJR,ISODG,NDSQ,NDJ,IPDBP,IUDBP,NBB,NFJS1,JJC,JDC,JMIC,JMPC,JLD
          COMMON JEXTN,MEXTN,LEXTN,JLC,NLDSI,IYOUNG,ISHEAR,IEXPAN,IDENS,
         1 PROFIL,
         4NAME,KXYZ,KJREL,JPLS,JMIN,MTYP,KPSI,MEMB,LOADS,MCDN,KS,KMKST,KSTDB
         5,KATKA,KPPLS,KPMNS,KUV,KPPRI,KR,KSAVE,KRAY,KDIAG,KOFDG,
         6 IOFDG,LOADN,MEGAO,JEXT,JINT,KUDBP,KMEGA,KPDBP,JTYP,MTYP1,KB
         7,MLOAD,JLOAD,KATR,LEXT,KYOUNG,KSHEAR,KEXPAN,KDENS
          EQUIVALENCE (U,IU,Y)
       10 DO 100 I = 1,JF
          DO 100 L = 1,JF
          K3 = K2+(L-1)*JF+I
          DO 100 J = 1,JF
          DO 100 K = 1,JF
          K4 = K1+(K-1)*JF+J
          GO TO (50,60),NT
C         TRANSPOSE THE PREMULTIPLIER
       50 U(K3) = U(K3)   +T(J,I)*U(K4)*T(K,L)
          GO TO  100
C         TRANSPOSE THE POSTMULTIPLIER
       60 U(K3)   = U(K3)+T(I,J)*U(K4)*T(L,K)
      100 CONTINUE
          RETURN
          END
*         FAP
*         ADDRESS ROUTINE FOR NEW BOOKKEEPING SYSTEM
*         R.D.LOGCHER  2/9/64
          COUNT   150
          LBL     ADRESS
          ENTRY   ADRESS
ADRESS SXA       END,1
          SXA     END+1,2
          CLA*    1,4                SAVE FOR UNCHANGED CALL SEQUENCE
          STO     N
          CLA*    2,4                DITTO
          STO     I
          CLA*    4,4                OPERATION NUMBER
          TNZ     BITC               TRANSFER NOT PLACING BIT
```

```
        TSX     COUNT,1
        ORS     U+1,2
END     AXT     **,1
        AXT     **,2
        TRA     5,4
BITC    SUB     TWO
        STO     TEST
        TZE     SEEK                    FIND ARRAY
        TPL     SIZEK                   FIND NUMBER OF ARRAYS IN COLUMN
        CLA     NFJS
        STO     TEMP
SEEK    TSX     COUNT,1
        ANA     U+1,2
        TZE     NPRES                   BIT AND ARRAY NOT PRESENT
        SXA     RNT,4
        CALL    ALOCAT,IOFDG,N
RNT     AXT     **,4
        CLA     IOFDG                   FIND ARRAY POSITION IN KOFDG
        ADD     N
        PDX     ,2
        CLA     U+1,2
        PDX     ,2
        CLA     U,2
        PDX     ,1                      NUMBER OF ARRAYS IN ROW
        PXA     ,2
        ADD     =1
        SSM
        ADD     TOP
        STA     LOOK
LOOK    CLA     **,1
        ANA     =0077777000000
        LAS     I                       FIND ARRAY POSITION
        TRA     *+2
        TRA     FOUND
        TIX     LOOK,1,1
        TSX     $EXIT,4
FOUND   CLA*    LOOK
        ALS     18
        SLW*    3,4
        TRA     END
NPRES   STZ*    3,4                     RETURN NOT FOUND INDICATION
        NZT     TEST                    RECYCLE FOR COLUMN SEARCH
        TRA     ENR
        CLA     TEMP
        SUB     N
        TZE     END
        TMI     END                     MORE ELEMENTS IN COLUMN
        CLA     N
        ADD     ONE
        STO     N                       UPDATE N
        STO*    1,4
        TRA     SEEK
ENR     SXA     TNR,4
        CALL    ALOCAT,IOFDG,N
TNR     AXT     **,4
        TRA     END
SIZEK   SUB     ONE
        TZE     *+4
        CLA     NJ
        STO     TEMP
```

```
            TRA     SEEK
            SXA     DNE,4
            CLA     I
            ADD     ONE
            STO     N
            AXT     0,4
     SZE    TSX     COUNT,1
            ANA     U+1,2
            TZE     *+2
            TXI     *+1,4,1                 ELEMENT PRESENT,INCREMENT COUNTER
            CLA     NFJS
            SUB     N
            TZE     *+5
            CLA     N
            ADD     ONE
            STO     N
            TRA     SZE
            PXD     ,4
     DNE    AXT     **,4
            STO*    3,4
            TRA     END
     *      CALCULATE BIT POSITION,DETERMINE WORD POSITION, AND READY FOR USE
     *      BIT COUNTER
     COUNT  CLA     I
            ADD     ONE
            XCA
            MPY     I
            LRS     36
            DVH     =2
            STQ     SUM
            CLA     I
            SUB     ONE
            XCA
            MPY     NJ
            ARS     1
            SUB     SUM
            STO     SUM
            CLA     N
            ARS     18
            ADD     SUM
            STO     SUM                     BIT COUNT,RIGHT JUST. IN WORD
     *      ARRAY COUNT
            SUB     =1
            LDQ     =0
            XCA
            DVH     =7200
            XCA                             REMAINDER IN MQ
            ADD     =1
            ALS     18                      ARRAY COUNT TO DECREMENT
            STO     M
            CLM
            DVH     =36
            STO     L                       BIT POSITION - 1
            XCA
            ADD     =1
            ALS     18
            STO     J                       WORD NUMBER
     *      IS ARRAY OF IFDT IN CORE
            CLA     IFDT
            TPL     TRN
```

```
        SXA     TRN-2,4
        CALL    ALOCAT,IFDT,ZERO
        AXT     **,4
        CLA     IFDT
TRN     ADD     M
        PDX     ,2
        CLA     U+1,2
        TPL     CONTN
        SXA     RET,4
        CALL    ALOCAT,IFDT,M
        CALL    RELEAS,IFDT,M
RET     AXT     **,4
        CLA     IFDT
        ADD     M
        PDX     ,2
        CLA     U+1,2
CONTN   ADD     J
        PDX     ,2
        CLA     L
        STA     *+2                        PREPARE BIT MASK
        CLA     =1
        ALS     **
        TRA     1,1
N       PZE
I
TEST
SUM
M
L
J
TEMP    PZE
ZERO    PZE
ONE     OCT     1000000
TWO     OCT     2000000
U       COMMON  119
TOP     COMMON  39
NJ      COMMON  9
NFJS    COMMON  65
IOFDG   COMMON  5
IFDT    COMMON  1
        END
*       LIST8
*       SYMBOL TABLE
*       LABEL
        SUBROUTINE STEP2
C  LOGICAL TRIPLE PRODUCT  (LA2R) (A//TKA//) (LA2RTRANSP)
C  NOTE IOFDG MUST BE OF SIZE((NJ-1)*NJ/2) FOR IT MUST INCLUDE SUPPORT PORTION
C  REVISION JAN 9 64  S M TO PROPERLY RELEAS ARRAYS
C   S M REVISION FOR NEW SOLVER BOOKKEEPING SYSTEM
        DIMENSION  Y(6,6),T(6,6),Q(6,6),U(36),IU(36),SYPA(40),FILL(20)
        COMMON U,T,Q,CHECK,NMAX,INORM,ISOLVE,ISCAN,IIII,IMOD,JJJJ,ICONT,
       1ISUCC,SYPA,NJ,NB,NDAT,ID,JF,NSQ,NCORD,IMETH,NLDS,NFJS,NSTV
       2,NMEMV,IPSI,NMR,NJR,ISODG,NDSQ,NDJ,IPB,IUPB,NBB ,NFJS1,FILL,
       3NAME,KXYZ,KJREL,JPLS,JMIN,MTYP,KPSI,MEMB,
       4LOADS, INPUT,KS,KMKST,KSTDB,KATKA,KPPLS,KPMIN,KUV,KPPRI,KR,KMK,
       5KV,K33,LA2R ,LA2RT,NSCR5,NSCR6,NSCR7,NSCR8,NSCR9,NSCR10,KDIAG,KOFD
       6G,IOFDG,LDNM,MEGAO,JEXT,JINT,KUDBP,KMEGA,KPDBP,JTYP,MTYP1
        EQUIVALENCE(U(1),IU(1),Y(1))
    6   CALL DEFINE(K33,NDSQ,0,1,1)
        CALL CLEAR(K33)
```

348

```
          CALL ALOCAT(K33)
          DO 55 K=1,NJR
          IKV=KV+K
          I=IU(IKV)
          DO 50 J=1,NJR
          IAP=KV+J
          IP=IU(IAP)
C      COMPUTE ADDRESS OF SUBMATRIX, IN SQUARE ARRAY.
C   INDICES OF TRIPLE PRODUCT ARE  (K,I)*(I,IP)*(IP,J)
          INS=I+NFJS
   1      IF(I-IP)14,11,13
C KDIAG AND KOFDG ARE SECOND LEVEL ARRAYS
    11 CALL ALOCAT(KDIAG,INS)
          NA2=KDIAG+INS
          N2=IU(NA2)
          GO TO 122
   13     IR=I
          IC=IP
          IND=0
          GO TO 16
   14     IR=IP
          IC=I
          IND=1
   16     CALL ADRESS(NFJS+IR,NFJS+IC,NA2,2)
   15     IF(NA2)50,50,120
  120     CALL ALOCAT(KOFDG,NA2)
          N2=KOFDG+NA2
          N2=IU(N2)
  122     N1=(K-1)*JF+LA2R
          N3=(J-1)*JF+LA2R
          N4=K33+(J-1)*NJR+K-1
C      COMPUTE THIS ELEMENT OF K33
          CALL MAPROD(N1,N2,N3,N4,1,JF,IND)
   56     IF(I-IP)140,110,140
  110     CALL RELEAS(KDIAG,INS)
          GO TO 50
  140     CALL RELEAS(KOFDG,NA2)
          CALL RELEAS(IOFDG,NFJS+IR)
C      FIRST 4 ARG. ARE U-ADDRESSES. LAST
C      TWO ARE SIZES OF ARRAYS.
   50     CONTINUE
   55     CONTINUE
C
C   END STEP 2
C      TRIPLE PROD. IS COMPUTED AND STORED.  IN K33.FULL SIZE (NJR*NJR)
          RETURN
          END
*      LIST8
*      LABLE
*      SYMBOL TABLE
          SUBROUTINE TTHETA(KJND,ID,JF)
          DIMENSION  U(72),Q(6,6)
          COMMON  U,Q
          CALL COPY(Q,Q,JF,0,0)
          PIMOD=57.2957795
  204     THA1=U(KJND+1)/PIMOD
          CT1=COSF(THA1)
          ST1=SINF(THA1)
          IF(ID-3)300,300,301
  300     THA2=0.0
```

```
      THA3=0.0
      GO TO 302
301   THA2=U(KJND+2)/PIMOD
      THA3=U(KJND+3)/PIMOD
302   CT2=COSF(THA2)
      ST2=SINF(THA2)
      CT3=COSF(THA3)
      ST3=SINF(THA3)
      IF(ID-3)304,303,304
304   I1=1
      I2=2
      I3=3
      GO TO 305
303   I1=2
      I2=3
      I3=1
305   Q(I1,I1)=CT1*CT2
      Q(I1,I2)=ST1*CT2
      Q(I2,I1)=-CT3*ST1
      Q(I2,I2)=CT3*CT1-ST1*ST2*ST3
      Q(I3,I3)=1.0
      IF(ID-4)306,307,307
307   Q(3,1)=-CT3*ST2*CT1+ST3*ST1
      Q(3,2)=-CT3*ST2*ST1-ST3*CT1
      Q(3,3)=CT3*CT2
      Q(1,3)=ST2
      Q(2,3)=ST3*CT2
      IF(ID-4)306,306,308
308   DO 309 I=1,3
      DO 309 J=1,3
309   Q(I+3,J+3)=Q(I,J)
C     Q NOW CONTAINS TCO//(ROTATION)
306   RETURN
      END
*     LIST8
*     LABEL
*     SYMBOL TABLE
      SUBROUTINE MAPROD(N1,N2,N3,N4,IS,JF,IND)
C     TRIPLE OR DOUBLE PROD. OF U-ARRAYS
C     N1*N2*N3=N4
C     ARRAYS ARE STORED COLUMN WISE
C  N2 IS SQUARE
C     IS=COLUMN ORDER OF N3=ROW ORDER OF
C  TRIPLE PROD   N2 IS NOT TRANSP      IND=0
C  TRIPLE PROD   N2 IS     TRANSP      IND=1
C  DOUBLE PROD   N2*N3    NOTHING TRANSP   IND=2
C  DOUBLE PROD N2*N3  N3 IS TRANSP IND=3
C    IN THIS CASE N3 MUST BE JF*JF   I.E.   JF=IS
C  N2*N3   N2 IS TRANSP    IND=4
C  1 IS TRIPLE PROD
      DIMENSION  U(36)
      COMMON  U
10    IF(IND-2)1,2,2
1     DO 100  I=1,IS
      DO 100  K=1,IS
      NA4=N4+(K-1)*IS+I
      DO 90   N=1,JF
      NA1=N1+(N-1)*IS+I
      DO 90   M=1,JF
      NA3=N3+(K-1)*JF+M
```

```
11      IF(IND)4,4,3
 4      IC=M
        IR=N
        GO TO 5
 3      IC=N
        IR=M
 5      NA2=N2+(IC-1)*JF+IR
90      U(NA4)=U(NA4)+U(NA1)*U(NA2)*U(NA3)
100     CONTINUE
        GO TO 1000
 2      IF(IND-3)6,6,7
 6      DO 200  N=1,JF
        DO 200   K=1,IS
        NA4=N4+(K-1)*JF+N
        DO 190  M=1,JF
        NA2=N2+(M-1)*JF+N
12      IF(IND-2)104,104,103
104     IC=K
        IR=M
        GO TO 105
103     IC=M
        IR=K
105     NA3=N3+(IC-1)*JF+IR
190     U(NA4)=U(NA4)+U(NA2)*U(NA3)
200     CONTINUE
        GO TO 1000
 7      DO 300   N=1,JF
        DO 300   K=1,IS
        NA4=N4+(K-1)*JF+N
        DO 290  M=1,JF
        NA3=N3+(K-1)*JF+M
        NA2=N2+(N-1)*JF+M
290     U(NA4)=U(NA4)+U(NA2)*U(NA3)
300     CONTINUE
1000    RETURN
        END
*       LIST8
*       LABEL
CMAIN5/6
        DIMENSION  Y(6,6),T(6,6),Q(6,6),U(36),IU(36),SYPA(40),FILL(6)
        COMMON U,T,Q,CHECK,NMAX,INORM,ISOLV,ISCAN,IIII,IMOD,ILINK,ICONT,
       1ISUCC,SYPA,NJ,NB,NDAT,ID,JF,NSQ,NCORD,IMETH,NLDS,NFJS,NSTV
       2,NMEMV,IPSI,NMR,NJR,ISODG,NDSQ,NDJ,SDJ,NPR,NBB,NFJS1,JJC,JDC,
       3JMIC,JMPC,JLD,JEXTN,MEXTN,LEXTN,JLC,NLDSI,IYOUNG,ISHER,IEXPAN,
       4IDENS,FILL,
       5NAME,KXYZ,KJREL,JPLS,JMIN,MTYP,KPSI,MEMB,
       6LOADS, INPUT,KS,KMKST,KSTOB,KATKA,KPPLS,KPMIN,KUV,KPPRI,KR,KMK,
       7KV,K33,LA2R ,LA2RT,NV    ,NTP    ,NSCR7,NSCR8,NSCR9,NSCR10,KDIAG,KOFD
       8G,IOFDG,LDNM,MEGAO,JEXT,JINT,KUDBP,KMEGA,KPDBP,JTYP,MTYP1,KB,MLOAD
       9,JLOAD,KATR,LEXT,KYOUNG,KSHER,KEXPAN,KDENS
        EQUIVALENCE(U(1),IU(1),Y(1))
        CALL SOLVER
        ISUCC=ISUCC
        GO TO (1,2),ISUCC
 1      ISOLV=10
        CALL CHAIN(6,A4)
 2      CALL CHAIN(1,A4)
        END
*       LIST8
*       LABEL
```

```
*       SYMBOL TABLE
        SUBROUTINE SOLVER
C  THIS SUBROUTINE SOLVES THE GOVERNIG JOINT EQUILIBRIUM EQUATIONS FOR
C THE DISPLACEMENTS OF THE FREE JOINTS
C    CALLED ONLY IF NFJS IS POS
C STRESS III
C  S MAUCH  NOV 15 63  (SOLVER AND BACKSU NOW ARE CNE SUBROUTINE)
C REVISION DEC 31 63  TO NOT REDEFINE KOFDG(0) EVERY TIME THERE IS A NEW ARRAY
        DIMENSION  Y(6,6),T(6,6),Q(6,6),U(36),IU(36),SYPA(40),FILL(9),
       1FIL(4),FILL1(5)
        COMMON U,T,Q,CHECK,NMAX,INORM,ISOLVE,ISCAN,IIII,IMOD,JJJJ,ICCNT,
       1ISUCC,SYPA,NJ,NB,NDAT,ID,JF,NSQ,NCORD,IMETH,NLDS,NFJS,NSTV
       2,NMEMV,IPSI,NMR,NJR,ISODG,NDSQ,NDJ,IPB,IUPB,NBB,NFJS1,FILL,NLDSI,F
       3IL,NBNEW,FILL1,NAME,KXYZ,KJREL,JPLS,JMIN,MTYP,KPSI,MEMB,
       4LOADS, INPUT,KS,KMKST,KSTDB,KATKA,KPPLS,KPMIN,KUV,KPPRI,KR,KMK,
       5KV,K33,LA2R ,LA2RT,NSCR5,NSCR6,NSCR7,NSCR8,NSCR9,NSCR10,KDIAG,KOFD
       6G,IOFDG,LDNM,MEGAO,JEXT,JINT,IFDT ,KMEGA,KPDBP,JTYP,MTYP1,IOFC
        EQUIVALENCE(U(1),IU(1),Y(1))
    3   IIII=2
C DEFINE TRANSPOSE OF IOFDG = IOFC
        CALL DEFINE(IOFC,0,NFJS,1,0,1)
        CALL ALOCAT(IOFC,0)
        CALL ALOCAT(KDIAG,0)
        CALL ALOCAT(KOFDG,0)
        CALL ALOCAT(IOFDG,0)
        CALL ALOCAT(KPPRI)
  203   DO 998 K=1,NFJS
C
C       ELIMINATE KTH ROW
C
C   LE IS CURRENT LENGTH OF COLUMN REPRESENTATION OF IOFDG COLUMN K
        LE=0
        CALL ALOCAT(KDIAG,K)
        IA=KDIAG+K
        IV=IU(IA)
        C=0.0
        DO 200  JQ=1,JF
        K1=72+(JQ-1)*JF
        DO 199 JW=1,JF
        LNK=K1+JW
  199   U(LNK)=0.0
        LKI=K1+JQ
  200   U(LKI)=1.0
  204   JQ=JF
        CALL BUGER(JF,JQ,JQ,IV,72,C,36)
        GO TO (205,2,2),JQ
    2   ISUCC=2
        CALL PRER2(8,K,JQ)
C PRER2 PRINTS 'SING MATRIX OR OVERFLOW)
        GO TO 1004
  205   K1=K+1
        IF(K-NFJS)1099,1002,1002
C1002 LAST SUBMATRIX INVERTED
C
C DETERMINE HOW MANY ELEMENTS ARE IN COLUMN K
C
 1099 CALL ADRESS(0,K,LTT,3)
        CALL DEFINE(IOFC,K,LTT,0,0,1)
C IF LTT   ROW K DOES NOT HAVE ANY CFF DIAG ELEMENTS
        IF(LTT)198,996,198
```

```
      198   CALL ALOCAT(IOFC,K)
            N=K1
     1100   CALL ADRESS(N,K,IANK,1)
C ADRESS WILL RETURN WITH A NEW N
C CHECK IF NO LEADING ELEMENTS BELOW IN COLUMN K
            IF(IANK)206,997,206
C 206 RECORD THE IOFC INFORMATION
      206   LE =LE+1
            JW=IOFC+K
            JW=IU(JW)+LE
            CALL PADP(IU(JW),0,N,IANK)
            CALL ALOCAT(KOFDG,IANK)
            CALL ALOCAT(KDIAG,N)
            LNK=KOFDG+IANK
            LNK=IU(LNK)
            IV=KDIAG+K
            IV=IU(IV)
            CALL MAPRDT(0,LNK,IV,0,JF,JF,3)
            CALL RELEAS(KOFDG,IANK)
C         RESULT IN Y=U(1).ANK*AKK(-1)
C FIRST MODIFY DIAG ELEMENT IN N TH ROW
      790   LNN=KDIAG+N
      801   CALL MAPRDT(0,0,LNK,IU(LNN),-JF,JF,-4)
            CALL RELEAS(KDIAG,N)
C  PROD   A(N,K)*(A(K,K)-1)*A(K,I)
C PRODUCT IS DIRECTLY SUBTRECTED FROM KDIAG(N)
C CHECK IF LEADING ELEMENT JUST FOUND IS IN ROW K1    (OR IF NO OFF DIAG ELEMENTS
C HAVE TO BE MODIFIED    THEN LEE=0)
      6     LEE=LE-1
            IF(LEE)991,991,901
C 901 GO ACROSS ROWS N AND K TAKING LINEAR COMB
C START OF I LOOP ACROSS ROW K
      901   DO 800 IM=1,LEE
            IND=IOFC+K
            IND=IU(IND)+IM
            CALL UPADP(IU(IND),D,I,IAKI)
            CALL ADRESS(N,I,IANI,2)
C    ADRESS DETERNINES IF THERE IS A BIT FOR ELEMENT ( N,I)
C IANI RETURN =0 IF NOT
            IF(IANI)212,802,212
C FOR 802 NEW KOFDG IS CREATED
C      802 MEANS THAT ATKA DOES NOT HAVE
C      THIS ELEMENT, BUT ATKA(-1) DOES
      802   NBB=NBB+1
            IANI=NBB
C  UPDATE IOFDG(N,I)=NBB
            IA=IOFDG+N
            IA=IU(IA)+1
            CALL UPADP(IU(IA),D,LCU,LDE)
            LCU=LCU+1
            IF(LDE-LCU)440,440,441
      440   LDE=XMINOF(N-LCU+1,LE/2+2)+LDE
            CALL DEFINE(IOFDG,N,LDE,0,1,1)
            IA=IOFDG+N
            IA=IU(IA)+1
C UPDATE IOFDG(N) FOR NEW LENGTH
      441   CALL PADP(IU(IA),0,LCU,LDE)
            CALL RELEAS(IOFDG,N)
C UPDATE IOFDG(N) FOR NEW ELEMENT
            IA=IA+LCU
```

```
      CALL PADP(IU(IA),0,I,NBB)
C UPDATE IFDT FOR NEW ELEMENT        I SMALLER THAN N
      CALL ADRESS(N,I,0,0)
C TEST WHETHER TO REDEFINE KOFDG(0)
C NBNEW IS PRESENT LENGHT OF KOFDG(0)          DEFINED
      IF(NBB-NBNEW)449,449,448
C     REDEFINE KOFDG FOR NEW SUBMATRIX
 448  NBNEW=NBNEW+3+XMINOF(300,(NB*NB)/(NJ*4))
      CALL DEFINE(KOFDG,0,NBNEW,1,1,1)
 449  CALL DEFINE(KOFDG,NBB,NSQ,0,0,1)
 212  CALL ALOCAT(KOFDG,IANI)
      CALL ALOCAT(KOFDG,IAKI)
      LKI=KOFDG+IAKI
      IA=KOFDG+IANI
      CALL MAPRDT(0,0,IU(LKI),IU(IA),-JF,JF,-4)
C MULTIPLIES AND SUBTRACTS DIRECTLY
C MODIFIED IAKI SUBTRACTED FROM IANI
 805  CALL RELEAS(KOFDG,IANI)
      CALL RELEAS(KOFDG,IAKI)
 800  CONTINUE
C
C     NEXT REDUCE LOADS IN KTH ROW
C
 991  DO 700  J=1,NLDSI
      KPPJ=KPPRI+(J-1)*NSTV
      ICK=KPPJ+JF*(K-1)
 213  ICN=KPPJ+JF*(N-1)
C ADD DIRECTLY TO ICN
      CALL MAPRDT(0,0,ICK,ICN,-1,JF,-3)
C
C     PUT REDUCED LOAD INTO ICN
C
 700  CONTINUE
 999  CALL RELEAS(IOFDG,N)
1000  N=N+1
      IF(N-NFJS)1100,1100,997
 997  CALL RELEAS(IOFC,K)
 996  CALL RELEAS(IOFDG,K,0,0)
 998  CALL RELEAS(KDIAG,K)
1002  CALL RELEAS(IOFDG,0)
      CALL RELEAS(IFDT,0)
C
C  NOW BACKSUBSTITUTION
C1002 MULT. LAST DIAG. ELEMENT(-1) WITH LOADS
C     STORE JOINT DISPLACEMENTS IN KPPRI
C
      DO 400 KK=1,NFJS
      K=NFJS-KK
      K1=K+1
      CALL ALOCAT(KDIAG,K1)
  1   IF(KK-1)20,20,21
  21  K2=K+2
      CALL ALOCAT(IOFC,K1)
      JW =IOFC+K1
C NOTE IOFC DOES NOT HAVE A LEAD WORD WITH CURRENT AND DEFINED LENGTH
C SO GET LENTGH FROM CODEWORD ADDRESS
      CALL UPADP(IU(JW),D,LCU,LDE)
      IF(LDE)299,20,299
 299  JW=IU(JW)
      DO 300 IM=1,LDE
```

```
      LE=JW+IM
      CALL UPADP(IU(LE),D,I,IAKI)
C I=ROW NUMBBER IN WICH KOFDG(IAKI) IS LOCATED I LARGER THAN K1
      CALL ALOCAT(KOFDG,IAKI)
      LKI=KOFDG+IAKI
 301  DO 310  J=1,NLDSI
      IA=KPPRI+(J-1)*NSTV
      ICI=IA+JF*(I-1)
      ICK=IA+JF*K
C
C   SUBTRACT PROD A(I,K1)*C(I) FROM C(K)
C
      CALL MAPRDT(0,IU(LKI),ICI,ICK,-1,JF,-5)
 310  CONTINUE
      CALL RELEAS(KOFDG,IAKI)
 300  CONTINUE
      CALL RELEAS(IOFC,K1,0,0)
C
C  FINAL MULT OF INVERSE DIAG SUBMATRIX WITH MODIFIED LOAD    ALL  LOAD CONDITICN
C
 20   DO 320  J=1,NLDSI
      ICK=KPPRI+(J-1)*NSTV+K*JF
      IA=KDIAG+K1
      IA=IU(IA)
C
C  REPLACE CISPLACEMENTS INTO KPPRI(K)
C
      CALL MAPRDT(0,IA,ICK,ICK,1,JF,3)
 320  CONTINUE
 400  CALL RELEAS(KDIAG,K1)
 1003 CALL RELEAS(KPPRI)
      CALL RELEAS(KDIAG,0)
      CALL RELEAS(KOFDG,0)
      CALL RELEAS(IOFC,0,0,0)
 1004 RETURN
      END
*     LIST8
*     LABEL
*     SYMBOL TABLE
      SUBROUTINE MAPRDT(N1,N2,N3,N4,IZ,KF,INP)
C IF IZ IS NEG PRODUCT IS ADDED TO OR SUBTRACTED FROM CONTENTS OF N4
C IF(INP) IS NEG. -PRODUCT IS PLACED INTO N4  NOTE N4 IS CLEARED FIRST
C IF IZ IS POS ) NOT OTHERWISE
C    TRIPLE OR DOUBLE PROD. OF U-ARRAYS
C    N1*N2*N3=N4
C    ARRAYS ARE STORED COLUMN WISE
C N2 IS SQUARE
C    IS=COLUMN ORDER OF N3=ROW ORDER OF
C TRIPLE PROD    N2 IS NOT TRANSP     IND=1
C TRIPLE PROD    N2 IS     TRANSP     IND=2
C DOUBLE PROD    N2*N3    NOTHING TRANSP    IND=3
C DOUBLE PROD N2*N3  N3 IS TRANSP IND=4
C   IN THIS CASE N3 MUST BE JF*JF    I.E.  JF=IS
C N2*N3    N2 IS TRANSP    IND=5
      DIMENSION U(72),Q(36)
      COMMON U,Q
      JF=KF
      IS=XABSF(IZ)
      IND=XABSF(INP)
      NSS=IS*JF
```

```
10     IF(IND-3)1,2,2
C  1 IS TRIPLE PROD
1      DO 100   I=1,IS
       DO 100   K=1,IS
       NA4=(K-1)*IS+I
       Q(NA4)=0.0
       DO 90  N=1,JF
       NA1=N1+(N-1)*IS+I
       DO 90   M=1,JF
       NA3=N3+(K-1)*JF+M
11     IF(IND-1)4,4,3
4      IC=M
       IR=N
       GO TO 5
3      IC=N
       IR=M
5      NA2=N2+(IC-1)*JF+IR
90     Q(NA4)=Q(NA4)+U(NA1)*U(NA2)*U(NA3)
100    CONTINUE
       GO TO 1000
C
C   DOUBLE PRODUCT
C
2      IF(IND-4)6,6,7
6      DO 200   N=1,JF
       DO 200   K=1,IS
       NA4=    (K-1)*JF+N
       Q(NA4)=0.0
       DO 190   M=1,JF
       NA2=N2+(M-1)*JF+N
12     IF(IND-3)104,104,103
104    IC=K
       IR=M
       GO TO 105
103    IC=M
       IR=K
105    NA3=N3+(IC-1)*JF+IR
190    Q(NA4)=Q(NA4)+U(NA2)*U(NA3)
200    CONTINUE
       GO TO 1000
7      DO 300   N=1,JF
       DO 300   K=1,IS
       NA4=    (K-1)*JF+N
       Q(NA4)=0.0
       DO 290   M=1,JF
       NA3=N3+(K-1)*JF+M
       NA2=N2+(N-1)*JF+M
290    Q(NA4)=Q(NA4)+U(NA2)*U(NA3)
300    CONTINUE
1000   IF(IZ)2000,1008,1008
1008   IF(INP)1001,1001,1002
1002   IF(N4-72)1005,1004,1005
1001   C=-1.0
       GO TO 1006
1005   C=1.0
1006   DO 1003   J=1,NSS
       NA4=N4+J
1003   U(NA4)=C*Q(J)
       GO TO 1004
2000   IF(INP)2001,2001,2002
```

```
      2001 C=-1.0
           GO TO 2006
      2002 C=1.0
      2006 DO 2003   J=1,NSS
           NA4=N4+J
      2003 U(NA4)=U(NA4)+C*Q(J)
      1004 RETURN
           END
    *      LIST8
    *      LABEL
CMAIN6/6
           DIMENSION  Y(6,6),T(6,6),Q(6,6),U(36),IU(36),SYPA(40),FILL(6)
           COMMON U,T,Q,CHECK,NMAX,INORM,ISOLV,ISCAN,IIII,IMOD,ILINK,ICONT,
          1ISUCC,SYPA,NJ,NB,NDAT,ID,JF,NSQ,NCORD,IMETH,NLDS,NFJS,NSTV
          2,NMEMV,IPSI,NMR,NJR,ISODG,NDSQ,NDJ,SDJ,NPR,NBB,NFJS1,JJC,JDC,
          3JMIC,JMPC,JLD,JEXTN,MEXTN,LEXTN,JLC,NLDSI,IYOUNG,ISHER,IEXPAN,
          4IDENS,FILL,
          5NAME,KXYZ,KJREL,JPLS,JMIN,MTYP,KPSI,MEMB,
          6LOADS, INPUT,KS,KMKST,KSTOB,KATKA,KPPLS,KPMIN,KUV,KPPRI,KR,KMK,
          7KV,K33,LA2R ,LA2RT,NV     ,NTP    ,NSCR7,NSCR8,NSCR9,NSCR10,KDIAG,KOFD
          8G,IOFDG,LDNM,MEGAO,JEXT,JINT,KUDBP,KMEGA,KPDBP,JTYP,MTYP1,KB,MLOAD
          9,JLOAD,KATR,LEXT,KYOUNG,KSHER,KEXPAN,KDENS
           EQUIVALENCE(U(1),IU(1),Y(1))
           CALL BAKSUB
           ISUCC=ISUCC
           GO TO (1,2),ISUCC
     1     ISOLV=11
     2     CALL CHAIN(1,A4)
           END
    *      LIST8
    *      LABEL
    *      SYMBOL TABLE
           SUBROUTINE BAKSUB
C          STRESS PROGRAM...STRUCTURAL ENGINEERING SYSTEMS SOLVER
C    VERS III   WITH COMBINATION LOADINGS 10/15  S MAUCH
C          USES AVECT TO BACKSUBSTITUTE AND
C          ANSOUT TO PRINT UNDER CONTROL OF IPRINT(JLD)
           DIMENSION U(32),IU(32),Y(32),T(6,6),JUNK(13),V(12),
          1 SYSFIL(27),PROFIL(6), KRAY(10)
           COMMON U,IU,Y,JLI,JLX,JLO,L,T,K1,K2,KU,KPMIN,KPPL,KUPR,JT,JTEST,
          1JUNK,KRR,KTEM,KSUP,V,
          4CHECK,NMAX,INORM,ISOLV,ISCAN,IIII,IMOD,JJJJ,ICONT,ISUCC,IMERG,
          5TOP,N1,NL,NT,NREQ,TN,LFILE,TOLER,IPRG,IRST,IRLD,IRPR,SYSFIL,
          8NJ,NB,NDAT,ID,JF,NSQ,NCORD,IMETH,NLDS,NFJS,NSTV,NMEMV,IPSI,NMR,
          9NJR,ISODG,NDSQ,NDJ,IPDBP,IUDBP,NBB,NFJS1,JJC,JDC,JMIC,JMPC,JLD
           COMMON JEXTN,MEXTN,LEXTN,JLC,NLDSI,IYOUNG,ISHEAR,IEXPAN,IDENS,
          1 PROFIL,
          4NAME,KXYZ,KJREL,JPLS,JMIN,MTYP,KPSI,MEMB,LOADS,MODN,KS,KMKST,KSTOB
          5,KATKA,KPPLS,KPMNS,KUV,KPPRI,KR,KSAVE,KRAY,KDIAG,KOFDG,
          6IOFDG,LOADN,MEGAO,JEXT,JINT,IFDT,KMEGA,KPDBP,JTYP,MTYP1,KB
          7,MLOAD,JLOAD,KATR,LEXT,KYOUNG,KSHEAR,KEXPAN,KDENS
           EQUIVALENCE (U,IU,Y)
C
      10 IF (NJR) 30,30,20
      20   IF(NFJS)255,32,255
      32   CALL ALOCAT(KPDBP)
C          NO FREE JOINTS, SUPPORT DISPLACEMENTS IN KPDBP FROM FOMOD NDJ*NLDSI
           GO TO 40
     255 CALL DEFSUP
           GO TO 40
```

```
   30    CALL DEFINE(KPDBP,NDJ,0,0,1)
C        NO JOINT RELEASES, KPDBP IS ZERO VECTOR NDJ IN LENGTH - THIS IS THE FIRST
C          DEFINITION OF KPDBP IF NJR=0
         CALL ALOCAT(KPDBP)
         CALL CLEAR(KPDBP)
   40    CALL ALOCAT(NAME)
         CALL RELEAS(KDIAG,0,0,0)
         CALL RELEAS(KOFDG,0,0,0)
         CALL RELEAS(IOFDG,0,0,0)
         CALL RELEAS(IFDT,0,0,0)
         IF (IPSI) 14,15,14
   14    CALL ALOCAT(KPSI)
   15    CALL ALOCAT(KS)
         CALL ALOCAT(KPPRI)
         CALL ALOCAT(KMKST,0)
         CALL ALOCAT(LOADS,0)
         CALL ALOCAT(KUV,0)
         CALL ALOCAT(KPPLS,0)
         CALL ALOCAT(KPMNS,0)
         CALL ALOCAT(JPLS)
         CALL ALOCAT(JMIN)
         CALL ALOCAT(LEXT)
         CALL ALOCAT(JINT)
         CALL ALOCAT(MTYP)
         IF(MODN)16,13,16
   16    CALL ALOCAT(MODN)
   13    CALL DEFINE(KR,0, NLDS,1,0,1)
         CALL ALOCAT(KR,0)
C        KR CONTAINS THE STATICS CHECKS FOR LOAD CONDITIONS
C        IF JLI=0, LOAD CONDITION IS DEPENDATN.  ALL CONDITIONS ON WHICH IT
C        DEPEND HAVE A SMALLER EXTERNAL NUMBER JLX,I.E. THEY HAVE BEEN
C        PROCESSED BY BAKSUB.
  100    DO 200 JLX=1,LEXTN
         CALL ALOCAT(LOADS,JLX)
         K1=LOADS+JLX
         JLO=IU(K1)+1
         CALL UPACW(U(JLO),K1,JUNK(2),JUNK(3),JUNK(4),JT)
 1775 IF(K1-1)200,18,18
C IF JLI-NLDSI IS POS WE HAVE A DEPENDANT LOADING COND.
   18    JLI=LEXT+JLX
         JLI=IU(JLI)
C JLI IS NOW INTERNAL LOADING NUMBER
C        ZERO THE KR MATRIX
   19    CALL DEFINE(KR,JLI,NJ*JF,0,0,1)
         CALL ALOCAT(KR,JLI)
         CALL ALOCAT(KUV,JLI)
         CALL ALOCAT(KPPLS,JLI)
         CALL ALOCAT(KPMNS,JLI)
         K1=KR+JLI
         CALL CLEAR(IU(K1))
         JSA=NLDSI+1-JLI
         IF(JSA)52,52,51
   52    CALL COMBLD
         GO TO 99
   51    CALL NEWADR
         IF(NFJS)39,38,39
   38    IF(NJR)39,101,39
C        DONT GO TO AVECT ONLY IF NO FREE JOINTS AND NO SUPPORT RELEASES
C     THEN MEMBER FORCES ARE FIXED END FORCES
   39    DO 503 L=1,MEXTN
```

```
        K1=MTYP+L
        IF(U(K1))37,503,37
C       CHECK FOR DELETED MEMBERS
37      K2=LFILE
        CALL ALOCAT(KMKST,L)
 31     IF(K2-LFILE)41,33,41
 41     CALL NEWADR
 33     CALL AVECT
503     CALL RELEAS(KMKST,L)
 101 CALL STATCK
C       RESTORE JLO
99      K1=LOADS+JLX
        JLO=IU(K1)+1
        JTEST=8
C I=1 FORCES   2=DISTORTIONS 3=DISPLACEMENTS   4=REACTIONS
 103 DO 110 J = 1,4
        IF(JT-JTEST) 109,105,105
 105    K1 = NAME+1
        K2 = NAME+12
        PRINT 300,(U(I),I=K1,K2)
 787    IF(MODN)106,107,106
 106    K1 = MODN+1
        K2 = MODN+12
        PRINT 302,(U(I),I=K1,K2)
 107    K1 = JLO+2
        K2=JLO+13
        PRINT 302 ,(U(I),I=K1,K2)
 108 CALL ANSOUT(J,0,1)
        JT = JT-JTEST
 109 JTEST = JTEST/2
 110 CONTINUE
        CALL RELEAS(KR,JLI)
        CALL RELEAS(KUV,JLI)
        CALL RELEAS(KPPLS,JLI)
        CALL RELEAS(KPMNS,JLI)
 190 CALL RELEAS(LOADS,JLX)
 200 CONTINUE
        CALL RELEAS(MODN)
        CALL RELEAS(KR,0)
        CALL RELEAS(LOADS,0)
        CALL RELEAS(KUV,0)
        CALL RELEAS(KPPLS,0)
        CALL RELEAS(KPMNS,0)
        CALL RELEAS(JPLS)
        CALL RELEAS(JMIN)
        CALL RELEAS(NAME)
        CALL RELEAS(JINT)
        CALL RELEAS(MTYP)
        CALL RELEAS(LEXT)
        CALL RELEAS(KPDBP)
        CALL RELEAS(KATKA)
        CALL RELEAS(KPPRI)
        CALL RELEAS(KMKST,0)
        CALL RELEAS(KS)
        CALL RELEAS(KPSI)
        RETURN
 300 FORMAT (1H1,3X,12A6)
 302 FORMAT(1H0,3X,12A6)
        END
*       LIST8
```

```
*        SYMBOL TABLE
*        LABEL
CAVECT
         SUBROUTINE   AVECT
C        DOES BACKSUBSTITUTION USING THE INCIDENCE TABEL (A)
C        KENNETH F. REINSCHMIDT, ROOM 1-255, EXT. 2117
C        STRESS...STRUCTURAL ENGINEERING SYSTEM SOLVER
C        VERSION III...23 AUGUST 1963
         DIMENSION U(32),IU(32),Y(32) ,T(6,6),          V(6),W(6),
        1 SYSFIL(27),PROFIL(6), KRAY(10)
         COMMON U,IU,Y,JLI,JLX,JLO,L,T,K1,K2,KU,KPMIN,KPPL,KUPR,JT,JTEST,
        1LMS,LPS,J,
        1JPS,JMS,LS,JL,JJ,K3,K4,K5,K6,K7,KRR,KTEM,KSUP,V,W,
        4CHECK,NMAX,INORM,ISOLV,ISCAN,IIII,IMOD,JJJJ,ICONT,ISUCC,IMERG,
        5TOP,N1,NL,NT,NREQ,TN,LFILE,TOLER,IPRG,IRST,IRLD,IRPR,SYSFIL,
        8NJ,NB,NDAT,ID,JF,NSQ,NCORD,IMETH,NLDS,NFJS,NSTV,NMEMV,IPSI,NMR,
        9NJR,ISODG,NDSQ,NDJ,IPDBP,IUDBP,NBB,NFJS1,JJC,JDC,JMIC,JMPC,JLD
         COMMON JEXTN,MEXTN,LEXTN,JLC,NLDSI,IYOUNG,ISHEAR,IEXPAN,IDENS,
        1 PROFIL,
        4NAME,KXYZ,KJREL,JPLS,JMIN,MTYP,KPSI,MEMB,LOADS,MODN,KS,KMKST,KSTDB
        5,KATKA,KPPLS,KPMNS,KUV,KPPRI,KR,KSAVE,KRAY,KDIAG,KOFDG,
        6 IOFDG,LOADN,MEGAO,JEXT,JINT,KUDBP,KMEGA,KPDBP,JTYP,MTYP1,KB
        7,MLOAD,JLOAD,KATR,LEXT,KYOUNG,KSHEAR,KEXPAN,KDENS
         EQUIVALENCE (U,IU,Y)
      30 LPS = JPLS+L
         LMS = JMIN+L
         LPS = IU(LPS)+JINT
         LMS = IU(LMS) + JINT
         LPS = IU(LPS)
         LMS = IU(LMS)
         JPS = JF*(LPS-1)
         JMS = JF*(LMS-1)
         LS = JF*(L-1)
         IF (LPS-NFJS) 50,50,55
C        PLUS NODE IS A FREE JOINT
      50 K1 = KUPR+JPS
         GO TO 60
C        PLUS NODE IS IN THE DATUM
      55 K1 = KSUP+JPS-NSTV
      60 IF (LMS-NFJS) 65,65,70
C        MINUS NODE IS FREE
      65 K2 = KUPR+JMS
         GO TO 75
C        MINUS NODE IS FIXED
      70 K2 = KSUP+JMS-NSTV
      75 K5 = KU+LS
C     GET TRANSLATION MATRIX, MINUS NODE TO PLUS NODE, IN GLOBAL
C     COORDINATES
         CALL TRAMAT(L,1)
C        TRANSLATE THE PLUS  END JOINT DISPLACEMENTS
         DO 100 I = 1,JF
         K3 = K2+I
C        LOAD IN MINUS END DEFLECTION
         V(I) =-U(K3)
         DO 100 J = 1,JF
         K4 = K1+J
C        SUBTRACT TRANSLATED PLUS  END DEFLECTION
C        TRANSPOSE TRANSLATION MATRIX FOR DISPLACEMENTS
         V(I) = V(I)+T(J,I)*U(K4)
     100 CONTINUE
```

```
C        GET ROTATION MATRIX, GLOBAL COORDINATES TO LOCAL
         CALL TRAMAT(L,2)
C        ROTATE MEMBER DISTORTION INTO MEMBER COORDINATES
         DO 200 I = 1,JF
         W(I) = 0.0
         K1 = K5+I
         DO 150 J = 1,JF
         W(I) = W(I)+T(I,J)*V(J)
  150 CONTINUE
C ADD TO EMB DIST POOL WITH DIFFERENCE IN JOINT DISPLACEMENTS
         U(K1) =  U(K1)+W(I)
  200 CONTINUE
C        NOW MULTIPLY BY STIFFNESS FOR FORCES
         K2 = KPMIN + LS
         DO 300 I = 1,JF
         V(I) = 0.0
         DO 250 J = 1,JF
         K1=KMKST+L
         K1=IU(K1)+(J -1)*JF+I
         V(I) = U(K1)*W(J) + V(I)
  250 CONTINUE
C        ADD FORCES TO P MINUS POOL
         K3 = K2+I
         U(K3) =  U(K3)+V(I)
  300 CONTINUE
C        FIND MEMBER LENGTH
         K1 = KS+(L-1)*(NCORD+1)+1
C        TRANSLATE ALONG MEMBER TO PLUS(LEFT) END
         GO TO (400,320,330,400,340),ID
C        PLANE FRAME
  320 V(3) = V(3)+U(K1)*V(2)
         GO TO 400
C        PLANE GRID
  330 V(3) = V(3)-U(K1)*V(1)
         GO TO 400
C        SPACE FRAME
  340 V(5) = V(5)-U(K1)*V(3)
         V(6) = V(6)+U(K1)*V(2)
C        ADD TO P PLUS POOL
  400 K2 = KPPL+LS
         DO 450 I = 1,JF
         K3 = K2+I
         U(K3) = U(K3)+V(I)
  450 CONTINUE
         RETURN
         END
*        LIST8
*        LABEL
*        SYMBOL TABLE
         SUBROUTINE COMBLD
C        SUBROUTINE COMBLD   S.MAUCH   11 OCTOBER 1963
C        THIS SUBROUTINE COMPUTES THE OUTPUT FOR A COMBINATION LOADING AND
C        STORES IT IN NLDSI+1 ST ARRAYS OF KPPLS,KPMNS,KUV,KR
         DIMENSION U(22),IU(22),Y(22 ),T(6,6),V(5),SPAC(2),
        1SYSFIL(27),PROFIL(6),KRAY(10)
         COMMON U,IU,Y,J10,J9,K10,K9,NK,NKPRI,NKPOP,NLSX,NBX,NBTJ,
        1JLI,JLX,JLO,L,T,SPAC,KU,KPMIN,KPPL,KUPR,JT,JTEST,KSA,NLSS,NBT,NLS,
        2FAC,K1,K2,K3,K4,K5,N2,N3,N4,KRR,N5,KSUP,L3,L4,L5,M2,M3,M4,M5,V,
        4CHECK,NMAX,INORM,ISOLV,ISCAN,IIII,IMOD,JJJ,ICONT,ISUCC,IMERG,
        5TOP,N1,NL,NT,NREQ,TN,LFILE,TOLER,IPRG,IRST,IRLD,IRPR,SYSFIL,
```

```
      8 NJ,NB,NDAT,ID,JF,NSQ,NCORD,IMETH,NLDS,NFJS,NSTV,NMEMV,IPSI,NMR,
      9NJR,ISODG,NDSQ,NDJ,IPDBP,IUDBP,NBB,NFJS1,JJC,JDC,JMIC,JMPC,JLD
       COMMON JEXTN,MEXTN,LEXTN,JLC,NLDSI,IYOUNG,ISHEAR,IEXPAN,IDENS,
      1PROFIL,
      4NAME,KXYZ,KJREL,JPLS,JMIN,MTYP,KPSI,MEMB,LOADS,MODN,KS,KMKST,KSTDB
      5,KATKA,KPPLS,KPMNS,KUV,KPPRI,KR,KSAVE,KRAY,KDIAG,KOFDG,
      6IOFDG,LOADN,MEGAO,JEXT,JINT,KUDBP,KMEGA,KPDBP,JTYP, MTYP1,KB
      7,MLOAD,JLOAD,KATR,LEXT,KYOUNG,KSHEAR,KEXPAN,KDENS
       EQUIVALENCE(U,IU,Y)
       NBTJ=NJ*JF
       CALL NEWADR
C
C CLEAR COMBINATION ARRAYS
C
       CALL CLEAR(KPMIN)
       CALL CLEAR(KPPL)
       CALL CLEAR(KU)
C
C CLEAR LAST 'SUBARRAY' OF KPPRI AND KPDBP
C
       DO 1 J=1,NSTV
       K1=KUPR+J
    1  U(K1)=0.0
       DO 2 J=1,NDJ
       K1=KSUP+J
    2  U(K1)=0.0
       K1=LOADS+JLX
       K1=IU(K1)+1
       CALL UPADP(U(K1),K2,NLS,K2)
C      NLS= NUMBER OF INDEPENDANT LOADINGS INVOLVED
       DO 10 I=1,NLS
       K1=LOADS+JLX
       K1=IU(K1)+13+2*I
       FAC=U(K1+1)
       K11=IU(K1)
C K11   1 IS NOW EXTERNAL LOAD NUMBER FOR THE I TH INDEPENDANT LOAD CONDITION
C      IN THE LIST FOR THE CURRENT DEPENDANT CONDITION  -  CONVERT TO INTERNAL
       K1=LEXT+K11
  801  K1=IU(K1)
       IF(K1)10,10,3
    3  CALL ALOCAT(KPMNS,K1)
       CALL ALOCAT(KPPLS,K1)
       CALL ALOCAT(KUV,K1)
       CALL ALOCAT(KR,K1)
       K2=KPMNS+K1
       K3=KPPLS+K1
       K4=KUV+K1
       K5=KR+K1
       CALL NEWADR
C      DO NOT TEST FOR DELETED MEMBERS
       DO 20 J=1,NMEMV
       L2=IU(K2)+J
       L3=IU(K3)+J
       L4=IU(K4)+J
       M2=KPMIN+J
       M3=KPPL+J
       M4=KU+J
       U(M2)=U(M2)+FAC*U(L2)
       U(M3)=U(M3)+FAC*U(L3)
   20  U(M4)=U(M4)+FAC*U(L4)
```

```
      DO 30 J=1,NBTJ
      L5=IU(K5)+J
      M5=KRR+J
   30 U(M5)=U(M5)+FAC  *U(L5)
C  DO COBINATION OF FREE AND SUPPORTS DISPLACEMENTS  ( IN KPPRI AND KPDBP)
C NOTE KPPRI AND KPDBP ARE FORST LEVEL ARRAYS (SHOULD BE MADE SECOND LEVEL TOO)
C
      NKPRI=KPPRI+(K1-1)*NSTV
      DO 70  J=1,NSTV
      NK=NKPRI+J
      K10=KUPR+J
   70 U(K10)=U(K10)+FAC*U(NK)
      IF(NJR)71,72,71
   71 NKPDP=KPDBP+(K1-1)*NDJ
      DO 80  J=1,NDJ
      NK=NKPDP+J
      J10=KSUP+J
   80 U(J10)=U(J10)+FAC*U(NK)
   72 CALL RELEAS(KPMNS,K1)
      CALL RELEAS(KPPLS,K1)
      CALL RELEAS(KUV,K1)
      CALL RELEAS(KR,K1)
   10 CONTINUE
   40 RETURN
      END
*     LIST8
*     LABEL
*     SYMBOL TABLE
      SUBROUTINE DEFSUP
C     STRESS...STRUCTURAL ENGINEERING SYSTEM SOLVER
C     VERSION II...29 JULY 1963
C     KENNETH F. REINSCHMIDT, ROOM 1-255, EXT. 2117
C
C     COMPUTES THE DEFLECTIONS OF THE SUPPORT JOINTS, IF THEY EXIST
C  PRODUCT MEGAO*A//TKA*KPPRI
C SUBSCRIPTS (L,I)*(I,J)*(J)
C
      DIMENSION U(36),IU(36),Y(6,6),T(6,6),JUNK(11),V(12),
     1 SYSFIL(27),PROFIL(6), KRAY(10)
      COMMON U,IU,Y,T,K1,K2,K3,K4,K5,K6,K7,K8,K9,ISTOP,INDAT,JUNK,
     1 KTEM,KSUP,V,
     4CHECK,NMAX,INORM,ISOLV,ISCAN,IIII,IMOD,JJJJ,ICONT,ISUCC,IMERG,
     5TOP,N1,NL,NT,NREQ,TN,LFILE,TOLER,IPRG,IRST,IRLD,IRPR,SYSFIL,
     8NJ,NB,NDAT,ID,JF,NSQ,NCORD,IMETH,NLDS,NFJS,NSTV,NMEMV,IPSI,NMR,
     9NJR,ISODG,NDSQ,NDJ,IPDBP,IUDBP,NBB,NFJS1,JJC,JDC,JMIC,JMPC,JLD
      COMMON JEXTN,MEXTN,LEXTN,JLC,NLDSI,IYOUNG,ISHEAR,IEXPAN,IDENS,
     1 PROFIL,
     4NAME,KXYZ,KJREL,JPLS,JMIN,MTYP,KPSI,MEMB,LOADS,MODN,KS,KMKST,KSTDB
     5,KATKA,KPPLS,KPMNS,KUV,KPPRI,KR,KSAVE,KRAY,KDIAG,KOFDG,
     6IOFDG,LOADN,MEGAO,JEXT,JINT,IFDT,KMEGA,KPDBP,JTYP,MTYP1,IOFC
     7,MLOAD,JLOAD,KATR,LEXT,KYOUNG,KSHEAR,KEXPAN,KDENS
      EQUIVALENCE (U,IU,Y)
      CALL ALOCAT(KPDBP)
      CALL ALOCAT(IOFDG,0)
    2 I=NFJS1
    3 II=I-NFJS
C CHECK IF MEGAO(II) IS NOT ALL ZERO
      K6=MEGAO+II
      IF(IU(K6))4,602,4
C   TEST IF IOFDG(I) CONTAINS ANY ELEMENTS
```

```
4      CALL ALOCAT(IOFDG,I)
       K6=IOFDG+I
       K6=IU(K6)+1
       CALL UPADP(IU(K6),D,LCU,LDE)
       IF(LCU)5,601,5
5      CALL ALOCAT(MEGAO,II)
       DO 600 JM=1,LCU
       K6=IOFDG+I
       K6=IU(K6)+1+JM
       CALL UPADP(IU(K6),D,J,K1)
       IF(J-NFJS)6,6,600
6      CALL ALOCAT(KOFDG,K1)
C   J= COLUMN ORDER OF KOFDG(K1)
       K2=KOFDG+K1
       K2=IU(K2)
       K6=MEGAO+II
 100 DO 500 K=1,NLDSI
       K3 = KPDBP+(K-1)*NDJ
       K4=KPPRI+(K-1)*NSTV+(J-1)*JF
       DO 500 L = 1,NDJ
       K5 = K3+L
       DO 500 M = 1,JF
       K7=IU(K6)+(M-1)*NDJ+L
       DO 500 N = 1,JF
       K8 = K2+(N-1)*JF+M
       K9 = K4+N
       U(K5) = U(K5)-U(K7)*U(K8)*U(K9)
 500 CONTINUE
 580 CALL RELEAS(KOFDG,K1)
600    CONTINUE
 601   CALL RELEAS(MEGAO,II,0,0)
602    CALL RELEAS(IOFDG,I,0,0)
       I=I+1
       IF(I-NJ)3,3,603
603    CALL RELEAS(KOFDG,0,0,0)
       CALL RELEAS(IOFDG,0,0,0)
       CALL RELEAS(MEGAO,0,0,0)
       RETURN
       END
*      LIST8
*      LABEL
*      SYMBOL TABLE
       SUBROUTINE NEWADR
C      STRESS PROGRAM...STRUCTURAL ENGINEERING SYSTEMS SOLVER
C   VERS III   WITH COMBINATION LOADINGS 10/15  S MAUCH
C      CALLED BY BAKSUB
       DIMENSION U(32),IU(32),Y(32),T(6,6),JUNK(13),V(12),
      1 SYSFIL(27),PROFIL(6), KRAY(10)
       COMMON U,IU,Y,JLI,JLX,JLO,L,T,K1,K2,KU,KPMIN,KPPL,KUPR,JT,JTEST,
      1JUNK,KRR,KTEM,KSUP,V,
      4CHECK,NMAX,INORM,ISOLV,ISCAN,IIII,IMOD,JJJJ,ICONT,ISUCC,IMERG,
      5TOP,N1,NL,NT,NREQ,TN,LFILE,TOLER,IPRG,IRST,IRLD,IRPR,SYSFIL,
      8NJ,NB,NDAT,ID,JF,NSQ,NCORD,IMETH,NLDS,NFJS,NSTV,NMEMV,IPSI,NMR,
      9NJR,ISODG,NDSQ,NDJ,IPDBP,IUDBP,NBB,NFJS1,JJC,JDC,JMIC,JMPC,JLD
       COMMON JEXTN,MEXTN,LEXTN,JLC,NLDSI,IYOUNG,ISHEAR,IEXPAN,IDENS,
      1 PROFIL,
      4NAME,KXYZ,KJREL,JPLS,JMIN,MTYP,KPSI,MEMB,LOADS,MCDN,KS,KMKST,KSTDB
      5,KATKA,KPPLS,KPMNS,KUV,KPPRI,KR,KSAVE,KRAY,KDIAG,KOFDG,
      6 IOFDG,LOADN,MEGAO,JEXT,JINI,KUDBP,KMEGA,KPDBP,JTYP,MTYP1,KB
      7,MLOAD,JLOAD,KATR,LEXT,KYOUNG,KSHEAR,KEXPAN,KDENS
```

```
      EQUIVALENCE (U,IU,Y)
   50 K1=KR+JLI
C     CONVERT TO INTERNAL LOADING NUMBERS
C     KRR= U ADDRESS OF JLI TH KR ARRAY
      KRR = IU(K1)
      K1 = KUV+JLI
      KU = IU(K1)
      K1 = KPPLS+JLI
      KPPL = IU(K1)
      K1 = KPMNS+JLI
      KPMIN = IU(K1)
      GO TO (4,2),IIII
    2 KTEM = KPPRI
      GO TO 60
    4 KTEM = KATKA
   60 KUPR = KTEM + NSTV*(JLI-1)
      IF (NJR) 90,90,80
C     SUPPORT DISPLACEMENTS
80    KSUP=KPDBP+(JLI-1)*NDJ
      GO TO 102
C     NO SUPPORT DISPLACEMENTS...USE KPDBP AS ZERO VECTOR
   90 KSUP = KPDBP
  102 RETURN
      END
*     LABEL
*     SYMBOL TABLE
*     LIST8
      SUBROUTINE STATCK
C     DOES A STATICS CHECK BY BACK COMPUTING THE SUPPORT REACTIONS
C     AND THE APPLIED JOINT LOADS
C     KENNETH F. REINSCHMIDT, ROOM 1-255, EXT. 2117
C     STRESS...STRUCTURAL ENGINEERING SYSTEM SOLVER
C     VERSION III...23 AUGUST 1963
      DIMENSION U(32),IU(32),Y(32) ,T(6,6),          V(6),W(6),
     1 SYSFIL(27),PROFIL(6), KRAY(10)
      COMMON U,IU,Y,JLI,JLX,JLO,L,T,K1,K2,KU,KPMIN,KPPL,KUPR,JT,JTEST,
     1LMS,LPS,LLS,
     1 JPS,JMS,LS,JL,JJ,K3,K4,K5,K6,K7,KRR,KTEM,KSUP,V,W,
     4CHECK,NMAX,INORM,ISOLV,ISCAN,IIII,IMOD,JJJJ,ICONT,ISUCC,IMERG,
     5TOP,N1,NL,NT,NREQ,TN,LFILE,TOLER,IPRG,IRST,IRLD,IRPR,SYSFIL,
     8NJ,NB,NDAT,ID,JF,NSQ,NCORD,IMETH,NLDS,NFJS,NSTV,NMEMV,IPSI,NMR,
     9NJR,ISODG,NDSQ,NDJ,IPDBP,IUDBP,NBB,NFJS1,JJC,JDC,JMIC,JMPC,JLD
      COMMON JEXTN,MEXTN,LEXTN,JLC,NLDSI,IYOUNG,ISHEAR,IEXPAN,IDENS,
     1 PROFIL,
     4NAME,KXYZ,KJREL,JPLS,JMIN,MTYP,KPSI,MEMB,LOADS,MODN,KS,KMKST,KSTDB
     5,KATKA,KPPLS,KPMNS,KUV,KPPRI,KR,KSAVE,KRAY,KDIAG,KOFDG,
     6 IOFDG,LOADN,MEGAO,JEXT,JINT,KUDBP,KMEGA,KPDBP,JTYP,MTYP1,KB
     7,MLOAD,JLOAD,KATR,LEXT,KYOUNG,KSHEAR,KEXPAN,KDENS
      EQUIVALENCE (U,IU,Y)
      DO 300 L = 1,MEXTN
      K1=MTYP+L
      IF ( U(K1)) 60,300,60
   60 LPS = JPLS+L
      LMS = JMIN+L
      LPS = IU(LPS)+JINT
      LMS = IU(LMS)+JINT
      JPS = JF*(IU(LPS)-1)
      JMS = JF*(IU(LMS)-1)
      LS = JF*(L-1)
      K1 = KPPL+LS
```

```
      K2 = KPMIN+LS
      K3 = KRR + JPS
      K4 = KRR + JMS
C     GET FORCE ROTATION MATRIX, GLOBAL TO LOCAL
      CALL TRAMAT(L,2)
      DO 200 I = 1,JF
      K6 = K3+I
C     ROTATE MEMBER END FORCES INTO GLOBAL COORDINATES FOR STATICS CHECK
      DO 100 J = 1,JF
      K5 = K1+J
      U(K6) = U(K6)+T(J,I)*U(K5)
  100 CONTINUE
      K6 = K4+I
      DO 150 J = 1,JF
      K5 = K2+J
      U(K6) = U(K6)-T(J,I)*U(K5)
  150 CONTINUE
  200 CONTINUE
  300 CONTINUE
      RETURN
      END
*     FAP
      COUNT    200                                                      CHAI0010
      SST                                                               CHAI0020
*     32K 709/7090 FORTRAN LIBRARY        9CHN                          CHAI0030
*     32K 709/7090 FORTRAN LIBRARY     MIT VERSION.  MARCH 28,1962  MICHAI0040
      TTL      MONITOR CHAIN ROUTINE / 9CHN FOR STRESS III PROCESSOR    CHAI0050
*     ALL CARDS SEQUENCE NUMBERED WITH MI ARE ONLY FOR MIT MONITOR
*     ALL CARDS SEQUENCE NUMBERED WITH STRES ARE FOR STRESS PROCESSOR
      LBL      STRCHN                                              STRESCHAI0060
*     MAY 16 1963 REVISED FOR TSS BACKGROUND                       MICHAI0070
*     AUG 22 1963 * REVISED FOR 7094                               MICHAI0071
*     NOV 20 1963 * REVISED FOR STRESS III                    STRESCHAI0072
      BCORE                                                         MICHAI0075
      ENTRY    CHAIN                                                    CHAI0080
CHAIN LTM                                                               CHAI0090
      EMTM                                           7094          MICHAI0092
      ENB      =0400000          DISABLE EVERY TRAP BUT CLOCK 7094  MICHAI0094
      CAL      1,4                                             ***MITCHAI0100
      STP      MADSW.                                          ***MITCHAI0110
      SXA      *+2,4                                                    CHAI0120
      XEC*     $(TES)                                                   CHAI0130
      AXT      **,4                                                     CHAI0140
      CLA*     1,4                                                      CHAI0150
      ZET      MADSW.                                          ***MITCHAI0160
      ALS      18                                              ***MITCHAI0170
      STD      CHWRD                                                    CHAI0180
      CLA*     2,4                                                      CHAI0190
      ZET      MADSW.                                          ***MITCHAI0200
      ALS      18                                              ***MITCHAI0210
      PDX      ,1               TAPE NO. TO IR A,B                      CHAI0220
      PDX      ,2                                                       CHAI0230
      AXT      (CLKL),4                                        ***MITCHAI0270
      SXA      (CTRP),4                                        ***MITCHAI0280
      PXA      0,1                                             ***MITCHAI0290
      PAX      0,2                                             ***MITCHAI0300
      AXT      1154,4           2202 (B2)                              CHAI0310
      SXA      XA3+1,4          SET ERROR                              CHAI0320
      TXL      CHA,1,1          LEAVE FOR                              CHAI0330
      TXH      CHA,1,3          CHANNEL A                              CHAI0340
```

```
        SXA     CHWRD,1                                                         CHAIO350
        TXH     CHAA,1,2                                                        CHAIO360
CHB     PXD     ,1              HERE FOR B2/B3                                MICHAIO370
        ALS     1                                                             MICHAIO372
        PDX     ,1                                                            MICHAIO374
        PXD     ,2                                                            MICHAIO376
        ALS     10                                                              CHAIO380
        ORS     XCI+18,1        SET FOR 2/3                                   MICHAIO390
        ALS     1                                                               CHAIO400
        ORS     XCI+14          SET B                                         MICHAIO410
        TRA     *+4                                                             CHAIO420
CHAA    TXI     *+1,4,1         SET FOR B3                                      CHAIO430
        SXA     XA3+1,4         AND GO SET                                      CHAIO440
        TXI     CHB,2,-1        ERRORS                                          CHAIO450
        CAL     CHB             PREFIX OF 4                                     CHAIO460
        LDQ     TCOB                                                            CHAIO470
        TRA     CHAB                                                            CHAIO480
CHA     TXI     *+1,4,-62       SET FOR                                         CHAIO490
        SXA     XA3+1,4         A4                                              CHAIO500
        TXI     *+1,4,-448      IRC = 1204 OCTAL                                CHAIO510
        AXT     4,2                                                             CHAIO520
        SXA     CHWRD,2                                                         CHAIO530
        AXT     2048,2                                                          CHAIO540
        PXD     0,2                                                             CHAIO550
        ORS     XCI+10          SET 4L                                        MICHAIO560
        ALS     1                                                               CHAIO570
        ORS     XCI+16          SET A                                         MICHAIO580
        CLM                                                                     CHAIO590
        LDQ     XXA             TCOA                                            CHAIO600
CHAB    SXA     CHD,4           AC=PREFIX +OR-,MQ=TCO  AC=2202,2203,1204 BSCHAIO610
        SXA     XLA,4           BSR                                             CHAIO620
        SXA     CHE+3,4         REW                                             CHAIO630
        TXI     *+1,4,16        SET IRC FOR BIN                                 CHAIO640
        SLQ     CHC+2                                                         MICHAIO645
        SXA     CHC+3,4         RTB                                           MICHAIO650
        STP     CHC+4           RCH                                           MICHAIO660
        SLQ     CHC+5           TCO                                           MICHAIO670
        STP     CHC+6           TRC                                           MICHAIO680
        STP     CHC+7           TEF                                           MICHAIO690
        SXA     XLC,4           RTB                                             CHAIO700
        STP     XLC+1           RCH                                             CHAIO710
        SLQ     XLD             TCO                                             CHAIO720
        STP     XLD+1           TRC                                             CHAIO730
        STP     XLD+2           SCH                                             CHAIO740
        STP     CHC-1           TEF                                             CHAIO750
        AXT     XL(0)-XLA+1,1 MOVE                                             CHAIO760
        AXT     0,2             LOADER TO UPPER                                 CHAIO770
        CLA     XLA,2                                                           CHAIO780
        STO     .XLA,2                                                          CHAIO790
        TXI     *+1,2,-1                                                        CHAIO800
        TIX     *-3,1,1                                                         CHAIO810
        TEFA    *+1                                                             CHAIO820
CHC     AXT     5,1                                                             CHAIO830
        CLA     CHWRD                                                           CHAIO840
        TCOB    *                                                             MICHAIO845
        RDS     **              2                                               CHAIO850
        RCHB    CHSEL1          3                                               CHAIO860
        TCOB    *               4                                               CHAIO870
        TRCB    CHD             5                                               CHAIO880
        TEFB    CHE             6                                               CHAIO890
```

```
        LXA     LBL,2                                                  CHAIO9CO
        TXH     *+2,2,4                                                CHAIO910
        TXH     *+3,2,1                                                CHAIO920
        AXT     4,2             ANY TAPE EXCEPT 2 OR 3 MAKE 4          CHAIO930
        SXA     LBL,2                                                  CHAIC940
        SUB     LBL             IS THIS THE CORRECT LINK               CHAIO950
        TNZ     CHC             NOT THIS ONE                           CHAIO960
        CLA     LKRCW                                                  CHAIO970
        STO     .XLZ            CONTROL WORD FOR READING LINK          CHAIC980
        STA     .XL(0)                                           MICHAIC985
        CAL     PROG                                                   CHAIO990
        ERA     XGO                                                    CHAI1000
        PDX     ,1                                                     CHAI1010
        TXH     .XXA,1,0        MAKE SURE 3RD WORD IS TRA TO SOMEWHERE CHAI1020
        TZE     .XXA            IF NOT, COMMENT BAD TAPE               CHAI1030
        CLA     PROG                                                   CHAI1040
        STO     .XGO            TRANSFER TO LINK                       CHAI1050
        LXD     .XLZ,1                                          STRESCHAI1051
        PXA     ,1                                              STRESCHAI1052
        ADD     =100                                            STRESCHAI1053
        STO     TEMP
        ZET     INORM                                           STRESCHAI1055
        TRA     *+5                                             STRESCHAI1056
        SUB     NL                                              STRESCHAI1057
        TMI     *+3                                             STRESCHAI1058
        STO     NREQ                                            STRESCHAI1059
        TSX     $REORG,4                                        STRESCHAI1060
        CLA     TEMP
        STO     NT
        AXT     5,1                                                    CHAI1061
        TRA     .XLC                                                   CHAI1070
TEMP    PZE
CHD     BSRB    **                                                     CHAI1080
        TIX     CHC+1,1,1                                              CHAI1C90
        TRA     .XXA                                                   CHAI1100
CHE     NZT     CHAIN                                                  CHAI1110
        TRA     NGD             SECONC EOF NO GOOD                     CHAI1120
        STZ     CHAIN                                                  CHAI1130
        REWA    **                                                     CHAI1140
        TRA     CHC                                                    CHAI1150
NGD     LXA     XA3+1,1.                                               CHAI1160
        SXD     COM1+3,1                                               CHAI1170
        TCOA    *                                               MICHAI1175
        WTDA    PRTAPE                                                 CHAI1180
        RCHA    CHSEL2                                                 CHAI1190
        TRA*    $EXIT                                                  CHAI1200
CHWRD   PZE                                                            CHAI1210
TCOB    TCOB    *                                                      CHAI1220
CHSEL1  IORT    LBL,,3                                                 CHAI1230
 LBL                                                                   CHAI1240
 LKRCW                                                                 CHAI1250
 PROG                                                                  CHAI1260
CHSEL2  IORT    COM1,,7                                                CHAI1270
COM1    BCI     7,1 LINK NOT ON TAPE  .  JOB TERMINATED.               CHAI1280
        REM                                                            CHAI1290
XLA     BSRA    **              REDUNDANCY, TRY AGAIN                  CHAI1300
        IOT                                                            CHAI1310
        NOP                                                            CHAI1320
        TIX     .XLC,1,1                                               CHAI1330
XXA     TCOA    .XXA                                                   CHAI1340
```

```
          WTDA      PRTAPE            COMMENT BAD TAPE                              CHAI1350
          RCHA      .XLX                                                           CHAI1360
XXA1      TCOA      .XXA1                                                         MICHAI1365
          WPDA                                                                     CHAI1370
          RCHA      .XCIL                                                          CHAI1390
XXA2      TCOA      .XXA2                                                         MICHAI1400
          REWA      1                                                             CHAI1470
          REWA      4                                                             CHAI1480
          REWB      1                                                             CHAI1490
          REWB      2                                                             CHAI1500
          REWB      3                                                             CHAI1510
Z1        HTR       .Z1+1                                                        MICHAI1520
          ENK                                                                      CHAI1530
          TQP       .XLB                                                           CHAI1540
          BSFA      2                 REPEAT JOB                                   CHAI1550
XLB       RTBA      1                 READ IN 1 TO CS                              CHAI1560
          RCHA      .XLY                                                           CHAI1570
Z2        TCOA      .Z2                                                          MICHAI1575
          RTBA      1                                                             CHAI1580
Z3        TCOA      .Z3                                                          MICHAI1585
          RTBA      1                                                             CHAI1590
          TRA       1                 GO TO SIGN ON                                CHAI1600
          REM                                                                      CHAI1610
XLC       RDS       **                READ IN LINK                                 CHAI1620
          RCHB      .XLZ                                                           CHAI1630
XLD       TCOB      .XLD                                                           CHAI1640
          TRCB      .XLA                                                           CHAI1650
          SCHB      .XSCH                                                          CHAI1660
          LAC       .XSCH,2                                                        CHAI1670
          CLA*      .XL(0)                                                       MICHAI1680
          STD       .XLD1                                                          CHAI1690
XLD1      TXI       .XLD1+1,2,**                                                   CHAI1700
          TXH       .XXA,2,0          WAS RECORD READ IN UP TO PROGRAM BREAK       CHAI1710
XGO       TRA       **                OK, GO TO THE LINK                           CHAI1720
          REM                                                                      CHAI1730
XCI       OCT       000100001200,0  9L                                          MICHAI1740
          OCT       000000400000,0  8L COL 6=1,2                                MICHAI1750
          OCT       000200000000,0  7L CHANNEL A,B                             MICHAI1760
          OCT       000000004004,0  6L COL 7=2,3,4                             MICHAI1770
          OCT       000441000440,0  5L TAPE NO.                                MICHAI1780
          OCT       041000000100,0  4L BAD XX DEPRESS                          MICHAI1790
          OCT       000000010000,0  3L KEY S TO RERUN                          MICHAI1800
          OCT       200032100002,0  2L                                         MICHAI1810
          OCT       100000000010,0  1L                                         MICHAI1820
          OCT       000030510100,0  OL                                         MICHAI1830
          OCT       000302005254,0  11L                                        MICHAI1840
          OCT       351441000402,0  12L                                        MICHAI1850
XA3       BCI       1,1 BAD                                                       CHAI1860
          BCI       1,                                                            CHAI1870
XLX       IORT      .XA3,,2                                                       CHAI1880
XCIL      IOCT      .XCI,,24                                                    MICHAI1890
XLY       IOCP      0,,3                                                          CHAI1910
          TCH       0                                                             CHAI1920
XLZ       IORT      **,,**                                                        CHAI1930
XSCH      PZE                                                                     CHAI1940
XL(0)     PZE                                                                     CHAI1950
          REM                                                                      CHAI1960
PRTAPE    EQU       3                                                             CHAI1970
.X        COMMON    0                                                             CHAI1980
.XLA      EQU       .X+2                                                          CHAI1990
```

```
.XXA    EQU     .XLA+4                                          CHAI2000
.XXA1   EQU     .XXA+3                                        MICHAI2010
.XXA2   EQU     .XXA1+3                                         CHAI2020
.Z1     EQU     .XXA2+6                                       MICHAI2025
.XLB    EQU     .Z1+4                                         MICHAI2030
.Z2     EQU     .XLB+2                                        MICHAI2033
.Z3     EQU     .Z2+2                                         MICHAI2036
.XLC    EQU     .Z3+3                                         MICHAI2040
.XLD    EQU     .XLC+2                                          CHAI2050
.XLD1   EQU     .XLD+6                                          CHAI2060
.XGO    EQU     .XLD1+2                                         CHAI2070
.XCI    EQU     .XGO+1                                          CHAI2080
.XA3    EQU     .XCI+24                                       MICHAI2090
.XLX    EQU     .XA3+2                                         CHAI21C0
.XCIL   EQU     .XLX+1                                          CHAI2110
.XLY    EQU     .XCIL+1                                       MICHAI2120
.XLZ    EQU     .XLY+2                                          CHAI2130
.XSCH   EQU     .XLZ+1                                          CHAI2140
.XL(O)  EQU     .XSCH+1                                         CHAI2150
MADSW.  PZE                                                ***MITCHAI2160
 U      COMMON  110                                        STRESCHAI2161
 INORM  COMMON  11                                         STRESCHAI2162
 NL     COMMON  1                                          STRESCHAI2163
 NT     COMMON  1                                          STRESCHAI2164
 NREQ   COMMON  1                                          STRESCHAI2165
        END                                                    CHAI2170
*       LIST8
*       LABEL
CFILES A4 CHAIN WITH BUFFER
        SUBROUTINE FILES(NOP,TN,NFILE,NCOUNT,ARRAY)
        COMMON U,ISCAN,ISUCC,FIL,LFILE,FIL2,NTAPE,NXREC
        DIMENSION ARRAY(100),U(112),FIL(7),FIL2(5),ISCAN(5)
        DIMENSION NTAPE(5),NXREC(5),BUF(254),LBUR(254)
        EQUIVALENCE (BUF,LBUR)
        XNTF(L)=((L-1)/20+1)-(4*((L-1)/ 80))
        XNRNF(L)=(L-1)/80 +1
        XLFNF(L)=((L-1)/20)*20+21
        XFIBF(L)=L-20*((L-1)/20)
  5     GO TO (10,20,30,30,19),NOP
  10    NTAPE(1)=5
        NTAPE(2)=6
        NTAPE(3)=7
        NTAPE(4)=9
        LFILE=1
        NBUF=0
        DO 11 I=1,4
        NT=NTAPE(I)
        NXREC(I)=1
  11    REWIND NT
  12    RETURN
C       DUMP BUFFER ON TAPE AND RETURN
  19    NS1=3
        IF(NBUF)12,12,234
  20    IF(NCOUNT-252)231,231,21
C       IS BUFFER EMPTY
  21    IF(NBUF)25,25,22
  22    NS1=1
C       EMPTY BUFFER
  234   NIT=XNTF(LFILE-1)
        NT=NTAPE(NIT)
```

```
        NIR=XNRNF(LFILE-1)
230     IF(NIR-NXREC(NIT))23,24,26
23      PRINT 1,LFILE
        GO TO 12
1       FORMAT(39H1ERROR IN TAPE WINDING FOR WRITING FILE,I6)
24      WRITE TAPE NT,(BUF(I),I=1,254)
        NBUF=0
        LFILE=XLFNF(LFILE-1)
        NXREC(NIT)=NXREC(NIT)+1
        GO TO (25,235,13),NS1
25      NIT=XNTF(LFILE)
        NT=NTAPE(NIT)
        NIR=XNRNF(LFILE)
        NS2=2
29      IF(NIR-NXREC(NIT))23,27,240
26      NS2=1
C       FORWARD SPACING ROUTINE
240     READ TAPE NT,NOLD
        NXREC(NIT)=NXREC(NIT)+1
        GO TO (230,29,37),NS2
C       WRITE ARRAY ON TAPE - IT EXCEEDS BUFFER SIZE
27      WRITE TAPE NT,LFILE,NCOUNT,(ARRAY(I),I=1,NCOUNT)
        NFILE=LFILE
        LFILE=XLFNF(LFILE)
        NXREC(NIT)=NXREC(NIT)+1
13      IF(65400-LFILE)14,12,12
14      PRINT 15
        INORM=1
        CALL CHAIN(1,4)
15      FORMAT(50H1TAPE CAPACITY EXCEEDED.  PROBLEM CANNOT CONTINUE.)
C       WILL ARRAY FIT IN BUFFER
231     IF(NBUF+NCOUNT-252)232,232,233
C       NO - EMPTY BUFFER
233     NS1=2
        GO TO 234
C       IS THIS THE FIRST FILE
232     IF(XFIBF(LFILE)-1)235,237,235
C       YES  -  DUMP BUFFER IF NON EMPTY AND -235- PUT ARRAY IN BUFFER
237     IF(NBUF)235,235,233
C       NO - 235 - PUT ARRAY IN BUFFER
235     IBUF=NBUF+2
        LBUR(IBUF-1)=LFILE
        LBUR(IBUF)=NCOUNT
        DO 236 I=1,NCOUNT
        I1=IBUF+I
236     BUF(I1)=ARRAY(I)
        NBUF=I1
        NFILE=LFILE
        LFILE=LFILE+1
        GO TO 12
30      LB=XFIBF(NFILE)-1
        IF(LBUR(1)-NFILE+LB)33,74,33
33      NIR=XNRNF(NFILE)
        NIT=XNTF(NFILE)
        NT=NTAPE(NIT)
        NS2=3
37      IF(NIR-NXREC(NIT))31,35,240
31      N=NXREC(NIT)-NIR
        IF(NXREC(NIT)/4-N)45,45,60
60      DO 32 I=1,N
```

```
32      BACKSPACE NT
35      NXREC(NIT)=NIR+1
        IF(NOP-3)22,36,40
45      REWIND NT
        NXREC(NIT)=1
        GO TO 37
36      IF(NCOUNT-252)70,70,38
38      READ TAPE NT,NOLD,N,(ARRAY(I),I=1,NCOUNT)
        IF(NFILE-NOLD)44,12,44
44      PRINT 2,NFILE
        ISUCC=2
        ISCAN=2
        CALL CHAIN(1,4)
70      READ TAPE NT,(BUF(I),I=1,254)
74      NBF=2
        IF(LB)44,77,78
78      DO 71 I=1,LB
71      NBF=NBF+2+LBUR(NBF)
77      IF(LBUR(NBF)-NCOUNT)75,72,72
72      DO 73 I=1,NCOUNT
        J=NBF+I
73      ARRAY(I)=BUF(J)
        GO TO 12
75      N=N+1
        DO 76 I=N,NCOUNT
76      ARRAY(I)=0.0
        NCOUNT=N-1
        GO TO 72
40      IF(XFIBF(NFILE)-1)44,41,70
41      READ TAPE NT,NOLD,N
        BACKSPACE NT
        IF(NCOUNT-N)38,38,42
42      N=N+1
        DO 43 I=N,NCOUNT
43      ARRAY(I)=0.0
        NCOUNT=N-1
        GO TO 38
2       FORMAT(25H1TAPE READING ERROR, FILE,I6)
        END
*       FAP
        COUNT   200                                                      CHAI0010
        SST                                                              CHAI0020
*       32K 709/7090 FORTRAN LIBRARY          9CHN                       CHAI0030
*       32K 709/7090 FORTRAN LIBRARY     MIT VERSION. MARCH 28,1962   MICHAI0040
        TTL     MONITOR CHAIN ROUTINE / 9CHN FOR STRESS III PROCESSOR   CHAI0050
*       ALL CARDS SEQUENCE NUMBERED WITH MI ARE ONLY FOR MIT MONITOR
*       ALL CARDS SEQUENCE NUMBERED WITH STRES ARE FOR STRESS PROCESSOR
        LBL     STATA5
*       MAY 16 1963 REVISED FOR TSS BACKGROUND                         MICHAI0070
*       AUG 22 1963 * REVISED FOR 7094                                  MICHAI0071
*       NOV 20 1963 * REVISED FOR STRESS III                        STRESCHAI0072
*       JAN 15 1964 * REVISED FOR STRESS III PROCESSOR TAPE ON A5  STRESCHAI0076
*       STARTER VERSION OF CHAIN FOR STARTING CHAIN TAPE ON A5
        BCORE                                                          MICHAI0075
        ENTRY   CHAIN                                                    CHAI0080
CHAIN LTM                                                                CHAI0090
        EMTM                                  7094                     MICHAI0092
        ENB     =0400000        DISABLE EVERY TRAP BUT CLOCK 7094     MICHAI0094
        CAL     1,4                                                  ***MITCHAI0100
        STP     MADSW.                                               ***MITCHAI0110
```

```
          SXA      *+2,4                                                    CHAIO120
          XEC*     $(TES)                                                   CHAIO130
          AXT      **,4                                                     CHAIO140
          CLA*     1,4                                                      CHAIO150
          ZET      MADSW.                                             ***MITCHAIO160
          ALS      18                                                ***MITCHAIO170
          STD      CHWRD                                                    CHAIO180
          AXT      (CLKL),4                                          ***MITCHAIO270
          SXA      (CTRP),4                                          ***MITCHAIO280
CHA       AXT      1093,4          SET FOR                                  CHAIO490
          SXA      XA3+1,4         A5                                       CHAIO500
          TXI      *+1,4,-448      IRC = 1205 OCTAL                         CHAIO510
          AXT      4,2                                                      CHAIO520
          SXA      CHWRD,2                                                  CHAIO530
          AXT      2048,2                                                   CHAIO540
          PXD      0,2                                                      CHAIO550
          ORS      XCI+10          SET 4L                                  MICHAIO560
          ALS      1                                                        CHAIO570
          ORS      XCI+16          SET A                                   MICHAIO580
          CLM                                                               CHAIO590
          LDQ      XXA             TCOA                                     CHAIO600
CHAB      SXA      CHD,4           AC=PREFIX +OR-,MQ=TCO  AC=2202,2203,1204 BSCHAIO610
          SXA      XLA,4           BSR                                      CHAIO620
          SXA      CHE+3,4         REW                                      CHAIO630
          TXI      *+1,4,16        SET IRC FOR BIN                          CHAIO640
          SLQ      CHC+2                                                   MICHAIO645
          SXA      CHC+3,4         RTB                                     MICHAIO650
          STP      CHC+4           RCH                                     MICHAIO660
          SLQ      CHC+5           TCO                                     MICHAIO670
          STP      CHC+6           TRC                                     MICHAIO680
          STP      CHC+7           TEF                                     MICHAIO690
          SXA      XLC,4           RTB                                      CHAIO700
          STP      XLC+1           RCH                                      CHAIO710
          SLQ      XLD             TCO                                      CHAIO720
          STP      XLD+1           TRC                                      CHAIO730
          STP      XLD+2           SCH                                      CHAIO740
          STP      CHC-1           TEF                                      CHAIO750
          AXT      XL(0)-XLA+1,1   MOVE                                     CHAIO760
          AXT      0,2             LOADER TO UPPER                          CHAIO770
          CLA      XLA,2                                                    CHAIO780
          STO      .XLA,2                                                   CHAIO790
          TXI      *+1,2,-1                                                 CHAIO800
          TIX      *-3,1,1                                                  CHAIO810
          TEFA     *+1                                                      CHAIO820
CHC       AXT      5,1                                                      CHAIO830
          CLA      CHWRD                                                    CHAIO840
          TCOB     *                                                       MICHAIO845
          RDS      **              2                                        CHAIO850
          RCHB     CHSEL1          3                                        CHAIO860
          TCOB     *               4                                        CHAIO870
          TRCB     CHD             5                                        CHAIO880
          TEFB     CHE             6                                        CHA1O890
          LXA      LBL,2                                                    CHAIO9C0
          TXH      *+2,2,4                                                  CHAIC910
          TXH      *+3,2,1                                                  CHAIC920
          AXT      4,2             ANY TAPE EXCEPT 2 OR 3 MAKE 4            CHAIU930
          SXA      LBL,2                                                    CHAIO940
          SUB      LBL             IS THIS THE CORRECT LINK                 CHAIC950
          TNZ      CHC             NOT THIS ONE                             CHAIC960
          CLA      LKRCW                                                    CHAIU970
```

```
          STO     .XLZ            CONTROL WORD FOR READING LINK          CHAIC980
          STA     .XL(0)                                                MICHAIC985
          CAL     PROG                                                  CHAI0990
          ERA     XGO                                                   CHAI1000
          PDX     ,1                                                    CHAI1010
          TXH     .XXA,1,0        MAKE SURE 3RD WORD IS TRA TO SOMEWHERE CHAI1020
          TZE     .XXA            IF NOT, COMMENT BAD TAPE              CHAI1030
          CLA     PROG                                                  CHAI1040
          STO     .XGO            TRANSFER TO LINK                      CHAI1050
          AXT     5,1                                                   CHAI1061
          TRA     .XLC                                                  CHAI1070
TEMP      PZE     .XLC                                                  CHAI1080
CHD       BSRB    **                                                    CHAI1090
          TIX     CHC+1,1,1                                             CHAI1100
          TRA     .XXA                                                  CHAI1110
CHE       NZT     CHAIN                                                 CHAI1120
          TRA     NGD             SECOND EOF NO GOOD                    CHAI1130
          STZ     CHAIN                                                 CHAI1140
          REWA    **                                                    CHAI1150
          TRA     CHC                                                   CHAI1160
NGD       LXA     XA3+1,1                                               CHAI1170
          SXD     COM1+3,1                                              MICHAI1175
          TCOA    *                                                     CHAI1180
          WTDA    PRTAPE                                                CHAI1190
          RCHA    CHSEL2                                                CHAI1200
          TRA*    $EXIT                                                 CHAI1210
CHWRD     PZE                                                           CHAI1220
TCOB      TCOB    *                                                     CHAI1230
CHSEL1    IORT    LBL,,3                                                CHAI1240
          LBL                                                           CHAI1250
          LKRCW                                                         CHAI1260
          PROG                                                          CHAI1270
CHSEL2    IORT    COM1,,7                                               CHAI1280
COM1      BCI     7,1 LINK NOT ON TAPE  .  JOB TERMINATED.             CHAI1290
          REM                                                           CHAI1300
XLA       BSRA    **              REDUNDANCY, TRY AGAIN                 CHAI1310
          IOT                                                           CHAI1320
          NOP                                                           CHAI1330
          TIX     .XLC,1,1                                              CHAI1340
XXA       TCOA    .XXA                                                  CHAI1350
          WTDA    PRTAPE          COMMENT BAD TAPE                      CHAI1360
          RCHA    .XLX                                                  MICHAI1365
XXA1      TCOA    .XXA1                                                 CHAI1370
          WPDA                                                          CHAI1390
          RCHA    .XCIL                                                 MICHAI1400
XXA2      TCOA    .XXA2                                                 CHAI1470
          REWA    1                                                     CHAI1480
          REWA    4                                                     CHAI1490
          REWB    1                                                     CHAI1500
          REWB    2                                                     CHAI1510
          REWB    3                                                     MICHAI1520
Z1        HTR     .Z1+1                                                 CHAI1530
          ENK                                                           CHAI1540
          TQP     .XLB            REPEAT JOB                            CHAI1550
          BSFA    2               READ IN 1 TO CS                       CHAI1560
XLB       RTBA    1                                                     CHAI1570
          RCHA    .XLY                                                  MICHAI1575
Z2        TCOA    .Z2                                                   CHAI1580
          RTBA    1                                                     MICHAI1585
Z3        TCOA    .Z3
```

```
        RTBA    1                                               CHAI1590
        TRA     1               GO TO SIGN ON                   CHAI1600
        REM                                                     CHAI1610
XLC     RDS     **              READ IN LINK                    CHAI1620
        RCHB    .XLZ                                            CHAI1630
XLD     TCOB    .XLD                                            CHAI1640
        TRCB    .XLA                                            CHAI1650
        SCHB    .XSCH                                           CHAI1660
        LAC     .XSCH,2                                         CHAI1670
        CLA*    .XL(0)                                        MICHAI1680
        STD     .XLD1                                           CHAI1690
XLD1    TXI     .XLD1+1,2,**                                    CHAI1700
        TXH     .XXA,2,0        WAS RECORD READ IN UP TO PROGRAM BREAK   CHAI1710
XGO     TRA     **              OK, GO TO THE LINK              CHAI1720
        REM                                                     CHAI1730
XCI     OCT     000100001200,0 9L                             MICHAI1740
        OCT     000000400000,0 8L COL 6=1,2                   MICHAI1750
        OCT     000200000000,0 7L CHANNEL A,B                 MICHAI1760
        OCT     000000004004,0 6L COL 7=2,3,4                 MICHAI1770
        OCT     000441000440,0 5L TAPE NO.                    MICHAI1780
        OCT     041000000100,0 4L BAD XX DEPRESS              MICHAI1790
        OCT     000000010000,0 3L KEY S TO RERUN              MICHAI1800
        OCT     200032100002,0 2L                             MICHAI1810
        OCT     100000000010,0 1L                             MICHAI1820
        OCT     000030510100,0 0L                             MICHAI1830
        OCT     000302005254,0 11L                            MICHAI1840
        OCT     351441000402,0 12L                            MICHAI1850
XA3     BCI     1,1 BAD                                         CHAI1860
        BCI     1,                                             CHAI1870
XLX     IORT    .XA3,,2                                        CHAI1880
XCIL    IOCT    .XCI,,24                                      MICHAI1890
XLY     IOCP    0,,3                                           CHAI1910
        TCH     0                                              CHAI1920
XLZ     IORT    **,,**                                         CHAI1930
XSCH    PZE                                                    CHAI1940
XL(0)   PZE                                                    CHAI1950
        REM                                                    CHAI1960
PRTAPE  EQU     3                                              CHAI1970
.X      COMMON  0                                              CHAI1980
.XLA    EQU     .X+2                                           CHAI1990
.XXA    EQU     .XLA+4                                         CHAI2000
.XXA1   EQU     .XXA+3                                        MICHAI2010
.XXA2   EQU     .XXA1+3                                        CHAI2020
.Z1     EQU     .XXA2+6                                       MICHAI2025
.XLB    EQU     .Z1+4                                         MICHAI2030
.Z2     EQU     .XLB+2                                        MICHAI2033
.Z3     EQU     .Z2+2                                         MICHAI2036
.XLC    EQU     .Z3+3                                         MICHAI2040
.XLD    EQU     .XLC+2                                         CHAI2050
.XLD1   EQU     .XLD+6                                         CHAI2060
.XGO    EQU     .XLD1+2                                        CHAI2070
.XCI    EQU     .XGO+1                                         CHAI2080
.XA3    EQU     .XCI+24                                       MICHAI2090
.XLX    EQU     .XA3+2                                         CHAI2100
.XCIL   EQU     .XLX+1                                         CHAI2110
.XLY    EQU     .XCIL+1                                       MICHAI2120
.XLZ    EQU     .XLY+2                                         CHAI2130
.XSCH   EQU     .XLZ+1                                         CHAI2140
.XL(0)  EQU     .XSCH+1                                        CHAI2150
MADSW.  PZE                                            ***MITCHAI2160
```

```
   U      COMMON   110                                                    STRESCHAI2161
  INORM   COMMON   11                                                     STRESCHAI2162
  NL      COMMON   1                                                      STRESCHAI2163
  NT      COMMON   1                                                      STRESCHAI2164
  NREQ    COMMON   1                                                      STRESCHAI2165
          END                                                             CHAI2170
*         FAP                                                             XDETC000
          COUNT    400                                                    XDET0010
*XSIMEQ        JULY 26, 1961      KORN                                    XDETC020
*SOLUTION OF SIMULTANEOUS LINEAR EQUATIONS / EVALUATION OF DETERMINANTS   XDETC030
*     REASSEMBLE XSIMEQ TO CORRECT ERROR INDUCED BY CHANGE TO FORTRAN     XDETC040
*         AUGUST 22,1962                                                  XDET0050
          LBL      XDET,X                                                 XDET0060
          ENTRY    XSIMEQ                                                 XDETC070
          ENTRY    XDETRM                                                 XDET0080
          REM                                                             XDET0090
          REM      XSIMEQ WILL SOLVE THE MATRIX EQUATION AX=B FOR THE     XDET0100
          REM      UNKNOWN MATRIX X AND WILL COMPUTE THE DETERMINANT OF A.XDET0110
          REM      FORTRAN STATEMENT TYPE                                 XDET0120
          REM                                                             XDET0130
          REM      XDETRM WILL COMPUTE THE DETERMINANT OF A.              XDET0140
          REM      FORTRAN STATEMENT TYPE        M=XDETRMF(N,LN,A,D)      XDETC150
          REM                                                             XDET0160
          REM      WHERE                                                  XDET0170
          REM                                                             XDET0180
          REM      N=FIXED POINT CONSTANT OR VARIABLE EQUAL TO MAXIMUM    XDET0190
          REM      POSSIBLE NUMBER OF ROWS (=COLUMNS) IN MATRIX A AS      XDET0200
          REM      SPECIFIED BY DIMENSION STATEMENT IN SOURCE PROGRAM     XDET0210
          REM                                                             XDET0220
          REM      LN=FIXED POINT CONSTANT OR VARIABLE EQUAL TO OBJECT    XDET0230
          REM      PROGRAM DIMENSION OF A                                 XDET0240
          REM                                                             XDET0250
          REM      LM=FIXED POINT CONSTANT OR VARIABLE EQUAL TO OBJECT    XDET0260
          REM      PROGRAM NUMBER OF CCLUMNS IN MATRIX B (NOT LARGER THAN N)XDET0270
          REM                                                             XDET0280
          REM      A=SOURCE PROGRAM FLOATING POINT VARIABLE USED TO       XDET0290
          REM      DESIGNATE THE ELEMENTS OF THE MATRIX A                 XDET03C0
          REM                                                             XDET0310
          REM      B=SOURCE PROGRAM FLOATING POINT VARIABLE USED TO       XDET0320
          REM      DESIGNATE THE ELEMENTS OF THE MATRIX B                 XDET0330
          REM                                                             XDET0340
          REM      D=SOURCE PROGRAM FLOATING POINT VARIABLE USED TO DENOTE XDET0350
          REM      SCALING FACTOR. AFTER EXECUTION D CONTAINS SCALED VALUE XDET0360
          REM      OF THE DETERMINANT                                     XDET0370
          REM                                                             XDET0380
          REM      E=SOURCE PROGRAM FIXED OR FLOATING POINT VARIABLE      XDET0390
          REM      DENOTING AT LEAST LN ERASABLE CELLS                    XDET0400
          REM                                                             XDET0410
          REM      M WILL BE SET TO                                       XDET0420
          REM                                                             XDET0430
          REM      1 IF SOLUTION IS SUCCESSFUL                            XDET0440
          REM      2 IF SPILL OCCURRED                                    XDET0450
          REM      3 IF MATRIX A IS SINGULAR.                             XDETC460
          REM                                                             XDETC470
          REM      LOCATE PIVOT AND RECORD I AND J                        XDETC480
   T1     LXD      AKK,1              INITIALIZE ELEMENT LOCATION INDEX    XDET0490
          SXD      AKQ,1                                                  XDET05C0
          LXD      K,2                INITIALIZE ROW INDEX                 XDET0510
          LXD      K,4                INITIALIZE COLUMN INDEX              XDET0520
          SXD      I,2                INITIALIZE MAXIMUM PIVOT ROW         XDETU530
```

```
        SXD     J,4                 INITIALIZE MAXIMUM PIVOT COLUMN           XDET0540
T7      PXD     0,0                                                           XDET0550
T8      ADM     0,1                 AC CONTAINS MAGNITUDE CURRENT MAXIMUM     XDET0560
        TXI     T10,1,1             NEXT ELEMENT                             XDET0570
T10     TXI     T11,2,1             NEXT ROW                                 XDET0580
T11     TXH     T17,2,LN            TRANSFER IF LAST ROW TESTED              XDET0590
T12     SBM     0,1                 TEST CURRENT ELEMENT                     XDET0600
        TPL     T8                  CURRENT MAXIMUM PIVOT HOLDS              XDET0610
        SXD     I,2                 CHANGE MAXIMUM PIVOT                     XDET0620
        SXD     J,4                                                          XDET0630
        TRA     T7                                                          XDET0640
T17     LXD     AKQ,1               KTH ELEMENT, CURRENT COLUMN              XDET0650
T18     TXI     T19,1,N                                                      XDET0660
T19     SXD     AKQ,1               KTH ELEMENT, NEXT COLUMN                 XDET0670
        LXD     K,2                 KTH ROW                                  XDET0680
        TXI     T20,4,1             NEXT COLUMN                              XDET0690
T20     TXL     T12,4,LN            EXIT IF LAST COLUMN TESTED               XDET0700
        REM                                                                  XDET0710
        REM     INTERCHANGE ROWS IF NECESSARY                                XDET0720
        REM                                                                  XDET0730
T21     CLA     I                                                            XDET0740
        SUB     K                                                            XDET0750
        TZE     T55                 NO ROW INTERCHANGE                       XDET0760
        ADD     AKK                                                          XDET0770
        PDX     0,2                 INITIALIZE ITH ROW INDEX                 XDET0780
        LXD     AKK,1               INITIALIZE KTH ROW INDEX                 XDET0790
        LXD     K,4                 INITIALIZE COLUMN INDEX                  XDET0800
T28     CLS     , D                 CHANGE SIGN OF                           XDET0810
T29     STO     , D                 DETERMINANT                              XDET0820
T30     LDQ     0,1                 INTERCHANGE                              XDET0830
        CLA     0,2                 KTH AND ITH                              XDET0840
        STO     0,1                 ROWS OF                                  XDET0850
        STQ     0,2                 MATRIX A                                 XDET0860
T34     TXI     T35,1,N             NEXT ELEMENT, KTH ROW                    XDET0870
T35     TXI     T36,2,N             NEXT ELEMENT, ITH ROW                    XDET0880
T36     TXI     T37,4,1             NEXT COLUMN                              XDET0890
T37     TXL     T30,4,LN                                                     XDET0900
T38     NOP                         TRANSFER TO T55 FOR XDETRM               XDET0910
        CLA     KM1                                                          XDET0920
        ADD     B                                                            XDET0930
        PDX     0,1                 INITIALIZE KTH ROW INDEX                 XDET0940
        SUB     K                                                            XDET0950
        ADD     I                                                            XDET0960
        PDX     0,2                 INITIALIZE ITH ROW INDEX                 XDET0970
        LXD     =01000000,4         INITIALIZE COLUMN INDEX                  XDET0980
T46     LDQ     0,1                 INTERCHANGE                              XDET0990
        CLA     0,2                 KTH AND ITH                              XDET1000
        STO     0,1                 ROWS OF                                  XDET1010
        STQ     0,2                 MATRIX B                                 XDET1020
T50     TXI     T51,1,N             NEXT ELEMENT, KTH ROW                    XDET1030
T51     TXI     T52,2,N             NEXT ELEMENT, ITH ROW                    XDET1040
T52     TXI     T53,4,1             NEXT COLUMN                              XDET1050
T53     TXL     T46,4,LM            EXIT IF LAST COLUMN PROCESSED            XDET1060
        REM                                                                  XDET1070
        REM     INTERCHANGE COLUMNS IF NECESSARY                             XDET1080
        REM                                                                  XDET1090
T55     CLA     J                                                            XDET1100
        SUB     K                                                            XDET1110
        TZE     T85                 NO COLUMN INTERCHANGE                    XDET1120
        ADD     KM1                                                          XDET1130
```

```
          LRS        35                                                      XDET1140
          MPY        N                                                       XDET1150
          ALS        17                                                      XDET1160
          ADD        A                                                       XDET1200
          PDX        0,2             INITIALIZE KTH COLUMN INDEX              XDET1210
          LXD        LN,4            INITIALIZE COMPLEMENTARY ROW INDEX       XDET1220
T68       CLS     ,  D               CHANGE SIGN OF                          XDET1230
T69       STO     ,  D                    DETERMINANT                        XDET1240
T70       LDQ        0,1             INTERCHANGE                             XDET1250
          CLA        0,2                 KTH AND JTH                         XDET1260
          STO        0,1                 COLUMNS OF                          XDET1270
          STQ        0,2                 MATRIX A                            XDET1280
          TXI        T75,1,1         NEXT ELEMENT, JTH COLUMN                XDET1290
T75       TXI        T76,2,1         NEXT ELEMENT, KTH COLUMN                XDET1300
T76       TIX        T70,4,1                                                 XDET1310
T77       NOP                        TRANSFER TO T85 FOR XDETRM              XDET1320
          LXD        J,1                                                     XDET1330
          LXD        K,2                                                     XDET1340
T80       CLA     ,1 E+1,1           INTERCHANGE                             XDET1350
T81       LDQ     ,2 E+1,2               JTH AND KTH                         XDET1360
T82       STO     ,2 E+1,2               ELEMENTS OF                         XDET1370
T83       STQ     ,1 E+1,1               ARRAY E                             XDET1380
          REM                                                                XDET1390
          REM        COMPUTE DETERMINANT                                     XDET1400
          REM                                                                XDET1410
T85       LXD        AKK,1                                                   XDET1420
          CLA        0,1             PIVOT ELEMENT                           XDET1430
          LRS        35                                                      XDET1450
          TZE        T251            MATRIX A SINGULAR                       XDET1440
T89       FMP     ,  D                                                       XDET1460
T90       STO     ,  D                                                       XDET1470
          REM                                                                XDET1480
          REM        ROW REDUCTION                                           XDET1490
          REM                                                                XDET1500
          LXD        KP1,1                                                   XDET1510
          SXD        E1,1            INITIALIZE ROW TO BE REDUCED            XDET1520
          LXD        AKK,1                                                   XDET1530
          SXD        E2,1                                                    XDET1540
          CLA        KM1                                                     XDET1550
          ADD        B                                                       XDET1560
          STO        E3                                                      XDET1570
T99       LXD        E3,1                                                    XDET1580
          TXI        T101,1,1                                                XDET1590
T101      SXD        E3,1            FIRST ELEMENT, CURRENT ROW, MATRIX B    XDET1600
          LXD        E2,1                                                    XDET1610
          TXI        T104,1,1                                                XDET1620
T104      SXD        E2,1            LEADING ELEMENT, CURRENT ROW, MATRIX A  XDET1630
          LXD        AKK,2                                                   XDET1640
          LXD        KP1,4           INITIALIZE COLUMN INDEX                 XDET1650
          CLA        0,1                                                     XDET1660
          TZE        T136            ROW NEEDS NO REDUCTION                  XDET1670
          FDP        0,2                                                     XDET1680
          STQ        G                                                       XDET1690
T111      TXI        T112,1,N                                                XDET17C0
T112      TXI        T113,2,N                                                XDET1710
T113      LDQ        G                                                       XDET1720
          FMP        0,2                                                     XDET1730
          CHS                                                                XDET1740
          FAD        0,1                                                     XDET1750
          STO        0,1             ELEMENT REDUCED                         XDET1760
```

```
T118   TXI      T119,1,N            NEXT ELEMENT, CURRENT ROW        XDET1770
T119   TXI      T120,2,N            NEXT ELEMENT, KTH ROW            XDET1780
T120   TXI      T121,4,1            NEXT COLUMN                      XDET1790
T121   TXL      T113,4,LN                                            XDET1800
T122   NOP                          TRANSFER TO T136 FOR XDETRM      XDET1810
       LXD      E3,1               BEGIN REDUCTION OF MATRIX B       XDET1820
       CLA      KM1                                                  XDET1830
       ADD      B                                                    XDET1840
       PDX      0,2                                                  XDET1850
       LXD      LM,4                                                 XDET1860
T128   LDQ      0,2                                                  XDET1870
       FMP      G                                                    XDET1880
       CHS                                                           XDET1890
       FAD      0,1                                                  XDET1900
       STO      0,1                ELEMENT REDUCED                   XDET1910
T133   TXI      T134,1,N           NEXT ELEMENT, CURRENT ROW         XDET1920
T134   TXI      T135,2,N           NEXT ELEMENT, KTH ROW             XDET1930
T135   TIX      T128,4,1                                             XDET1940
T136   LXD      E1,1                                                 XDET1950
       TXI      T138,1,1                                             XDET1960
T138   SXD      E1,1               NEXT ROW TO BE REDUCED            XDET1970
T139   TXL      T99,1,LN                                             XDET1980
       LXD      KP1,1                                                XDET1990
       TXI      T142,1,1                                             XDET2000
T142   TXH      T156,1,LN          REDUCTION COMPLETE                XDET2010
       SXD      KP1,1              K+1                               XDET2020
       TIX      T145,1,1                                             XDET2030
T145   SXD      K,1                K                                 XDET2040
       TIX      T147,1,1                                             XDET2050
T147   SXD      KM1,1              K-1                               XDET2060
       CLA      KM1N                                                 XDET2070
       ADD      N                                                    XDET2080
       STO      KM1N               (K-1)N                            XDET2090
       CLA      AKK                                                  XDET2100
       ADD      N                                                    XDET2110
       ADD      =01000000                                            XDET2120
       STO      AKK                                                  XDET2130
       TRA      T1                 BEGIN NEW STAGE                   XDET2140
T156   CLA      AKK                                                  XDET2150
       ADD      N                                                    XDET2160
       ADD      =01000000                                            XDET2170
       PDX      0,1                                                  XDET2180
       CLA      0,1                LAST PIVOT                        XDET2190
       TZE      T251               MATRIX A SINGULAR                 XDET2200
       LRS      35                                                   XDET2210
T163   FMP   ,  D                  FINAL VALUE OF                    XDET2220
T164   STO   ,  D                       DETERMINANT                  XDET2230
T165   NOP                         THRU FOR XDETRM                   XDET2240
       REM                                                           XDET2250
       REM      BACK SUBSTITUTION                                    XDET2260
       REM                                                           XDET2270
       SXD      AKK,1                                                XDET2280
       CLA      LN                                                   XDET2290
       SUB      =01000000                                            XDET2300
       ADD      B                                                    XDET2310
       STD      E3                                                   XDET2320
       LXD      LM,1                                                 XDET2330
       SXD      E1,1                                                 XDET2340
T174   LXD      LN,1                                                 XDET2350
       SXD      E4,1                                                 XDET2360
```

```
        LXD     AKK,1                                                    XDET2370
        SXD     E2,1                                                     XDET2380
        LXD     E3,2                                                     XDET2390
        CLA     0,2                                                      XDET2400
        FDP     0,1                                                      XDET2410
        STQ     0,2                                                      XDET2420
T182    LXD     E2,1                                                     XDET2430
        TXI     T184,1,-1                                                XDET2440
T184    SXD     E2,1              LAST ELEMENT, CURRENT ROW, MATRIX A    XDET2450
        STZ     G                                                        XDET2460
        LXD     E4,4                                                     XDET2470
        TNX     T204,4,1          ROW TO BE PROCESSED                    XDET2480
        SXD     E4,4                                                     XDET2490
        LXD     E3,2                                                     XDET2500
T191    LDQ     0,1                                                      XDET2510
        FMP     0,2                                                      XDET2520
        FAD     G                                                        XDET2530
        STO     G                                                        XDET2540
T195    TXI     T196,1, -N                                               XDET2550
T196    TXI     T197,2,-1                                                XDET2560
T197    TXI     T198,4,1                                                 XDET2570
T198    TXL     T191,4,LN-1                                              XDET2580
        CLA     0,2                                                      XDET2590
        FSB     G                                                        XDET2600
        FDP     0,1               VALUE OF UNKNOWN                       XDET2610
        STQ     0,2                                                      XDET2620
        TRA     T182                                                     XDET2630
T204    LXD     E3,2                                                     XDET2640
T205    TXI     T206,2,N                                                 XDET2650
T206    SXD     E3,2              LAST ROW, NEXT COLUMN, MATRIX B        XDET2660
        LXD     E1,2                                                     XDET2670
        TNX     T212,2,1                                                 XDET2680
T209    SXD     E1,2              NUMBER OF REMAINING COLUMNS            XDET2690
        TRA     T174              USE (LM-E1+1)TH COLUMN OF B            XDET2700
        REM                                                              XDET2710
        REM     REARRANGEMENT AND PERMANENT STORAGE ASSIGNMENT          XDET2720
        REM                                                              XDET2730
T212    CLA     A                                                        XDET2740
        STO     E1                                                       XDET2750
        CLA     =01000000                                                XDET2760
        STO     E2                                                       XDET2770
T216    LXD     =0,1                                                     XDET2780
T217    CLA     E2                                                       XDET2790
T218    SUB     ,1 E,1                                                   XDET2800
        TZE     T221                                                     XDET2810
        TXI     T217,1,1                                                 XDET2820
T221    PXD     0,1                                                      XDET2830
        ADD     B                                                        XDET2840
        PDX     0,1                                                      XDET2850
        LXD     E1,2                                                     XDET2860
        LXD     LM,4                                                     XDET2870
T226    CLA     0,1                                                      XDET2880
        STO     0,2                                                      XDET2890
T228    TXI     T229,1,N                                                 XDET2900
T229    TXI     T230,2,N                                                 XDET2910
T230    TIX     T226,4,1                                                 XDET2920
        LXD     LN,4                                                     XDET2930
        TNX     T242,4,1          THRU WITH XSIMEQ                       XDET2940
        SXD     LN,4                                                     XDET2950
        CLA     E1                                                       XDET2960
```

```
        ADD     =01000000                                                     XDET2970
        STO     E1              FIRST ELEMENT, NEXT ROW, MATRIX A             XDET2980
        CLA     E2                                                            XDET2990
        ADD     =01000000                                                     XDET3000
        STO     E2              NEXT ROW                                      XDET3010
        TRA     T216                                                          XDET3020
        REM                                                                   XDET3030
        REM     FINAL RESULTS                                                 XDET3040
        REM                                                                   XDET3050
T242    CLA     =01000000       SOLUTION SUCCESSFUL                           XDET3060
T243    LXD     REG12,1         RESTORE INDEX REGISTERS                       XDET3070
        LXA     REG12,2                                                       XDET3080
        LXD     T1-2,4                                                        XDET3090
        LDQ     SAVE            RESTORE LOCATION 8                            XDET3100
        STQ     8                                                             XDET3110
        TRA     1,4             EXIT FROM SUBROUTINE                          XDET3120
T249    CLA     =02000000       SPILL                                        XDET3130
        TRA     T243                                                         XDET3140
T251    CLA     =03000000       MATRIX A SINGULAR                            XDET3150
        TRA     T243                                                         XDET3160
        REM                                                                   XDET3170
        REM     ENTRY POINTS                                                  XDET3180
        REM                                                                   XDET3190
T254    STO     N               ENTRY FOR XSIMEQ                             XDET3200
        STQ     LN                                                            XDET3210
        CLA     =076100000000   OCTAL CODE FOR NOP                           XDET3220
        STO     T38                                                           XDET3230
        STO     T77                                                           XDET3240
        STO     T122                                                          XDET3250
        STO     T165                                                          XDET3260
        STO     T298                                                          XDET3270
        SXD     T1-2,4                                                        XDET3280
        CAL     -1,4            CHECH WORD PRECEEDING THE TSX               MXDET3290
        ARS     15              TO SEE IF IT WAS AN SXD                     MXDET3300
        ERA     =04634004                                                   MXDET3310
        TNZ     *+2                                                           XDET3320
        TXI     *+1,4,1         HAS SXD  REDUCE IR 4  BY ONE                 XDET3330
        CLA     -9,4            FOURTH ARGUMENT (A)                          XDET3340
        ALS     18                                                            XDET3350
        STD     A                                                             XDET3360
        CLA     -3,4            SEVENTH ARGUMENT (E)                         XDET3370
        STA     T218                                                          XDET3380
        STA     T301                                                          XDET3390
        ADD     =1                                                            XDET3400
        STA     T80                                                           XDET3410
        STA     T81                                                           XDET3420
        STA     T82                                                           XDET3430
        STA     T83                                                           XDET3440
        CLA     COMMON-2        THIRD ARGUMENT (LM)                          XDET3450
        STO     LM                                                            XDET3460
        CLA     -7,4            FIFTH ARGUMENT (B)                           XDET3470
        ALS     18                                                            XDET3480
        STD     B                                                             XDET3490
        CLA     -5,4            SIXTH ARGUMENT (D)                           XDET3500
T280    STA     T28                                                           XDET3510
        STA     T29                                                           XDET3520
        STA     T68                                                           XDET3530
        STA     T69                                                           XDET3540
        STA     T89                                                           XDET3550
        STA     T90                                                           XDET3560
```

```
        STA     T163                                            XDET3570
        STA     T164                                            XDET3580
        CLA     LN                                              XDET3590
        STD     T11                                             XDET3600
        STD     T20                                             XDET3610
        STD     T37                                             XDET3620
        STD     T121                                            XDET3630
        STD     T139                                            XDET3640
        STD     T142                                            XDET3650
        SUB     =01000000                                       XDET3660
        STD     T198                                            XDET3670
        STD     T304                                            XDET3680
T298    NOP             TRANSFER TO T305 FOR XDETRM             XDET3690
        LXD     =0,4                                            XDET3700
        CLA     =01000000                                       XDET3710
T301    STO     ,4 E,4          FILL ARRAY E                    XDET3720
        ADD     =01000000                                       XDET3730
        TXI     T304,4,1                                        XDET3740
T304    TXL     T301,4,LN-1                                     XDET3750
T305    SXD     REG12,1                                         XDET3760
        SXA     REG12,2                                         XDET3770
        LDC     A,4                                             XDET3780
        SXD     A,4                                             XDET3790
        SXD     AKK,4                                           XDET3800
        LDC     B,4                                             XDET3810
        SXD     B,4                                             XDET3820
        CLA     =01000000                                       XDET3830
        STO     K                                               XDET3840
        ADD     =01000000                                       XDET3850
        STO     KP1                                             XDET3860
        STZ     KM1                                             XDET3870
        STZ     KM1N                                            XDET3880
        CLA     LM                                              XDET3890
        STD     T53                                             XDET3900
        CLA     N                                               XDET3910
        STD     T18                                             XDET3920
        STD     T34                                             XDET3930
        STD     T35                                             XDET3940
        STD     T50                                             XDET3950
        STD     T51                                             XDET3960
        STD     T111                                            XDET3970
        STD     T112                                            XDET3980
        STD     T118                                            XDET3990
        STD     T119                                            XDET4000
        STD     T133                                            XDET4010
        STD     T134                                            XDET4020
        STD     T205                                            XDET4030
        STD     T228                                            XDET4040
        STD     T229                                            XDET4050
        LDC     N,4                                             XDET4060
        SXD     T195,4                                          XDET4070
        CLA     8                                               XDET4080
        STO     SAVE                                            XDET4090
        CLA     SPILL                                           XDET4100
        STO     8                                               XDET4110
        TRA     T1              ENTRY FOR XDETRM                XDET4120
T343    STO     N                                               XDET4130
        STQ     LN                                              XDET4140
        SXD     T1-2,4                                          XDET4150
        CLA     TRA1                                            XDET4160
```

```
        STO     T38                                                    XDET4170
        CLA     TRA2                                                   XDET4180
        STO     T77                                                    XDET4190
        CLA     TRA3                                                   XDET4200
        STO     T122                                                   XDET4210
        CLA     TRA4                                                   XDET4220
        STO     T165                                                   XDET4230
        CLA     TRA5                                                   XDET4240
        STO     T298                                                   XDET4250
        CLA     -5,4            THIRD ARGUMENT (A)                     XDET4260
        ALS     18                                                     XDET4270
        STD     A                                                      XDET4280
        CLA     -3,4            FOURTH ARGUMENT (D)                    XDET4290
        TRA     T280                                                   XDET4300
        REM                                                            XDET4310
TRA1    TRA     T55                                                    XDET4320
TRA2    TRA     T85                                                    XDET4330
TRA3    TRA     T136                                                   XDET4340
TRA4    TRA     T242                                                   XDET4350
TRA5    TRA     T305                                                   XDET4360
A       PZE                     -A                                     XDET4370
AKK     PZE                     -A+(K-1)(N+1)                          XDET4380
AKQ     PZE                                                            XDET4390
B       PZE                     -B                                     XDET4400
E1      PZE                                                            XDET4410
E2      PZE                                                            XDET4420
E3      PZE                                                            XDET4430
E4      PZE                                                            XDET4440
G       PZE                                                            XDET4450
I       PZE                                                            XDET4460
J       PZE                                                            XDET4470
K       PZE                     STAGE OF REDUCTION                     XDET4480
KM1     PZE                     K-1                                    XDET4490
KM1N    PZE                     (K-1)N                                 XDET4500
KP1     PZE                     K+1                                    XDET4510
LM      PZE                                                            XDET4520
LN      PZE                                                            XDET4530
N       PZE                                                            XDET4540
REG12   PZE                                                            XDET4550
SAVE    PZE                     CONTENTS OF LOCATION 8                 XDET4560
SPILL   TRA     TEST            MODIFIED TREATMENT OF UNDERFLOWS       XDET4580
TMP     PZE                     CONTENTS OF INDICATORS                 XDET4590
TEST    STI     TMP                                                    XDET4600
        LDI     0                                                      XDET4610
        LFT     4                                                      XDET4620
        TRA     OVER            SKIPPED IF UNDERFLOW                   XDET4630
        LFT     2                                                      XDET4640
        CLM                     SKIPPED IF ONLY MQ UNDERFLOW           XDET4650
        XCA                                                            XDET4660
        LFT     1                                                      XDET4670
        CLM                     SKIPPED IF ONLY AC UNDERFLOW           XDET4680
        XCA                                                            XDET4690
        LDI     TMP                                                    XDET4700
        TRA*    0                                                      XDET4710
OVER    LDI     TMP                                                    XDET4720
        TRA     T249                                                   XDET4730
XSIMEQ  SYN     T254                                                   XDET4740
XDETRM  SYN     T343                                                   XDET4750
        COMMON  -206                                                   XDET4760
COMMON  COMMON  1                                                      XDET4770


        END                                                            XDET4780
            END OF FILE
```

Appendix C

SEQUENTIAL LISTING OF PARAMETERS AND CODEWORDS

SEQUENTIAL LISTING OF PARAMETERS AND CODEWORDS

Common Location	Name	Refer to Section
1. System parameters		
77305	CHECK	5.2.1
77304	NMAX	5.2.2
77303	INORM	5.2.1
77302	ISOLV	5.2.1
77301	ISCAN	5.2.1
77277	IMOD	5.2.1
77276	NXFIL	5.2.3
77275	ICONT	5.2.1
77274	ISUCC	5.2.1
77272	TOP	5.2.2
77271	N1	5.2.2
77270	NL	5.2.2
77267	NT	5.2.2
77266	NREQ	5.2.2
77265	TN	5.2.3
77264	LFILE	5.2.3
77263	TOLER	5.2.4
77262	IPRG	5.2.1
77261	IRST	5.2.1
77260	IRLD	5.2.1
77257	IRPR	5.2.1
77256	NTAPE	5.2.3
77251	NXFILE	5.2.3

Common Location	Name	Refer to Section
2. Problem parameters		
77273	NJ	5.3.2
77222	NB	5.3.2
77221	NDAT	5.3.2
77220	ID	5.3.1
77217	JF	5.3.2
77216	NSQ	5.3.4
77215	NCORD	5.3.2
77214	IMETH	5.3.1
77213	NLDS	5.3.2
77212	NFJS	5.3.2
77211	NSTV	5.3.4
77210	NMEMV	5.3.4
77207	IPSI	5.3.6
77206	NMR	5.3.2
77205	NJR	5.3.2
77204	ISODG	5.3.7
77203	NDSQ	5.3.4
77202	NDJ	5.3.4
77201	IXX	5.3.7
77200	NPR	5.3.7
77177	NBB	5.3.7
77176	NFJSI	5.3.4
77175	JJC	5.3.5
77174	JDC	5.3.5
77173	JMIC	5.3.5
77172	JMPC	5.3.5
77171	JLD	5.3.5
77170	JEXTN	5.3.3
77167	MEXTN	5.3.3
77166	LEXTN	5.3.3

Common Location	Name	Refer to Section
77165	JLC	5.3.5
77164	NLDSI	5.3.2
77163	IYOUNG	5.3.6
77162	ISHEAR	5.3.6
77161	IEXPAN	5.3.6
77160	IDENS	5.3.6
77157	NBNEW	5.3.7
77156	NLDG	5.3.6

3. Codewords

Common Location	Name	Refer to Section
77151	NAME	5.4.1
77150	KXYZ	5.4.2
77147	KJREL	5.4.2
77146	JPLS	5.4.3
77145	JMIN	5.4.3
77144	MTYP	5.4.3
77143	KPSI	5.4.4
77142	MEMB	5.4.3
77141	LOADS	5.4.5
77140	MODN	5.4.1
77137	KS	5.4.6
77136	KMKST	5.4.6
77135	KSTDB	5.4.6
77133	KPPLS	5.4.8
77132	KPMNS	5.4.8
77131	KUV	5.4.8
77130	KPPRI	5.4.8
77127	KR	5.4.8
77126	KSAVE	5.4.9
77125 } 77114 }	KSRTCH	5.4.9
77113	KDIAG	5.4.7

Common Location	Name	Refer to Section
77112	KOFDG	5.4.7
77111	IOFDG	5.4.7
77107	MEGA0	5.4.7
77106	JEXT	5.4.2
77105	JINT	5.4.2
77104	IFDT	5.4.7
77103	KMEGA	5.4.6
77102	KPDBP	5.4.8
77101	JTYP	5.4.2
77100	MTYP1	5.4.3
77077	KB	5.4.7
77076	MLOAD	5.4.5
77075	JLOAD	5.4.5
77074	KATR	5.4.2
77073	LINT	5.4.5
77072	KYOUNG	5.4.4
77071	KSHEAR	5.4.4
77070	KEXPAN	5.4.4
77067	KDENS	5.4.4

DATE DUE

GAYLORD			PRINTED IN U.S.A.